Electric Motor Handbook

Electric Motor Handbook

H. Wayne Beaty

James L. Kirtley, Jr.

McGraw-Hill Book Company

New York   San Francisco   Washington, D.C.   Auckland   Bogotá
Caracas   Lisbon   London   Madrid   Mexico City   Milan
Montreal   New Delhi   San Juan   Singapore
Sydney   Tokyo   Toronto

McGraw-Hill

A Division of The **McGraw·Hill** *Companies*

1 2 3 4 5 6 7 8 9 0   DOC/DOC   9 0 3 2 1 0 9 8

ISBN 0-07-035971-7

The sponsoring editor for this book was Larry Hager and the production supervisor was Tina Cameron. It was set in Century Schoolbook by Pro-Image Corporation.

Printed and bound by R. R. Donnelley & Sons Company.

This book is printed on recycled, acid-free paper containing a minimum of 50% recycled, de-inked fiber.

Contents

Preface

Electric motors have been an important element of our industrial and commercial economy for over a century. Hundreds of books and thousands of scholarly works have been written about electric motors over the years. So it is fair to ask, "Why are we writing yet another book about motors?" The answer to this lies in an observation that one might make about the existing literature.

Virtually all books on electric machinery can be categorized as being introductory (that is textbooks), theoretical (principally having to do with design) or practical books about application. The introductory books typically do not concern the reader with any hard numbers. They simply try to promote a basic understanding of how motors and generators operate. Books on motor design are full of formulae useful to a machine designer who has a motor specification to fill, but say little if anything about application. Finally, in our experience, applications-oriented books tend to view the machine as a 'black box', with little disclosure about what is inside and where the real limits of the machine might lie.

This book is for the applications engineer, as opposed to the machine designer. It embodies the premise that effective application of electric motors requires an understanding of the internal mechanisms and physical principles on which the motor operates. For this reason it incorporates quite a lot of "theory" of operation. However, it also incorporates much practical knowledge, such as recommendations for machine type in certain applications and discussion of standards. It also incorporates a chapter on motor drives, as much application of electric motors will also incorporate power electronic controls, and a chapter on motor noise, an increasingly important consideration in many products.

Much of the theoretical material in this book had as its origin a set of notes for a Summer Professional Institute subject at the Massachusetts Institute of Technology called "Design of Electric Motors, Generators and Drive Systems".

Despite the long existence of electric motors and their importance in the modern world, more applications are found continually for them. As motors are more closely integrated into products and systems, and as drive electronics and motors become more closely integrated, a firm understanding of how the motors work, including their internal limits, production of noise and heat, and physical and electrical parameters, will be increasingly important for the applications engineer.

Electric Motor Handbook

Electric Motors

J. Kirtley

1.1 Electric Motors

Electric motors provide the driving power for a large and still increasing part of our modern industrial economy. The range of sizes and types of motors is large and the number and diversity of applications continues to expand. The computer on which this book is typed, for example, has several electric motors inside, in the cooling fan and in the disk drives. There is even a little motor that is used to eject the removable disk from its drive.

All around us there are electrical devices that move things around. Just about everything in one's life that whine, whirrs or clicks does so because an electric motor caused the motion.

At the small end of the power scale are motors that drive the hands in wristwatches, a job that was formerly done by a mechanical spring mechanism. At the large end of the power scale are motors, rated in the hundreds of megawatts (MW), that pump water uphill for energy storage. Somewhat smaller motors, rated in the range of 12 to 15 MW, have taken over the job of propulsion for cruise ships—a job formerly done by steam engines or very large, low speed diesel engines.

The flexibility of electric motors and generators and the possibility of transmitting electric power from place to place makes the use of electric motors in many drive mechanisms attractive. Even in situations in which the prime mover is aboard a vehicle, as in diesel-electric locomotives or passenger ships, electric transmission has displaced most mechanical or hydraulic transmission. As well, because electric power can be delivered over sliding contacts, stationary power plants can provide motive power for rail vehicles. The final drive is, of course, an electric motor.

The expansion of the use of electric motors' industrial, commercial and consumer applications is not at an end. New forms of energy storage systems, hybrid electric passenger vehicles, and other applications not yet envisioned will require electric motors, in some cases motors that have not yet been invented.

This book provides a basic and in-depth explanation for the operation of several different classes of electric motor. It also contains information about motor standards and application. The book is mostly concerned with application of motors, rather than on design or production. It takes, however, the point of view that good application of a motor must rely on understanding of its operation.

1.2   Types of Motor

It is important to remember at the outset that electric motors operate through the interaction of magnetic flux and electric current, or flow of charge. They develop *force* because a charge moving in a magnetic field produces a force which happens to be orthogonal to the motion of the charge and to the magnetic field. Electric machines also produce a voltage if the conductor in which current can flow moves through the magnetic field. Describing the interaction in a electric motor requires both phenomena, since the energy conversion typified by *torque* times *rotational speed* must also be characterized by *current* times *back voltage*.

Electric motors are broadly classified into two categories: AC and DC. Within those categories there are subdivisions. Recently, with the development of economical and reliable power electronic components, the classifications have become less rigorous and many other types of motor have appeared. However, it is probably best to start with the existing classifications of motor.

1.2.1   DC motors

DC motors, as the name implies, operate with terminal voltage and current that is "direct", or substantially constant. While it is possible to produce a "true DC" machine in a form usually called "acyclic", with homopolar geometry, such machines have very low terminal voltage and consequently high terminal current relative to their power rating. Thus all application of DC motors have employed a mechanical switch or commutator to turn the terminal current, which is constant or DC, into alternating current in the armature of the machine.

DC motors have usually been applied in two broad types of application. One of these categories is when the power source is itself DC. This is why motors in automobiles are all DC, from the motors that drive fans for engine cooling and passenger compartment ventilation to the engine starter motor.

A second reason for using DC motors is that their torque-speed characteristic has, historically, been easier to tailor than that of all AC motor categories. This is why most traction and servo motors have been DC machines. For example, motors for driving rail vehicles were, until recently, exclusively DC machines.

The mechanical commutator and associated brushes are problematical for a number of reasons, and because of this, the advent of cheaper high power semiconductors have led to applications of AC machines in situations formerly dominated by DC machines. For example, induction motors are seeing increased application in railroad traction applications. The class of machine known as "brushless DC" is actually a synchronous machine coupled with a set of semiconductor switches controlled by rotor position. Such machines have characteristics similar to commutator machines.

1.2.2 AC motors

Electric motors designed to operate with alternating current (AC) supplies are themselves broadly categorized into two classes: induction and synchronous. There are many variations of synchronous machines.

AC motors work by setting up a magnetic field pattern that rotates with respect to the stator and then employing electromagnetic forces to entrain the rotor in the rotating magnetic field pattern. Synchronous machines typically have a magnetic field which is stationary with respect to the rotor and which therefore rotate at the same speed as the stator magnetic field. In induction motors, the magnetic field is, as the name implies, induced by motion of the rotor through the stator magnetic field.

Induction motors are probably the most numerous in today's economy. Induction machines are simple, rugged and usually are cheap to produce. They dominate in applications at power levels from fractional horsepower (a few hundred watts) to hundreds of horsepower (perhaps half a megawatt) where rotational speeds required do not have to vary.

Synchronous motors are not as widely used as induction machines because their rotors are more complex and they require exciters. However, synchronous motors are used in large industrial applications in situations where their ability to provide leading power factor helps to support or stabilize voltage and to improve overall power factor. Also, in ratings higher than several hundred horsepower, synchronous machines are often more efficient than induction machines and so very large synchronous machines are sometimes chosen over induction motors.

Operated against a fixed frequency AC source, both synchronous and induction motors run at (nearly) fixed speed. However, when coupled with an adjustable frequency AC source, both classes of machine can form adjustable speed drives. There are some important distinctions based on method of control:

Brushless DC motors: permanent magnet synchronous machines coupled with switching mechanisms controlled by rotor position. They have characteristics similar to permanent magnet commutator machines.

Adjustable speed drives: synchronous or induction motors coupled to inverters that generate variable frequency. The speed of the motor is proportional to the frequency.

Vector control: also called *field oriented control,* is used to produce high performance servomechanisms by predicting the location of internal flux and then injecting current to interact optimally with that flux.

Universal motors are commutator machines, similar to DC machines, but are adapted to operation with AC terminal voltage. These machines are economically very important as large numbers are made for consumer appliances. They can achieve high shaft speed, and thus relatively high power per unit weight or volume, and therefore are economical on a watt-per-unit-cost basis. They are widely used in appliances such as vacuum cleaners and kitchen appliances.

Variable reluctance machines, (VRMs) also called *switched reluctance machines,* are mechanically very simple, operating by the principle that, under the influence of current excitation, magnetic circuits are pulled in a direction that increases inductance. They are somewhat akin to synchronous machines in that they operate at a speed that is proportional to frequency. However, they typically must operate with switching power electronics, as their performance is poor when operating against a sinusoidal supply. VRMs have not yet seen wide application, but their use is growing because of the simplicity of the rotor and its consequent ability to operate at high speeds and in hostile environments.

1.3   Description of the Rest of the Book

The book is organized as follows:

Chapter 2   contains a more complete description of the terminology of electric motors and more fully categorizes the machine types.

Chapter 3   contains the analytical principles used to describe electric motors and their operation, including loss mechanisms which limit machine efficiency and power density. This includes the elementary physics of electromechanical interactions employing the concepts of stored energy and co-energy; field-based force descriptions employing the "Maxwell Stress Tensor"; analytical methods for estimating loss densities in linear materials and in saturating iron; and empirical ways of describing losses in steel laminations.

Chapter 4   discusses induction machines. In this chapter, the elementary theory of the induction machine is derived and used to explain torque-speed curves. Practical aspects of induction motors, including different classes of motors and standards are described. Ways of controlling induction motors using adjustable frequency are presented, along with their limitations. Finally, single-phase motors are described and an analytic framework for their analysis is presented.

Chapter 5   concerns wound-field synchronous motors. It opens with a description of the synchronous motor. Analytical descriptions of synchronous motors and models for dynamic performance estimation and simulation are included. Standards and ways of testing synchronous motors are also examined.

Chapter 6   discusses "Brushless DC Motors". It includes a description of motor morphology, an analytic framework for brushless motors and a description of how they are operated.

Chapter 7   examines conventional, commutator type DC machines. It presents an analytical framework and a description of operation. It also contains nomenclature and a description of applicable standards.

Chapter 8   investigates other types of electric motors, including several types which do not fit into the conventional categories but which are nevertheless important, including types such as universal motors. This chapter also contains a section on high performance "high torque" motors.

Chapter 9   discusses the acoustic signature production in electric motors.

Chapter 10   explores the power-electronics systems that make up the other half of an electromechanical drive system.

Terminology and Definitions

N. Ghai

2.1 Types of Motor

There are many ways in which electric motors may be categorized or classified. Some of these are presented below and in Fig. 2.1.

2.1.1 AC and DC

One way of classifying electric motors is by the type of power they consume. Using this approach, we may state that all electric motors fall into one or the other of the two categories, viz., AC or DC. AC motors are those that run on alternating current or AC power, and DC motors are those that run on direct current, or DC power.

2.1.2 Synchronous and induction

Alternating current motors again fall into two distinct categories, synchronous or induction. Synchronous motors run at a fixed speed, irrespective of the load they carry. Their speed of operation is given by the relationship

$$\text{Speed in r/min} = 120 \times f/P$$

where f is the system frequency in Hz and P is the number of poles for which the stator is wound. The speed given by the above relationship is called the synchronous speed, and hence the name synchronous motor. The induction motor, on the other hand, runs very close to but less than the synchronous speed. The difference between the synchronous speed and the actual speed is called the slip speed. The slip speed

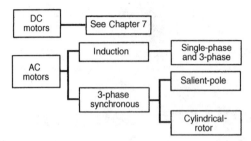

Figure 2.1 Classification of AC and DC motors.

of any induction motor is a function of its design and of desired performance. Further, for a given motor, the slip speed and the running speed vary with the load. The running speed decreases as the load on the motor is increased.

2.1.3 Salient-pole and cylindrical-rotor

Synchronous motors fall into two broad categories defined by their method of construction. These are salient-pole motors and cylindrical-rotor motors. High-speed motors, those running at 3600 r/min with 60 Hz supply, are of the cylindrical-rotor construction for mechanical strength reasons, whereas slower speed motors, those running at 1800 r/min and slower, are mostly of the salient-pole type.

2.1.4 Single-phase and three-phase motors

All AC motors may also be classified as single-phase and multiphase motors, depending on whether they are intended to run on single-phase supply or on multiphase supply. Since the distribution systems are universally of the three-phase type, multiphase motors are almost always of the three-phase type. Single-phase motors are limited by the power they can produce, and are generally available in sizes up to only a few horsepower, and in the induction motor variety only. Synchronous motors are usually available in three-phase configurations only.

2.1.5 Other variations

Many variations of the basic induction and synchronous motors are available. These include but are not limited to the *synchronous-induction* motor, which is essentially a wound-rotor-induction motor supplied with DC power to its rotor winding to make it run at synchronous speed; the *permanent-magnet* motor in which the field excitation is provided by permanent magnets; the *reluctance* motor in which the

TABLE 2.1   Operating Temperatures for Insulation System Classes

Class	A	E	B	F	H
Total operating temperature °C	105	120	130	155	180

TABLE 2.2   Allowable Temperature Rises

Insulation class	A	E	B	F	H
Temperature rise °C	55	65	80	105	125

surface of the rotor of a squirrel-cage induction motor is shaped to form salient-pole structures causing the motor to run up to speed as an induction motor and pull into synchronism by reluctance action and operate at synchronous speed; and the *ac-commutator motor* or *universal motor,* which possesses the wide speed range and higher starting torque advantages of DC motor, to name a few. One could also include here single-phase induction motor variations based on the method of starting used — the *split-phase* motor, the *capacitor-start* motor, the *resistance-start* motor, and the *shaded-pole* motor.

2.2   Insulation System Classes

The classification of winding insulation systems is based on their operating temperature capabilities. These classes are designated by the letters A, E, B, F, and H. The operating temperatures for these insulation classes are shown in Table 2.1.

These temperatures represent the maximum allowable operating temperature of the winding at which, if the motor were operated in a clean, dry, free-from-impurities environment at up to 40 hours per week, an operation life of 10 to 20 years could be expected, before the insulation deterioration due to heat destroys its capability to withstand the applied voltage.

The temperatures in the Table 2.1 are the maximum temperatures existing in the winding, or the hot spot temperatures, and are not the average winding temperatures. It is generally assumed that in a well-designed motor, the hot spot is approximately 10°C higher than the average winding temperature. This yields the allowable temperature rises (average, or rises by resistance) in an ambient temperature not exceeding 40°C, that one finds in standards. These are shown in Table 2.2.

Class A insulation is obsolete, and no longer in use. Class E insulation is not used in the United States, but is common in Europe. Class B is the most commonly specified insulation. Class F is slowly winning favor, although for larger motors in the United States, the users tend to specify class F systems with class B temperature rises to improve the life expectancy of the windings. Class H systems are widely specified in synchronous generators up to 5 mW in size.

2.3 Codes and Standards

Both national and international standards exist for electric motors. For the most part, these apply to general purpose motors. However, in the United States, some definite purpose standards also exist which are industry or application specific. Examples of the latter are the IEEE 841, which applies to medium size motors for petroleum and chemical applications, American Petroleum Institute standards API 541 (large induction motors) and API 546 (large synchronous motors), both for petroleum and chemical industry applications, and the American National Standards Institute standard ANSI C50.41 for large induction motors for generating station applications.

In the United States, in general, the Institute of Electrical and Electronics Engineers (IEEE) writes standards for motor testing and test methods, and the National Electrical Manufacturers Association (NEMA) writes standards for motor performance. In the international field, the International Electrotechnical Commission (IEC), which is a voluntary association of countries, writes all standards applicable to electric motors. U.S. and international standards that apply to electric motors are:

- NEMA MG1-1993, Rev 4, "Motors and Generators."
- IEEE Std 112-1996, "IEEE Standard Test Procedure for Polyphase Induction Motors and Generators."
- IEEE Std 115-1983, "IEEE Guide: Test Procedures for Synchronous Machines."
- IEEE Std 522-1992, "IEEE Guide for Testing Turn-to-Turn Insulation on Form-Wound Stator Coils for Alternating Current Rotating Electric Machines."
- IEC 34-1, 1996, 10th ed., "Rotating Electrical Machines, Part 1: Rating and Performance."
- IEC 34-1, Amendment 1, 1997, "Rotating Electrical Machines, Part 1: Rating and Performance."
- IEC 34-2, 1972, "Rotating Electrical Machines, Part 2: Methods of Determining Losses and Efficiency of Rotating Electrical Machinery from Tests."

- IEC 34-2, Amendment 1, 1995 and Amendment 2, 1996, "Rotating Electrical Machines, Part 2: Methods of Determining Losses and Efficiency of Rotating Electrical Machinery from Tests."

- IEC 34-5, 1991, "Rotating Electrical Machines, Part 5: Classification of Degrees of Protection Provided by Enclosures of Rotating Electrical Machines (IP Code)."

- IEC 34-6, 1991, "Rotating Electrical Machines, Part 6: Methods of Cooling (IC Code)."

- IEC 34-9, 1990 and 2/979/FDIS, 1997, "Rotating Electrical Machines, Part 9, "Noise Limits."

- IEC 34-12, 1980, "Rotating Electrical Machines, Part 12: Starting Performance of Single-speed, Three-phase Cage Induction Motors for Voltages up to and Including 600 Volts."

- IEC 34-14, 1990 and 2/940/FDIS, 1996, "Rotating Electrical Machines, Part 14: Mechanical Vibration of Certain Machines with Shaft Heights 56 mm and Larger."

- IEC 34-15, 1995, "Rotating Electric Machines, Part 15: Impulse Voltage Withstand Levels of Rotating AC Machines with Form-wound Coils."

- IEC 38, 1983, "IEC Standard Voltages."

- IEC 72-1, 1991, "Dimension and Output Series for Rotating Electrical Machines."

3

Fundamentals of Electromagnetic Forces and Loss Mechanisms

J. Kirtley

3.1 Introduction

This chapter covers some of the fundamental processes involved in electric machinery. In the section on energy conversion processes are examined the two major ways of estimating electromagnetic forces: those involving thermodynamic arguments (conservation of energy), and field methods (Maxwell's Stress Tensor). In between these two explications is a bit of description of electric machinery, primarily there to motivate the description of field based force calculating methods.

The section dealing with losses is really about eddy currents in both linear and nonlinear materials and about semi-empirical ways of handling iron losses and exciting currents in machines.

3.2 Energy Conversion Process

In a motor, the energy conversion process (see Fig. 3.1) can be thought of in simple terms. In "steady state", electric power input to the machine is just the sum of electric power inputs to the different phase terminals

$$P_e = \sum_i v_i i_i$$

Mechanical power is torque times speed

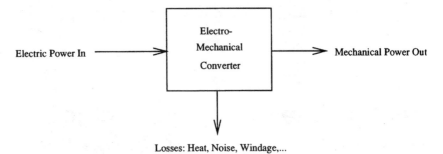

Figure 3.1 Energy conversion process.

$$P_m = T\Omega$$

and the sum of the losses is the difference

$$P_d = P_e - P_m$$

It will sometimes be convenient to employ the fact that, in most machines, dissipation is small enough to approximate mechanical power with electrical power. In fact, there are many situations in which the loss mechanism is known well enough that it can be idealized away. The "thermodynamic" arguments for force density take advantage of this and employ a "conservative" or lossless energy conversion system.

3.2.1 Energy approach to electromagnetic forces

To start, consider some electromechanical system which has two sets of "terminals", electrical and mechanical, as shown in Fig. 3.2. If the system stores energy in magnetic fields, the energy stored depends on the *state* of the system, defined by, in this case, two of the identifiable variables: flux (λ), current (i) and mechanical position (x). In fact, with only a little reflection, you should be able to convince yourself that this state is a single-valued function of two variables and that the

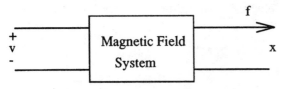

Figure 3.2 Conservative magnetic field system.

energy stored is independent of how the system was brought to this state.

Now, all electromechanical converters have loss mechanisms and so are not themselves conservative. However, the magnetic field system that produces force is, in principle, conservative in the sense that its state and stored energy can be described by only two variables. The "history" of the system is not important.

It is possible to choose the variables in such a way that electrical power *into* this conservative system is

$$P^e = vi = i\frac{d\lambda}{dt}$$

Similarly, mechanical power *out of* the system is

$$P^m = f^e \frac{dx}{dt}$$

The difference between these two is the rate of change of energy stored in the system

$$\frac{dW_m}{dt} = P^e - P^m$$

It is then possible to compute the change in energy required to take the system from one state to another by

$$W_m(a) - W_m(b) = \int_b^a id\lambda - f^e\,dx$$

where the two states of the system are described by $a = (\lambda_a, x_a)$ and $b = (\lambda_b, x_b)$.

If the energy stored in the system is described by two-state variables, λ and x, the *total differential* of stored energy is

$$dW_m = \frac{\partial W_m}{\partial \lambda} d\lambda + \frac{\partial W_m}{\partial x} dx$$

and it is also

$$dW_m = id\lambda - f^e\,dx$$

so that we can make a direct equivalence between the derivatives and

$$f^e = -\frac{\partial W_m}{\partial x}$$

This generalizes in the case of multiple electrical terminals and/or multiple mechanical terminals. For example, a situation with multiple electrical terminals will have

$$dW_m = \sum_k i_k d\lambda_k - f^e \, dx$$

In the case of rotary, as opposed to linear, motion has in place of force f^e and displacement x, torque T^e and angular displacement θ.

In many cases, we might consider a system which is electrically *linear*, in which case inductance is a function only of the mechanical position x

$$\lambda(x) = L(x)i$$

In this case, assuming that the energy integral is carried out from $\lambda = 0$ (so that the part of the integral carried out over x is zero)

$$W_m = \int_0^\lambda \frac{1}{L(x)} \lambda d\lambda = \frac{1}{2} \frac{\lambda^2}{L(x)}$$

This makes

$$f^e = -\frac{1}{2} \lambda^2 \frac{\partial}{\partial x} \frac{1}{L(x)}$$

Note that this is numerically equivalent to

$$f^e = -\frac{1}{2} i^2 \frac{\partial}{\partial x} L(x)$$

This is true *only* in the case of a linear system. Note that substituting $L(x)i = \lambda$ too early in the derivation produces erroneous results: in the case of a linear system, it is a sign error, but in the case of a nonlinear system, it is just wrong.

3.2.2 Co-energy

Often, systems are described in terms of inductance rather than its reciprocal, so that current, rather than flux, appears to be the relevant variable. It is convenient to derive a new energy variable, *co-energy*, by

$$W'_m = \sum_i \lambda_i i_i - W_m$$

and in this case, it is quite easy to show that the energy differential is (for a single mechanical variable) simply

$$dW'_m = \sum_k \lambda_k di_k + f^e \, dx$$

so that force produced is

$$f_e = \frac{\partial W'_m}{\partial x}$$

Consider a simple electric machine example in which there is a single winding on a rotor (call it the *field* winding) and a polyphase armature. Suppose the rotor is round so that we can describe the flux linkages as

$$\lambda_a = L_a i_a + L_{ab} i_b + L_{ab} i_c + M \cos(p\theta) i_f$$

$$\lambda_b = L_{ab} i_a + L_a i_b + L_{ab} i_c + M \cos\left(p\theta - \frac{2\pi}{3}\right) i_f$$

$$\lambda_c = L_{ab} i_a + L_{ab} i_b + L_a i_c + M \cos\left(p\theta + \frac{2\pi}{3}\right) i_f$$

$$\lambda_f = M \cos(p\theta) i_a + M \cos\left(p\theta - \frac{2\pi}{3}\right) i_b + M \cos\left(p\theta + \frac{2\pi}{3}\right) + L_f i_f$$

Now, this system can be simply described in terms of co-energy. With multiple excitation it is important to exercise some care in taking the co-energy integral (to ensure that it is taken over a valid path in the multi-dimensional space). In this case, there are actually five dimensions, but only four are important since the rotor can be positioned with all currents at zero so there is no contribution to co-energy from setting rotor position. Suppose the rotor is at some angle θ and that the four currents have values i_{a0}, i_{b0}, i_{c0} and i_{f0}. One of many correct path integrals to take would be

$$W'_m = \int_0^{i_{a0}} L_a i_a di_a$$

$$+ \int_0^{i_{b0}} (L_{ab} i_{a0} + L_a i_b) \, di_b$$

$$+ \int_0^{i_{c0}} (L_{ab} i_{a0} + L_{ab} i_{b0} + L_a i_c) \, di_c$$

$$+ \int_0^{i_{f0}} \left(M \cos (p\theta) i_{a0} + M \cos \left(p\theta - \frac{2\pi}{3} \right) i_{b0} \right.$$

$$\left. + M \cos \left(p\theta + \frac{2\pi}{3} \right) i_{c0} + L_f i_f \right) di_f$$

The result is

$$W'_m = \frac{1}{2} L_a(i_{a0}^2 + i_{b0}^2 + i_{c0}^2) + L_{ab}(i_{a0}i_{b0} + i_{a0}i_{c0} + i_{c0}i_{b0})$$

$$+ Mi_{f0} \left(i_{a0} \cos (p\theta) + i_{b0} \cos \left(p\theta - \frac{2\pi}{3} \right) \right.$$

$$\left. + i_{c0} \cos \left(p\theta + \frac{2\pi}{3} \right) \right) + \frac{1}{2} L_f i_{f0}^2$$

If the rotor is round so that there is no variation of the stator inductances with rotor position θ, torque is easily given by

$$T_e = \frac{\partial W'_m}{\partial \theta}$$

$$= -pMi_{f0} \left(i_{a0} \sin(p\theta) + i_{b0} \sin \left(p\theta - \frac{2\pi}{3} \right) + i_{c0} \sin \left(p\theta + \frac{2\pi}{3} \right) \right)$$

3.2.3 Generalization to continuous media

Consider a system with not just a multiplicity of circuits, but a continuum of current-carrying paths. In that case, we could identify the co-energy as

$$W'_m = \int_{area} \int \lambda(\vec{a}) \, d\vec{J} \cdot d\vec{a}$$

where that area is chosen to cut all of the current carrying conductors. This area can be picked to be perpendicular to each of the current filaments since the divergence of current is zero. The flux λ is calculated over a path that coincides with each current filament (such paths exist since current has zero divergence). Then the flux is

$$\lambda(\vec{a}) = \int \vec{B} \cdot d\vec{n}$$

Now, if the vector potential $\vec{A}$ for which the magnetic flux density is

$$\vec{B} = \nabla \times \vec{A}$$

the flux linked by any one of the current filaments is

$$\lambda(\vec{a}) = \oint \vec{A} \cdot d\vec{\ell}$$

where $d\vec{\ell}$ is the path around the current filament. This implies directly that the co-energy is

$$W'_m = \int_{\text{area}} \int_J \oint \vec{A} \cdot d\vec{\ell} d\vec{J} \cdot d\vec{a}$$

Now it is possible to make $d\vec{\ell}$ coincide with $d\vec{a}$ and be parallel to the current filaments, so that

$$W'_m = \int_{\text{vol}} \vec{A} \cdot d\vec{J} \, dv$$

3.2.4 Permanent magnets

Permanent magnets are becoming an even more important element in electric machine systems. Often systems with permanent magnets are approached in a relatively ad-hoc way, made equivalent to a current that produces the same MMF as the magnet itself.

The constitutive relationship for a permanent magnet relates the magnetic flux density $\vec{B}$ to magnetic field $\vec{H}$ and the property of the magnet itself, the *magnetization* $\vec{M}$

$$\vec{B} = \mu_0(\vec{H} + \vec{M})$$

Now, the effect of the magnetization is to act as if there were a current (called an *amperian current*) with density

$$\vec{J}* = \nabla \times \vec{M}$$

Note that this amperian current "acts" just like ordinary current in making magnetic flux density. Magnetic co-energy is

$$W'_m = \int_{\text{vol}} \vec{A} \cdot \nabla \times d\vec{M} \, dv$$

Next, note the vector identity $\nabla \cdot (\vec{C} \times \vec{D}) = \vec{D} \cdot (\nabla \times \vec{C}) - \vec{C} \cdot (\nabla \times \vec{D})$. Now

$$W'_m = \int_{vol} -\nabla \cdot (\vec{A} \times d\vec{M}) \, dv + \int_{vol} (\nabla \times \vec{A}) \cdot d\vec{M} \, dv$$

Then, noting that $\vec{B} = \nabla \times \vec{A}$

$$W'_m = -\oint \vec{A} \times d\vec{M} \, d\vec{s} + \int_{vol} \vec{B} \cdot d\vec{M} \, dv$$

The first of these integrals (closed surface) vanishes if it is taken over a surface just outside the magnet, where $\vec{M}$ is zero. Thus the magnetic co-energy in a system with only a permanent magnet source is

$$W'_m = \int_{vol} \vec{B} \cdot d\vec{M} \, dv$$

Adding current carrying coils to such a system is done in the obvious way.

3.2.5   Electric machine description

Actually, this description shows a conventional induction motor. This is a very common type of electric machine and will serve as a reference point. Most other electric machines operate in a fashion which is the same as the induction machine or which differ in ways which are easy to reference to the induction machine.

Consider the simplified machine drawing shown in Fig. 3.3. Most machines, but not all, have essentially this morphology. The rotor of the machine is mounted on a shaft which is supported on some sort of bearing(s). Usually, but not always, the rotor is inside. Although this rotor is round, this does not always need to be the case. Rotor

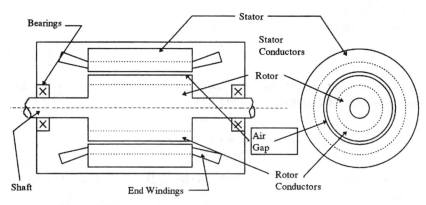

Figure 3.3 Form of electric machine.

conductors are shown, but sometimes the rotor has permanent magnets either fastened to it or inside, and sometimes (as in Variable Reluctance Machines), it is just an oddly shaped piece of steel. The stator is, in this drawing, on the outside and has windings. With most machines, the stator winding is the armature, or electrical power input element. (In dc and Universal motors, this is reversed, with the armature contained on the rotor.)

In most electrical machines, the rotor and the stator are made of highly magnetically-permeable materials: steel or magnetic iron. In many common machines such as induction motors, the rotor and stator are both made up of thin sheets of silicon steel. Punched into those sheets are slots which contain the rotor and stator conductors.

Figure 3.4 is a picture of part of an induction machine distorted so that the air-gap is straightened out (as if the machine had infinite radius). This is actually a convenient way of drawing the machine and, we will find, leads to useful methods of analysis.

What is important to note for now is that the machine has an air gap g which is relatively small (that is, the gap dimension is much less than the machine radius r). The air-gap also has a physical length ℓ. The electric machine works by producing a shear stress in the air-gap (with of course side effects such as production of "back voltage"). It is possible to define the average air-gap shear stress τ. Total developed torque is force over the surface area times moment (which is rotor radius)

$$T = 2\pi r^2 \ell \langle \tau \rangle$$

Power transferred by this device is just torque times speed, which is the same as force times surface velocity, since surface velocity is $u = r\Omega$

$$P_m = \Omega T = 2\pi r \ell \langle \tau \rangle u$$

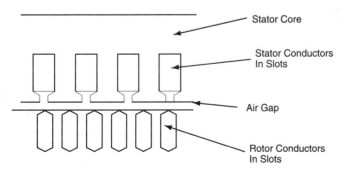

Figure 3.4 Windings in slots.

If active rotor volume is $V_r = \pi r^2 \ell$, the ratio of torque to volume is just

$$\frac{T}{V_r} = 2\langle \tau \rangle$$

Now, determining what can be done in a volume of machine involves two things. First, it is clear that the calculated volume is not the whole machine volume, since it does not include the stator. The actual estimate of total machine volume from the rotor volume is actually quite complex and detailed. Second, estimate the value of the useful average shear stress. Suppose both the radial flux density Br and the stator surface current density Kz are sinusoidal flux waves of the form

$$B_r = \sqrt{2}B_0 \cos (p\theta - \omega t)$$

$$K_z = \sqrt{2}K_0 \cos (p\theta - \omega t)$$

Note that this assumes these two quantities are exactly in phase, or oriented to ideally produce torque, and this will produce an "optimistic" estimate. Then the average value of surface traction is

$$\langle \tau \rangle = \frac{1}{2\pi} \int_0^{2\pi} B_r k_z \, d\theta = B_0 K_0$$

This actually makes some sense in view of the empirically derived Lorentz Force Law: Given a (vector) current density and a (vector) flux density, in the absence of magnetic materials (those with permeability different from that of free space), the observed force on a conductor is

$$\vec{F} = \vec{J} \times \vec{B}$$

where $\vec{J}$ is the vector describing current density (A/m^2) and $\vec{B}$ is the magnetic flux density (T). This is actually enough to describe the forces we see in many machines, but since electric machines have permeable magnetic material and since magnetic fields produce forces on permeable material even in the absence of macroscopic currents, it is necessary to observe how force appears on such material. A suitable empirical expression for force density is

$$\vec{F} = \vec{J} \times \vec{B} - \frac{1}{2}(\vec{H} \cdot \vec{H}) \nabla \mu$$

where $\vec{H}$ is the magnetic field intensity and μ is the permeability.

Now, note that current density is the curl of magnetic field intensity, so that

$$\vec{F} = (\nabla \times \vec{H}) \times \mu\vec{H} - \frac{1}{2} (\vec{H} \cdot \vec{H}) \nabla\mu$$

$$= \mu(\nabla \times \vec{H}) \times \vec{H} - \frac{1}{2} (\vec{H} \cdot \vec{H}) \nabla\mu$$

And, since

$$(\nabla \times \vec{H}) \times \vec{H} = (\vec{H} \cdot \nabla) \vec{H} - \frac{1}{2} \nabla(\vec{H} \cdot \vec{H})$$

force density is

$$\vec{F} = \mu(\vec{H} \cdot \nabla)\vec{H} - \frac{1}{2} \mu\nabla(\vec{H} \cdot \vec{H}) - \frac{1}{2} (\vec{H} \cdot \vec{H}) \nabla\mu$$

$$= \mu(\vec{H} \cdot \nabla)\vec{H} - \nabla \left(\frac{1}{2} \mu(\vec{H} \cdot \vec{H}) \right)$$

This expression can be written by components: the component of force in the i'th dimension is

$$F_i = \mu \sum_k \left(H_k \frac{\partial}{\partial x_k} \right) H_i - \frac{\partial}{\partial x_i} \left(\frac{1}{2} \mu \sum_k H_k^2 \right)$$

Now, the divergence of magnetic flux density is

$$\nabla \cdot \vec{B} = \sum_k \frac{\partial}{\partial x_k} \mu H_k = 0$$

and

$$\mu \sum_k \left(H_k \frac{\partial}{\partial x_k} \right) H_i = \sum_k \frac{\partial}{\partial x_k} \mu H_k H_i - H_i \sum_k \frac{\partial}{\partial x_k} \mu H_k$$

but since the last term is zero, the force density is

$$F_k = \frac{\partial}{\partial x_i} \left(\mu H_i H_k - \frac{\mu}{2} \delta_{ik} \sum_n H_n^2 \right)$$

where the Kroneker delta $\delta_{ik} = 1$ if $i = k$, 0 otherwise. Note that this force density is in the form of the divergence of a tensor

$$F_k = \frac{\partial}{\partial x_i} T_{ik}$$

or

$$\vec{F} = \nabla \cdot \underline{\underline{T}}$$

In this case, force on some object that can be surrounded by a closed surface can be found by using the divergence theorem

$$\vec{f} = \int_{\text{vol}} \vec{F} \, dv = \int_{\text{vol}} \nabla \cdot \underline{\underline{T}} \, dv = \oiint \underline{\underline{T}} \cdot \vec{n} \, da$$

or, if the surface traction is $\tau_i = \Sigma_k T_{ik} n_k$, where n is the surface normal vector, then the total force in direction i is just

$$\vec{f} = \oint_s \tau_i da = \oint \sum_k T_{ik} n_k da$$

The interpretation of all of this is less difficult than the notation suggests. This field description of forces gives a simple picture of surface traction, the force per unit area on a surface. Integrate this traction over the area of some body to get the whole force on the body.

Note one more thing about this notation. Sometimes when subscripts are repeated as they are here, the summation symbol is omitted. Thus $\tau_i = \Sigma_k T_{ik} n_k = T_{ik} n_k$.

Now, in the case of a circular cylinder and torque, one can compute the circumferential force by noting that the normal vector to the cylinder is just the radial unit vector, and then the circumferential traction must simply be

$$\tau_s = \mu_0 H_r H_\theta$$

Assuming that there are no fields inside the surface of the rotor, simply integrating this over the surface gives azimuthal force, and then multiplying by radius (moment arm) gives torque. The last step is to note that, if the rotor is made of highly permeable material, the azimuthal magnetic field is equal to surface current density.

3.3 Surface Impedance of Uniform Conductors

The objective of this section is to describe the calculation of the surface impedance presented by a layer of conductive material. Two problems are considered here. The first considers a layer of *linear* material backed up by an infinitely permeable surface. This is approximately the situation presented by, for example, surface-mounted permanent magnets and is probably a decent approximation to the conduction mechanism that would be responsible for loss due to asynchronous harmonics in these machines. It is also appropriate for use in estimating losses in solid-rotor induction machines and in the poles of turbogenerators. The second problem concerns saturating ferromagnetic material.

3.3.1 Linear case

The situation and coordinate system are shown in Fig. 3.5. The conductive layer is of thickness T and has conductivity σ and permeability μ_0. To keep the mathematical expressions within bounds, assume rectilinear geometry. This assumption will present errors which are small to the extent that curvature of the problem is small compared with the wavenumbers encountered. Presume that the situation is excited, as it would be in an electric machine, by a current sheet of the form $K_z = Re\{\underline{K}e^{j(\omega t - kx)}\}$

In the conducting material, the diffusion equation must be satisfied

$$\nabla^2 \overline{H} = \mu_0 \sigma \frac{\partial \overline{H}}{\partial t}$$

In view of the boundary condition at the back surface of the material, taking that point to be $y = 0$, a general solution for the magnetic field in the material is

$$H_x = Re\{A \sinh \alpha y e^{j(\omega t - kx)}\}$$

$$H_y = Re\left\{j \frac{k}{\alpha} A \cosh \alpha y e^{j(\omega t - kx)}\right\}$$

where the coefficient α satisfies

$$\alpha^2 = j\omega\mu_0\sigma + k^2$$

and note that the coefficients above are chosen so that $\overline{H}$ has no divergence.

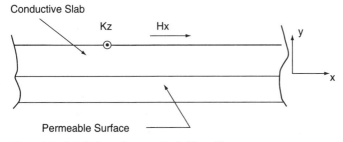

Figure 3.5 Axial view of magnetic field problem.

Note that if k is small (that is, if the wavelength of the excitation is large), this spatial coefficient α becomes

$$\alpha = \frac{1 + j}{\delta}$$

where the skin depth is

$$\delta = \sqrt{\frac{2}{\omega \mu_0 \sigma}}$$

Faraday's law

$$\nabla \times \overline{\mathbf{E}} = -\frac{\partial \overline{B}}{\partial t}$$

gives

$$\underline{E}_z = -\mu_0 \frac{\omega}{k} \underline{H}_y$$

Now, the "surface current" is just

$$\underline{K}_s = -\underline{H}_x$$

so that the equivalent surface impedance is

$$\underline{Z} = \frac{\underline{E}_z}{-\underline{H}_x} = j\mu_0 \frac{\omega}{\alpha} \coth \alpha T$$

A pair of limits are interesting here. Assuming that the wavelength is long so that k is negligible, then if αT is *small* (i.e. thin material)

$$\underline{Z} \to j\mu_0 \frac{\omega}{\alpha^2 T} = \frac{1}{\sigma T}$$

On the other hand, as $\alpha T \to \infty$

$$\underline{Z} \to \frac{1 + j}{\sigma \delta}$$

Next, it is necessary to transfer this surface impedance across the air-gap of a machine. So, assume a new coordinate system in which the surface of impedance $\underline{Z}_s$ is located at $y = 0$, and determine the impedance $\underline{Z} = -\underline{E}_z/\underline{H}_x$ at $y = g$.

In the gap, there is no current, so magnetic field can be expressed as the gradient of a scalar potential which obeys Laplace's equation

$$\overline{H} = -\nabla \psi$$

and

$$\nabla^2 \psi = 0$$

Ignoring a common factor of $e^{j(\omega t - kx)}$, $\overline{H}$ in the gap as

$$\underline{H}_x = jk(\underline{\psi}_+ e^{ky} + \underline{\psi}_- e^{-ky})$$

$$\underline{H}_y = -k(\underline{\psi}_+ e^{ky} - \underline{\psi}_- e^{-ky})$$

At the surface of the rotor

$$\underline{E}_z = -\underline{H}_x \underline{Z}_s$$

or

$$-\omega \mu_0 (\underline{\psi}_+ - \underline{\psi}_-) = jk\underline{Z}_s(\underline{\psi}_+ + \underline{\psi}_-)$$

and then, at the surface of the stator

$$\underline{Z} = -\frac{\underline{E}_z}{\underline{H}_x} = j\mu_0 \frac{\omega}{k} \frac{\underline{\psi}_+ e^{kg} - \underline{\psi}_- e^{-kg}}{\underline{\psi}_+ e^{kg} + \underline{\psi}_- e^{-kg}}$$

A bit of manipulation is required to obtain

$$\underline{Z} = j\mu_0 \frac{\omega}{k} \left\{ \frac{e^{kg}(\omega \mu_0 - jk\underline{Z}_s) - e^{-kg}(\omega \mu_0 + jk\underline{Z}_s)}{e^{kg}(\omega \mu_0 - jk\underline{Z}_s) + e^{-kg}(\omega \mu_0 + jk\underline{Z}_s)} \right\}$$

It is useful to note that, in the limit of $\underline{Z}_s \rightarrow \infty$, this expression approaches the *gap impedance*

$$\underline{Z}_g = j \frac{\omega \mu_0}{k^2 g}$$

and, if the gap is small enough that $kg \rightarrow 0$

$$\underline{Z} \rightarrow \underline{Z}_g \| \underline{Z}_s$$

3.3.2 Iron

Electric machines employ ferromagnetic materials to carry magnetic flux from and to appropriate places within the machine. Such materials have properties which are interesting, useful and problematical, and the designers of electric machines must understand these materials. The purpose of this note is to introduce the most salient properties of the kinds of magnetic materials used in electric machines.

For materials which exhibit *magnetization,* flux density is something other than $\vec{B} = \mu_0 \vec{H}$. Generally, materials are *hard* or *soft.* Hard materials are those in which the magnetization tends to be permanent, while soft materials are used in magnetic circuits of electric machines and transformers. They are related even though their uses are widely disparate.

3.3.2.1 Magnetization. It is possible to relate, in all materials, magnetic flux density to magnetic field intensity with a constitutive relationship of the form

$$\vec{B} = \mu_0 (\vec{H} + \vec{M})$$

where magnetic field intensity H and magnetization M are the two important properties. Now, linear-magnetic material magnetization is a simple linear function of magnetic field

$$\vec{M} = \chi_m \vec{H}$$

so that the flux density is also a linear function

$$\vec{B} = \mu_0 (1 + \chi_m) \vec{H}$$

Note that in the most general case, the magnetic susceptibility χ_m might be a tensor, leading to flux density being non-colinear with magnetic field intensity. But such a relationship would still be linear. Generally, this sort of complexity does not have a major effect on electric machines.

3.3.2.2 Saturation and hysteresis. In useful magnetic materials, this nice relationship is not correct and a more general view is taken. The microscopic picture is not dealt with here, except to note that the magnetization is due to the alignment of groups of magnetic dipoles — the groups often called *domaines.* There are only so many magnetic dipoles available in any given material, so that once the flux density is high enough, the material is said to saturate, and the relationship between magnetic flux density and magnetic field intensity is nonlinear.

Shown in Fig. 3.6, for example, is a "saturation curve" for a magnetic sheet steel that is sometimes used in electric machinery. Note the magnetic field intensity is on a logarithmic scale. If this were plotted on linear coordinates, the saturation would appear to be quite abrupt.

At this point, it is appropriate to note that the units used in magnetic field analysis are not always the same, nor even consistent. In

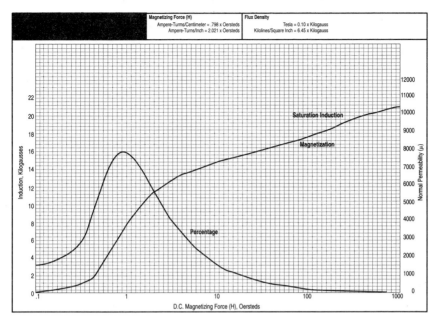

Figure 3.6 Saturation curve: Commercial M-19 silicon iron. Source: United States Steel, Applications handbook "Nonoriented Sheet Steel for Magnetic Applications."

almost all systems, the unit of flux is the weber (W), which is the same as a volt-second. In SI, the unit of flux density is the tesla (T), but many people refer to the gauss (G), which has its origin in CGS. 10,000 G = 1 T. There is an English system measure of flux density generally called kilo-lines per square inch, in which the unit of flux is the line. 10^8 lines is equal to a weber. Thus, a Tesla is 64.5 kilo-lines per square inch.

The SI and CGS units of flux density are easy to reconcile, but the units of magnetic field are a bit harder. In SI, H has dimensions of amperes/meter (or ampere-turns per meter). Often, however, magnetic field is represented as Oersteds (Oe). One Oe is the same as the magnetic field required to produce one gauss in free space. So 79.577 A/m is one Oe.

In most useful magnetic materials, the magnetic domains tend to be somewhat "sticky", and a more-than-incremental magnetic field is required to get them to move. This leads to the property called "hysteresis", both useful and problematical in many magnetic systems.

Hysteresis loops take many forms: a generalized picture of one is shown in Fig. 3.7. Salient features of the hysteresis curve are the remanent magnetization B_r and the coercive field H_c. Note that the actual loop that will be traced out is a function of field amplitude and history. Thus, there are many other "minor loops" that might be traced

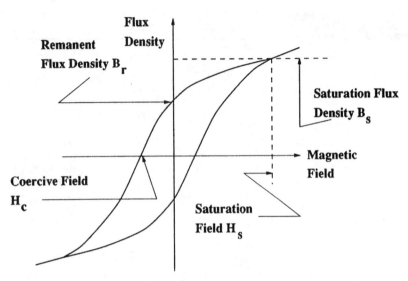

Figure 3.7 Hysteresis curve nomenclature.

out by the B-H characteristic of a piece of material, depending on just what the fields and fluxes have done and are doing.

Now, hysteresis is important for two reasons. First, it represents the mechanism for "trapping" magnetic flux in a piece of material to form a permanent magnet. We will have more to say about that anon. Second, hysteresis is a loss mechanism. To show this, consider some arbitrary chunk of material for which one can characterize an MMF and a flux

$$F = NI = \int \vec{H} \cdot d\vec{l}$$

$$\Phi = \int \frac{V}{N} dt = \iint_{\text{Area}} \vec{B} \cdot d\vec{A}$$

Energy input to the chunk of material over some period of time is

$$w = \int VI \, dt = \int F \, d\Phi = \int_t \int \vec{H} \cdot d\vec{l} \iint d\vec{B} \cdot d\vec{A} \, dt$$

Now, imagine carrying out the second (double) integral over a continuous set of surfaces which are perpendicular to the magnetic field H. (This IS possible!.) The energy becomes

$$w = \int_t \iiint \vec{H} \cdot d\vec{B} d\text{vol} \, dt$$

and, done over a complete cycle of some input waveform, that is

$$w = \iiint_{vol} W_m d\text{vol}$$

$$W_m = \oint_t \vec{H} \cdot d\vec{B}$$

That last expression simply expresses the area of the hysteresis loop for the particular cycle.

Generally, most electric machine applications use magnetic material characterized as "soft", having as narrow a hysteresis loop, and therefore as low a hysteretic loss as possible. At the other end of the spectrum are "hard" magnetic materials which are used to make permanent magnets. The terminology comes from steel, in which soft, annealed steel material tends to have narrow loops and hardened steel tends to have wider loops. However, permanent magnet technology has advanced to the point where the coercive forces, possible in even cheap ceramic magnets, far exceed those of the hardest steels.

3.3.2.3 Conduction, eddy currents and laminations.
Steel, being a metal, is an electrical conductor. Thus, when time-varying, magnetic fields pass through it, they cause eddy currents to flow, and of course those produce dissipation. In fact, for almost all applications involving "soft" iron, eddy currents are the dominant source of loss. To reduce the eddy current loss, magnetic circuits of transformers and electric machines are almost invariably laminated, or made up of relatively thin sheets of steel. To further reduce losses, the steel is alloyed with elements (often silicon) which poison the electrical conductivity.

There are several approaches to estimating the loss due to eddy currents in steel sheets and in the surface of solid iron, and it is worthwhile to look at a few of them. It should be noted that this is a "hard" problem, since the behavior of the material itself is difficult to characterize.

3.3.2.4 Complete penetration case.
Consider the problem of a stack of laminations. In particular, consider one sheet in the stack represented in Fig. 3.8. It has thickness t and conductivity σ. Assume that the "skin depth" is much greater than the sheet thickness so that magnetic field penetrates the sheet completely. Further, assume that the applied magnetic flux density is parallel to the surface of the sheets

$$\vec{B} = \vec{i}_z \text{Re}\{\sqrt{2}B_0 e^{j\omega t}\}$$

Faraday's law determines the electric field and therefore current density in the sheet. If the problem is uniform in the x- and z- directions

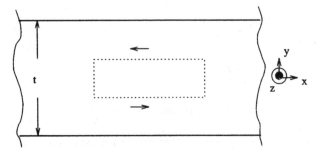

Figure 3.8 Lamination section for loss calculation.

$$\frac{\partial \underline{E}_x}{\partial y} = -j\omega_0 B_0$$

Note also that, unless there is some net transport current in the x-direction, E must be anti-symmetric about the center of the sheet. Thus, if the origin of y is in the center, electric field and current are

$$\underline{E}_x = -j\omega B_0 y$$

$$\underline{J}_x = -j\omega B_0 \sigma y$$

Local power dissipated is

$$P(y) = \omega^2 B_0^2 \sigma y^2 = \frac{|\underline{J}|^2}{\sigma}$$

To find average power dissipated, we integrate over the thickness of the lamination

$$\langle P \rangle = \frac{2}{t} \int_0^{1/2} P(y) \, dy = \frac{2}{t} \omega^2 B_0^2 \sigma \int_0^{1/2} y^2 \, dy = \frac{1}{12} \omega^2 B_0^2 t^2 \sigma$$

Pay attention to the orders of the various terms here: power is proportional to the square of flux density and to the square of frequency. It is also proportional to the square of the lamination thickness (this is average volume power dissipation).

As an aside, consider a simple magnetic circuit made of this material, with some length l and area A, so that volume of material is lA. Flux lined by a coil of N turns would be

$$\Lambda = N\Phi = NAB_0$$

and voltage is of course just $V = j\omega L$. Total power dissipated in this core would be

$$P_c = A\ell \, \frac{1}{12} \, \omega^2 B_0^2 t^2 \sigma = \frac{V^2}{R_c}$$

where the equivalent core resistance is now

$$R_c = \frac{A}{\ell} \frac{12 N_2}{\sigma t^2}$$

3.3.2.5 Eddy currents in saturating iron.

Although the same geometry holds for this pattern, consider the one-dimensional problem ($k \to 0$). The problem was worked by Agarwal[1] and MacLean.[2] They assumed that the magnetic field at the surface of the flat slab of material was sinusoidal in time and of high enough amplitude to saturate the material. This is true if the material has high permeability and the magnetic field is strong. What happens is that the impressed magnetic field saturates a region of material near the surface, leading to a magnetic flux density parallel to the surface. The depth of the region affected changes with time, and there is a separating surface (in the flat problem, this is a plane) that moves away from the top surface in response to the change in the magnetic field. An electric field is developed to move the surface, and that magnetic field drives eddy currents in the material.

Assume that the material has a perfectly rectangular magnetization curves as shown in Fig. 3.9, so that flux density in the x-direction is

$$B_x = B_0 \, \text{sign}(H_x)$$

The flux per unit width (in the z-direction) is

$$\Phi = \int_0^{-\infty} B_x \, dy$$

and Faraday's law becomes

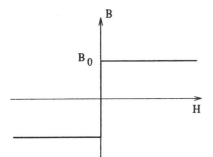

Figure 3.9 Idealized saturating characteristic.

$$E_z = \frac{\partial \Phi}{\partial t}$$

while Ampere's law in conjunction with Ohm's law is

$$\frac{\partial H_x}{\partial y} = \sigma E_z$$

MacLean[2] suggested a solution to this set in which there is a "separating surface" at depth ζ below the surface, as shown in Fig. 3.10. At any given time

$$H_x = H_s(t) \left(1 + \frac{y}{\zeta} \right)$$

$$J_z = \sigma E_z = \frac{H_s}{\zeta}$$

That is, in the region between the separating surface and the top of the material, electric field E_z is uniform and magnetic field H_x is a linear function of depth, falling from its impressed value at the surface to zero at the separating surface. Now, electric field is produced by the rate of change of flux which is

$$E_z = \frac{\partial \Phi}{\partial t} = 2B_x \frac{\partial \zeta}{\partial t}$$

Eliminating E, we have

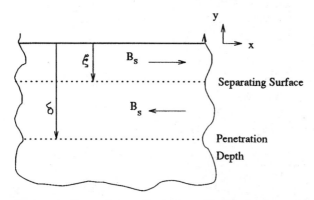

Figure 3.10 Separating surface and penetration depth.

$$2\zeta \frac{\partial \zeta}{\partial t} = \frac{H_s}{\sigma B_x}$$

and then, if the impressed magnetic field is sinusoidal, this becomes

$$\frac{d\zeta^2}{dt} = \frac{H_0}{\sigma B_0} |\sin \omega t|$$

This is easy to solve, assuming that $\zeta = 0$ at $t = 0$

$$\zeta = \sqrt{\frac{2H_0}{\omega \sigma B_0}} \sin \frac{\omega t}{2}$$

Now, the surface always moves in the downward direction, so at each half cycle, a new surface is created. The old one just stops moving at a maximum position, or penetration depth

$$\delta = \sqrt{\frac{2H_0}{\omega \sigma B_0}}$$

This penetration depth is analogous to the "skin depth" of the linear theory. However, it is an absolute penetration depth.

The resulting electric field is

$$E_z = \frac{2H_0}{\sigma \delta} \cos \frac{\omega t}{2} \qquad 0 < \omega t < \pi$$

This may be Fourier analyzed. Noting that if the impressed magnetic field is sinusoidal, only the time fundamental component of electric field is important, leading to

$$E_z = \frac{8}{3\pi} \frac{H_0}{\sigma \delta} (\cos \omega t + 2 \sin \omega t + \cdots)$$

Complex surface impedance is the ratio between the complex amplitude of electric and magnetic field, which becomes

$$\underline{Z}_s = \frac{\underline{E}_z}{\underline{H}_x} = \frac{8}{3\pi} \frac{1}{\sigma \delta} (2 + j)$$

Thus, in practical applications, nonlinear iron surfaces are treated in the same way as linear-conductive surfaces — by establishing a skin depth and assuming that current flows within that skin depth of the surface. The resistance is modified by the factor of $^{16}/_{3\pi}$ and the "power factor" of this surface is about 89% (as opposed to a linear surface, where the "power factor" is about 71%).

Agarwal[1] suggests using a value for B_0 of about 75% of the saturation flux density of the steel.

3.3.2.6 Semi-empirical method of handling iron loss. Neither of the models described so far are fully satisfactory in specifying the behavior of laminated iron, because losses are a combination of eddy current and hysteresis losses. The rather simple model employed for eddy currents is precise because of its assumption of abrupt saturation. The hysteresis model, while precise, would require an empirical determination of the size of the hysteresis loops anyway. So, one must often resort to empirical loss data. Manufacturers of lamination-steel sheets will publish data, usually in the form of curves, for many of their products. Here are a few ways of looking at the data.

A low frequency flux density vs. magnetic field ("saturation") curve was shown in Fig. 3.6. Included with that was a measure of the incremental permeability

$$\mu' = \frac{dB}{dH}$$

In *some* machine applications, either the "total" inductance (ratio of flux to MMF) or "incremental" inductance (slope of the flux to MMF curve) is required. In the limit of low frequency, these numbers may be useful.

For designing electric machines, however, a second way of looking at steel may be more useful. This is to measure the real and reactive power as a function of magnetic flux density and (sometimes) frequency. In principal, this data is immediately useful. In any well-designed electric machine the flux density in the core is distributed fairly uniformly and is not strongly affected by eddy currents, etc., in the core. Under such circumstances, one can determine the flux density in each part of the core. With that information, one can go to the published empirical data for real and reactive power and determine core loss and reactive power requirements.

Figure 3.11 shows core loss and "apparent" power per unit mass as a function of (RMS) induction for 29-gage, fully processed M-19 steel. The two left-hand curves are the most useful. "P" denotes real power while "P_a" denotes "apparent power". The use of this data is quite straightforward. If the flux density in a machine is estimated for each part of the machine and the mass of steel calculated, then with the help of this chart, a total core loss and apparent power can be estimated. Then the effect of the core may be approximated with a pair of elements in parallel with the terminals, with

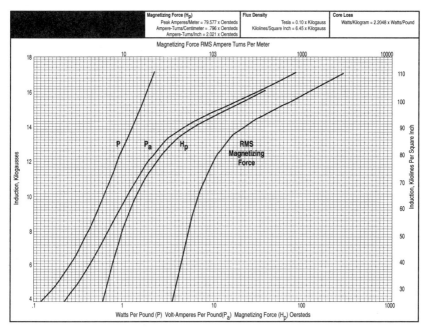

Figure 3.11 Real and apparent loss: M-19, fully processed, 29 gage. Source: United States Steel, Applications handbook "Nonoriented Sheet Steel for Magnetic Applications."

$$R_c = \frac{q|V|^2}{P}$$

$$X_c = \frac{q|V|^2}{Q}$$

$$Q = \sqrt{P_a^2 - P^2}$$

where q is the number of machine phases and V is *phase* voltage. Note that this picture is, strictly speaking, only valid for the voltage and frequency for which the flux density was calculated. But it will be approximately true for small excursions in either voltage or frequency and therefore useful for estimating voltage drop due to exciting current and such matters. In design program applications, these parameters can be pre-calculated repeatedly if necessary.

"Looking up" this data is a bit awkward for design studies, so it is often convenient to do a "curve fit" to the published data. There are a large number of possible ways of doing this. One method that has been found to work reasonably well for silicon iron is an "exponential fit"

$$P \approx P_0 \left(\frac{B}{B_0}\right)^{\epsilon_B} \left(\frac{f}{f_0}\right)^{\epsilon_F}$$

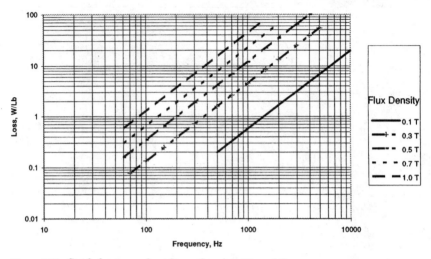

Figure 3.12 Steel sheet core loss fit vs. flux density and frequency.

This fit is appropriate if the data appears on a log-log plot to lie in approximately straight lines. Figure 3.12 shows such a fit for the same steel sheet as the other figures.

For "apparent power", the same sort of method can be used. It appears, however, that the simple exponential fit which works well for real power is inadequate, at least if relatively high inductions are to be used. This is because, as the steel saturates, the reactive component of exciting current rises rapidly. In some cases, a "double exponential" fit

$$VA \approx VA_0 \left(\frac{B}{B_0}\right)^{\epsilon_0} + VA_1 \left(\frac{B}{B_0}\right)^{\epsilon_1}$$

TABLE 3.1 Exponential Fit Parameters for Two-Sided Steel Sheets, 29 Ga., Fully Processed

		M-19	M-36
Base Flux Density	B_0	1 T	1 T
Base Frequency	f_0	60 Hz	60 Hz
Base Power (w/lb)	P_0	0.59	0.67
Flux Exponent	ϵ_B	1.88	1.86
Frequency Exponent	ϵ_F	1.53	1.48
Base Apparent Power 1	VA_0	1.08	1.33
Base Apparent Power 2	VA_1	.0144	.0119
Flux Exponent	ϵ_0	1.70	2.01
Flux Exponent	ϵ_1	16.1	17.2

gives adequate results. To first order the reactive component of exciting current will be linear in frequency. Parameters for two commonly used electrical sheet steels are shown in Table 3.1.

3.4 References

1. Agarwal, P.D. "Eddy-Current Losses in Solid and Laminated Iron." *Trans. AIEE* 78. (1959):169–171.
2. MacLean, W. "Theory of Strong Electromagnetic Waves in Massive Iron." *Journal of Applied Physics* 25, no. 10 (October 1954).

Chapter

4

Induction Motors

J. Kirtley and N. Ghai

4.1 Theory of the Polyphase Inductor Motor

4.1.1 Principle of operation

An induction motor is an electric transformer whose magnetic circuit is separated by an air gap into two relatively movable portions, one carrying the primary and the other the secondary winding. Alternating current supplied to the primary winding from an electric power system induces an opposing current in the secondary winding, when the latter is short-circuited or closed through an external impedance. Relative motion between the primary and secondary structures is produced by the electromagnetic forces corresponding to the power thus transferred across the air gap by induction. The essential feature which distinguishes the induction machine from other types of electric motors is that the secondary currents are created solely by induction, as in a transformer, instead of being supplied by a dc exciter or other external power source, as in synchronous and dc machines.

Induction motors are classified as squirrel-cage motors and wound-rotor motors. The secondary windings on the rotors of squirrel-cage motors are assembled from conductor bars short-circuited by end rings or are cast in place from a conductive alloy. The secondary windings of wound-rotor motors are wound with discrete conductors with the same number of poles as the primary winding on the stator. The rotor windings are terminated on slip rings on the motor shaft. The windings can be short-circuited by brushes bearing on the slip rings, or they can be connected to resistors or solid-state converters for starting and speed control.

41

4.1.2 Construction features

The normal structure of an induction motor consists of a cylindrical rotor carrying the secondary winding in slots on its outer periphery and an encircling annular core of laminated steel carrying the primary winding in slots on its inner periphery. The primary winding is commonly arranged for three-phase power supply, with three sets of exactly similar multipolar coil groups spaced one-third of a pole pitch apart. The superposition of the three stationary, but alternating, magnetic fields produced by the three-phase windings produces a sinusoidally distributed magnetic field revolving in synchronism with the power-supply frequency, the time of travel of the field crest from one phase winding to the next being fixed by the time interval between the reaching of their crest values by the corresponding phase currents. The direction of rotation is fixed by the time sequence of the currents in successive phase belts and so many be reversed by reversing the connections of one phase of a two- or three-phase motor.

Figure 4.1 shows the cross section of a typical polyphase induction motor, having in this case a three-phase four-pole primary winding with 36 stator and 28 rotor slots. The primary winding is composed of 36 identical coils, each spanning eight teeth, one less than the nine teeth in one pole pitch. The winding is therefore said to have ⁸⁄₉ pitch. As there are three primary slots per pole per phase, phase A comprises four equally spaced "phase belts," each consisting of three consecutive coils connected in series. Owing to the short pitch, the top and bottom coil sides of each phase overlap the next phase on either side. The

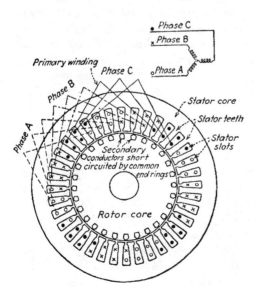

Figure 4.1 Section of squirrel-cage induction motor, 3-phase, 4-pole, 8/9-pitch stator winding.

rotor, or secondary, winding consists merely of 28 identical copper or cast-aluminum bars solidly connected to conducting end rings on each end, thus forming a "squirrel-cage" structure. Both rotor and stator cores are usually built on silicon-steel laminations, with partly closed slots, to obtain the greatest possible peripheral area for carrying magnetic flux across the air gap.

4.1.3   The revolving field

The key to understanding the induction motor is a thorough comprehension of the revolving magnetic field.

The rectangular wave in Fig. 4.2 represents the mmf, or field distribution, produced by a single full-pitch coil, carrying H At. The air gap between stator and rotor is assumed to be uniform, and the effects of slot openings are neglected. To calculate the resultant field produced by the entire winding, it is most convenient to analyze the field of each single coil into its space-harmonic components, as indicated in Fig. 4.2 or expressed by the following equation:

$$H(x) = \frac{4H}{\pi} \left(\sin x + \frac{1}{3} \sin 3x + \frac{1}{5} \sin 5x + \frac{1}{7} \sin 7x + \cdots \right) \quad (4.1)$$

When two such fields produced by coils in adjacent slots are superposed, the two fundamental sine-wave components will be displaced by the slot angle θ, the third-harmonic components by the angle 3θ, the fifth harmonics by the angle 5θ, etc. Thus, the higher space-harmonic components in the resultant field are relatively much reduced as compared with the fundamental. By this effect of distributing the winding in several slots for each phase belt, and because of the further reductions due to fractional pitch and to phase connections, the space-harmonic fields in a normal motor are reduced to negligible values, leaving only the fundamental sine-wave components to be considered in determining the operating characteristics.

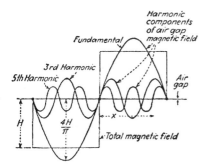

Figure 4.2 Magnetic field produced by a single coil.

The alternating current flowing in the winding of each phase therefore produces a sine-wave distribution of magnetic flux around the periphery, stationary in space but varying sinusoidally in time in synchronism with the supply frequencies. Referring to Fig. 4.3, the field of phase A at an angular distance x from the phase axis may be represented as an alternating phasor $I \cos x \cos \omega t$ but may equally well be considered as the resultant of two phasors constant in magnitude but revolving in opposite directions as synchronous speed

$$I \cos x \cos \omega t = \frac{I}{2} [\cos(x - \omega t) + \cos(x + \omega t)] \qquad (4.2)$$

Each of the right-hand terms in this equation represents a sine-wave field revolving at the uniform rate of one pole pitch, or 180 electrical degrees, in the time of each half cycle of the supply frequency. The synchronous speed N_s of a motor is therefore given by

$$N_s = \frac{120f}{P} \qquad \text{r/min} \qquad (4.3)$$

where f = line frequency in hertz and P = number of poles of the winding.

Considering phase A alone (Fig. 4.4), two revolving fields will coincide along the phase center line at the instant its current is a maxi-

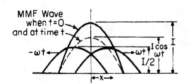

Figure 4.3 Resolution of alternating wave into two constant-magnitude waves revolving in opposite directions.

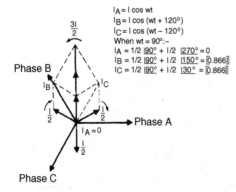

$I_A = I \cos wt$
$I_B = I \cos (wt + 120°)$
$I_C = I \cos (wt - 120°)$
When wt = 90°:-
$I_A = 1/2 \underline{|90°} + 1/2 \underline{|270°} = 0$
$I_B = 1/2 \underline{|90°} + 1/2 \underline{|150°} = \boxed{0.866|}$
$I_C = 1/2 \underline{|90°} + 1/2 \underline{|30°} = \boxed{0.866|}$

Figure 4.4 Resolution of alternating emf of each phase into oppositely revolving constant-magnitude components shown at instant when phase A current is zero ($\omega\tau = 90°$).

mum. One-third of a cycle later, each will have traveled 120 electrical degrees, one forward and the other backward, the former lining up with the axis of phase B and the latter with the axis of phase C. But at this moment, the current in phase B is a maximum, so that the forward-revolving B field coincides with the forward A field, and these two continue to revolve together. The backward B field is 240° behind the backward A field, and these two remain at this angle, as they continue to revolve. After another third of a cycle, the forward A and B fields will reach the phase C axis, at the same moment that phase C current becomes a maximum. Hence, the forward fields of all three phases are directly additive, and together they create a constant-magnitude sine-wave-shaped synchronously revolving field with a crest value ³⁄₂ the maximum instantaneous value of the alternating field due to one phase alone. The backward-revolving fields of the three phases are separated by 120°, and their resultant is therefore zero so long as the three-phase currents are balanced in both magnitude and phase.

If a two-phase motor is considered, it will have two 90° phase belts per pole instead of three 60° phase belts, and a similar analysis shows that it will have a forward-revolving constant-magnitude field with a crest value equal to the peak value of one phase alone and will have zero backward-revolving fundamental field. A single-phase motor will have equal forward and backward fields and so will have no tendency to start unless one of the fields is suppressed or modified in some way.

While the space-harmonic-field components are usually negligible in standard motors, it is important to the designer to recognize that there will always be residual harmonic-field values which may cause torque irregularities and extra losses if they are not minimized by an adequate number of slots and correct winding distribution. An analysis similar to that given for the fundamental field shows that in all cases the harmonic fields corresponding to the number of primary slots (seventh and 19th in a nine-slot-per-pole motor) are important and that the fifth and seventh harmonics on three-phase, or third and fifth on two-phase, may also be important.

The third-harmonic fields and all multiples of the third are zero in a three-phase motor, since the mmf's of the three phases are 120° apart for both backward and forward components of all of them. Finally, therefore, a three-phase motor has the following distinct fields:

a. The fundamental field with P poles revolving forward at speed N_s.

b. A fifth-harmonic field with five P poles revolving backward at speed $N_s/5$.

c. A seventh-harmonic field with seven P poles revolving forward at speed $N_s/7$.

d. Similar 13th, 19th, 25th, etc., forward-revolving and 11th, 17th, 23rd, etc., backward revolving harmonic fields.

Figure 4.5 shows a test speed-torque curve obtained on a two-phase squirrel-cage induction motor with straight (unspiraled) slots. The torque dips due to three of the forward-revolving fields are clearly indicated.

4.1.4 Torque, slip, and rotor impedance

When the rotor is stationary, the revolving magnetic field cuts the short-circuited secondary conductors at synchronous speed and induces in them line-frequency currents. To supply the secondary IR voltage drop, there must be a component of voltage in time phase with the secondary current, and the secondary current, therefore, must lag in space position behind the revolving air-gap field. A torque is then produced corresponding to the product of the air-gap field by the secondary current times the sine of the angle of their space-phase displacement.

At standstill, the secondary current is equal to the air-gap voltage divided by the secondary impedance at line frequency, or

$$I_2 = \frac{E_2}{Z_2} = \frac{E_2}{R_2 + jX_2} \tag{4.4}$$

where R_2 = effective secondary resistance and X_2 = secondary leakage reactance at primary frequency.

The speed at which the magnetic field cuts the secondary conductors is equal to the difference between the synchronous speed and the actual rotor speed. The ratio of the speed of the field relative to the rotor, to synchronous speed, is called the slip s

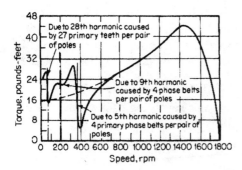

Figure 4.5 Speed-torque curve of 2-phase motor showing harmonic torque.

$$s = \frac{N_s - N}{N_s}$$

or $$N = (1 - s)N_s \qquad (4.5)$$

where N = actual and N_s = synchronous rotor speed.

As the rotor speeds up, with a given air-gap field, the secondary induced voltage and frequency both decrease in proportion to s. Thus, the secondary voltage becomes sE_2, and the secondary impedance $R_2 + jsX_2$, or

$$I_2 = \frac{sE_2}{R_2 + jsX_2} = \frac{E_2}{(R_2/s) + jX_2} \qquad (4.6)$$

The only way that the primary is affected by a change in the rotor speed, therefore, is that the secondary resistance as viewed from the primary varies inversely with the slip.

In practice, the effective secondary resistance and reactance, or R_2 and X_2, change with the secondary frequency, owing to the varying "skin effect," or current shifting into the outer portion of the conductors, when the frequency is high. This effect is employed to make the resistance, and therefore the torque, higher at starting and low motor speeds, by providing a double cage, or deep-bar construction, as shown in Fig. 4.6. The leakage flux between the outer and inner bars makes the inner-bar reactance high, so that most of the current must flow in the outer bars or at the top of a deep-bar at standstill, when frequency is high. At full speed, the secondary frequency is very low, and most of the current flows in the inner bars, or all over the cross section of a deep bar, owing to their lower resistance.

4.1.5 Analysis of induction motors

Induction motors are analyzed by three methods: (1) circle diagram; (2) equivalent circuit; (3) coupled-circuit, generalized machine. The

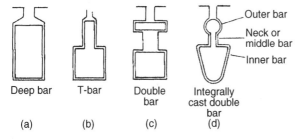

Deep bar   T-bar   Double   Integrally
                      bar   cast double
                               bar
(a)   (b)   (c)   (d)

Figure 4.6 Alternating forms of squirrel-cage rotor bars.

first two methods are used for steady-state conditions; the third method is used for transient conditions. The circle diagram is convenient for visualizing overall performance but is too inaccurate for detailed calculations and design. The magnetizing current is not constant, but decreases with load because of the primary impedance drop. All of the circuit constants vary over the operating range due to magnetic saturation and skin effect. The equivalent circuit method predominates for analysis and design under steady-state conditions. The impedances can be adjusted to fit the conditions at each calculation point.

4.1.6 Circle diagram

The voltage-current relations of the polyphase induction machine are roughly indicated by the circuit of Fig. 4.7. The magnetizing current I_M proportional to the voltage and lagging 90° in phase is nearly constant over the operating range, while the load current varies inversely with the sum of primary and secondary impedances. As the slip s increases, the load current and its angle of lag behind the voltage both increase, following a nearly circular locus. Thus, the circle diagram (Fig. 4.8) provides a picture of the motor behavior.

The data needed to construct the diagram are the magnitude of the no-load current ON and of the blocked-rotor current OS and their phase angles with reference to the line voltage OE. A circle with its

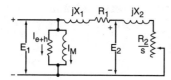

Figure 4.7 Equivalent circuit for circle diagram.

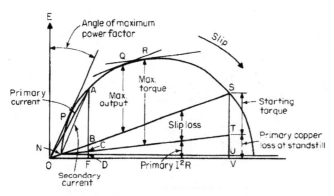

Figure 4.8 Circle diagram of polyphase induction motor.

center on the line NU at right angles to OE is drawn to pass through N and S. Each line on the diagram can be measured directly in amperes, but it also represents voltamperes or power, when multiplied by the phase voltage times number of phases. The line VS drawn parallel to OE represents the total motor power input with blocked rotor, and on the same scale VT represents the corresponding primary I^2R loss. Then ST represents the power input to the rotor at standstill, which, divided by the synchronous speed, gives the starting torque.

At any load point A, OA is the primary current, NA the secondary current, and AF the motor power input. The motor output power is AB, the torque × (synchronous speed) is AC, the secondary I^2R loss is BC, primary I^2R loss CD, and no-load copper loss plus core loss DF. The maximum power-factor point is P, located by drawing a tangent to the circle from O. The maximum output and maximum torque points are similarly located at Q and R by tangent lines parallel to NS and NT, respectively.

The diameter of the circle is equal to the voltage divided by the standstill reactance or to the blocked-rotor current value on the assumption of zero resistance in both windings. The maximum torque of the motor, measured in kilowatts at synchronous speed, is equal to a little less than the radius of the circle multiplied by the voltage OE.

4.1.7 Equivalent circuit

Figure 4.9 shows the polyphase motor circuit usually employed for accurate work. The advantages of this circuit over the circle-diagram method are that it facilitates the derivation of simple formulas, charts, or computer programs for calculating torque, power factor, and other motor characteristics and that it enables impedance changes due to saturation or multiple squirrel cages to be readily taken into account.

Inspection of the circuit reveals several simple relationships which are useful for estimating purposes. The maximum current occurs at standstill and is somewhat less than E/X. Maximum torque occurs

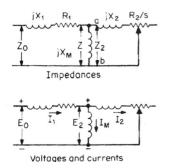

Figure 4.9 Equivalent circuit of polyphase induction motor.

when $s = R_2/X$, approximately, at which point the current is roughly 70% of the standstill current. Hence, the maximum torque is approximately equal to $E^2/2X$. This gives the basic rule that the percent maximum torque of a low-slip polyphase motor at a constant impressed voltage is about half the percent starting current.

By choosing the value of R_2, the slip at which maximum torque occurs can be fixed at any desired value. The maximum-torque value itself is affected, not by changes in R_2, but only by changes in X and to a slight degree by changes in X_M.

The magnetizing reactance X_M is usually eight or more times as great as X, while R_1 and R_2 are usually much smaller than X, except in the case of special motors designed for frequent-starting service.

Definitions of Equivalent-Circuit Constants

Unless otherwise noted, all quantities except watts, torque, and power output are per phase for two-phase motors and per phase Y for three-phase motors.

E_0 = impressed voltage (volts) = line voltage $\div \sqrt{3}$ for three-phase motors
I_1 = primary current (amperes)
I_2 = secondary current in primary terms (amperes)
I_M = magnetizing current (amperes)
R_1 = primary resistance (ohms)
R_2 = secondary resistance in primary terms (ohms)
R_0 = resistance at primary terminals (ohms)
X_1 = primary leakage reactance (ohms)
X_2 = secondary leakage reactance (ohms)
$X = X_1 + X_2$
X_0 = reactance at primary terminals (ohms)
X_M = magnetizing reactance (ohms)
Z_1 = primary impedance (ohms)
Z_2 = secondary impedance in primary terms (ohms)
Z_0 = impedance at primary terminals (ohms)
Z = combined secondary and magnetizing impedance (ohms)
s = slip (expressed as a fraction of synchronous speed)
N = synchronous speed (revolutions per minute)
m = number of phases
f = rated frequency (hertz)
f_t = frequency used in locked-rotor test
T = torque (foot-pounds)
W_0 = watts input
W_H = core loss (watts)
W_F = friction and windage (watts)
W_{RL} = running light watts input
W_s = stray-load loss (watts)

The equivalent circuit of Fig. 4.9 shows that the total power P_{g1} transferred across the air-gap from the stator is

$$P_{g1} = mI_2^2 \frac{R_2}{s} \qquad (4.7)$$

The total rotor copper loss is evidently

$$\text{Rotor copper loss} = mI_2^2R_2 \qquad (4.8)$$

The internal mechanical power P developed by the motor is therefore

$$P = P_{g1} - \text{rotor copper loss} = mI_2^2\frac{R_2}{s} - mI_2^2R_2$$

$$= mI_2^2R_2\frac{1-s}{s}$$

$$= (1-s)P_{g1} \qquad (4.9)$$

We see, then, that of the total power delivered to the motor, the fraction $1-s$ is converted to mechanical power and the fraction s is dissipated as rotor-circuit copper loss. The internal mechanical power per stator phase is equal to the power absorbed by the resistance $R_2(1-s)/s$. The internal electromagnetic torque T corresponding to the internal power P can be obtained by recalling that mechanical power equals torque times angular velocity. Thus, when ω_s is the synchronous angular velocity of the rotor in mechanical radians per second

$$P = (1-s)\omega_s T \qquad (4.10)$$

with T in newton-meters. By use of Eq. (4.9),

$$T = \frac{1}{\omega_s} mI_2^2\frac{R_2}{s} \qquad (4.11)$$

For T in foot-pounds and N_s in revolutions per minute

$$T = \frac{7.04}{N_s} mI_2^2\frac{R_2}{s} \qquad (4.12)$$

4.1.8 Torque and power

Considerable simplification results from application of Thévenin's network theorem to the induction-motor equivalent circuit. Thévenin's theorem permits the replacement of any network of linear circuit elements and constant phasor voltage sources, as viewed from two terminals by a single phasor voltage source E in series with a single impedance Z. The voltage E is that appearing across terminals a and b of the original network when these terminals are open-circuited; the impedance Z is that viewed from the same terminals when all voltage sources within the network are short-circuited. For application to the induction-motor equivalent circuit, points a and b are taken as those so designated in Fig. 4.9. The equivalent circuit then assumes the forms given in Fig. 4.10. So far as phenomena to the right of points a and b are concerned, the circuits of Figs. 4.9 and 4.10 are identical when the voltage V_{1a} and the impedance $R_1 + jX_1$ have the proper values. According to Thévenin's theorem, the equivalent source voltage V_{1a} is the voltage that would appear across terminals a and b of Fig. 4.9 with the rotor circuits open and is

$$V_{1a} = E_0 - I_0(R_1 + jX_1) = E_0 \frac{jX_M}{R_1 + jX_{11}} \qquad (4.13)$$

where I_M is the zero-load exciting current and

$$X_{11} = X_1 + X_M$$

is the self-reactance of the stator per phase and very nearly equals the reactive component of the zero-load motor impedance. For most induction motors, negligible error results from neglecting the stator resistance in Eq. (4.13). The Thévenin equivalent stator impedance $R_1 + jX_1$ is the impedance between terminals a and b of Fig. 4.9, viewed toward the source with the source voltage short-circuited, and therefore is

$$\overline{R}_1 + j\overline{X}_1 = R_1 + jX_1 \qquad \text{in parallel with } jX_M$$

From the Thévenin equivalent circuit (Fig. 4.10) and the torque expression (Eq. 4.11), it can be seen that

$$T = \frac{1}{\omega_2} \frac{mV_{1a}^2(R_2/s)}{(\overline{R}_1 + R_2/s)^2 + (\overline{X}_1 + X_2)^2} \qquad (4.14)$$

The slip at maximum torque, $s_{\max T}$, is obtained by differentiating Eq. (4.14) with respect to s and equating to zero

$$s_{\max T} = \frac{R_2}{\sqrt{\overline{R}_1^2 + (\overline{X}_1 + X_2)^2}}$$

The corresponding maximum torque is

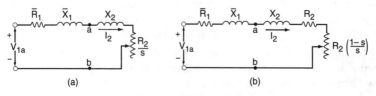

Figure 4.10 Induction-motor equivalent circuits simplified by Thévenin's theorem.

$$T_{max} = \frac{1}{\omega_s} \frac{0.5mV_{1a}^2}{R_1 + \sqrt{R_1^2 + (\overline{X}_1 + X_2)^2}}$$

4.1.9 Service factor

General-purpose fractional- and integral-horsepower motors are given a "service factor," which allows the motor to deliver greater than rated horsepower, without damaging its insulation system. The motor is operated at rated voltage and frequency. The standard service factors are 1.4 for motors rated $\frac{1}{20}$ to $\frac{1}{8}$ hp; 1.35 for $\frac{1}{6}$ to $\frac{1}{3}$ hp; 1.25 for $\frac{1}{2}$ hp to the frame size for 1 hp at 3600 r/min. For all larger motors through 200 hp, the service factor is 1.15. For 250 to 500 hp, the service factor is 1.0.

4.1.10 Efficiency and power factor

Typical full-load efficiencies and power factors of standard Design B squirrel cage induction motors are given in Figs. 4.11, and 4.12, respectively. The efficiencies of Design A motors are generally slightly lower, and those of Design D motors considerably lower. The power

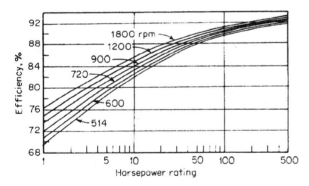

Figure 4.11 Typical full-load efficiencies of Design B squirrel-cage motors.

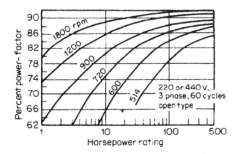

Figure 4.12 Typical full-load power factor of Design B squirrel-cage motors.

factors of Design A squirrel-cage induction motors are slightly higher, and those of Design C are slightly lower. Energy-efficient motors are those whose design is optimized to reduce losses. Comparative efficiencies of standard and energy-efficient motors of NEMA Design B are shown in Fig. 4.13.

4.1.11 Full-load current

With the efficiency and power factor of a three-phase motor known, its full-load current may be calculated from the formula

$$\text{Full-load current} = \frac{746 \times \text{hp rating}}{1.73 \times \text{efficiency} \times \text{pf} \times \text{voltage}}$$

where the efficiency and power factor are expressed as decimals.

4.1.12 Torques and starting currents

Starting and breakdown torques of common Design A, B, and C squirrel-cage induction motors are given in Table 4.1. The maximum breakdown torque for wound-rotor motors is 200% of full-load torque. The starting torque and starting current of wound-rotor motors vary with the amount of external resistance in the secondary circuit.

The starting kVA of a squirrel-cage motor is indicated by a code letter stamped on the nameplate. Table 4.2 gives the corresponding

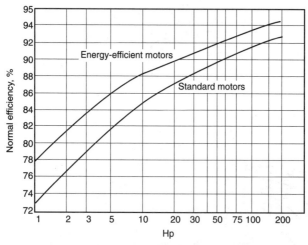

Figure 4.13 Nominal efficiencies for NEMA Design B 4-pole motors, 1800 r/min; standard vs. energy-efficient motors.

TABLE 4.1   Torques—Polyphase Induction Motors
(Percent of full-load torque)

Rpm	3,600		1,800				1,200				900				720	
Torque	LR	BD	LR	LR	BD	BD	LR	LR	BD	BD	LR	LR	BD	BD	LR	BD
Design	AB	B	AB	C	B	C	AB	C	B	C	AB	C	B	C	AB	B
½ hp	...	...	...	...	...	...	...	...	...	...	150	...	250	...	150	200
¾ hp	...	...	...	...	...	...	175	...	275	...	150	...	250	...	150	200
1 hp	...	...	275	...	300	...	175	...	275	...	150	...	250	...	150	200
1½ hp	175	275	265	...	300	...	175	...	275	...	150	...	250	...	150	200
2 hp	175	250	250	...	275	...	175	...	250	...	150	...	225	...	145	200
3 hp	175	250	250	...	275	...	175	250	250	225	150	225	225	200	135	200
5 hp	150	225	185	250	225	200	160	250	225	200	130	225	225	200	130	200
7½ hp	150	215	175	250	215	190	150	225	215	190	125	200	215	190	120	200
10 hp	150	200	175	250	200	190	140	225	200	190	125	200	200	190	120	200
15 hp	150	200	165	225	200	190	135	200	200	190	125	200	200	190	120	200
20 hp	150	200	150	200	200	190	135	200	200	190	125	200	200	190	120	200
25 hp	150	200	150	200	200	190	135	200	200	190	125	200	200	190	120	200
30 hp	150	200	150	200	200	190	135	200	200	190	125	200	200	190	120	200
40–200 hp	*	200	*	200	200	190	*	200	200	190	125	200	200	190	120	200

NOTE. LR = locked-rotor torque; BD = breakdown torque; A, B, and C refer to Design A, etc.
*Progressively lower values for these larger ratings.

TABLE 4.2   Locked-Rotor kVA for Code-Letter Motors

Code Letter*	Kva per Hp, with Locked Rotor	Code Letter*	Kva per Hp, with Locked Rotor
A	0–3.14	L	9.0–9.99
B	3.15–3.54	M	10.0–11.19
C	3.55–3.99	N	11.2–12.49
D	4.0–4.49	P	12.5–13.99
E	4.5–4.99	R	14.0–15.99
F	5.0–5.59	S	16.0–17.99
G	5.6–6.29	T	18.0–19.99
H	6.3–7.09	U	20.0–22.39
J	7.1–7.99	V	22.4 and up
K	8.0–8.99		

*National Electrical Code.

kVA for each code letter, and the locked-motor current can be determined from

$$\text{Locked-rotor current} = \frac{\text{kVA/hp} \times \text{hp} \times 1000}{k \times \text{line volts}}$$

where $k = 1$ for single-phase, and $k = 1.73$ for three-phase.

Maximum locked-rotor current for Design B, C, and D three-phase motors has been standardized as shown in Table 4.3 for 230 V. The

TABLE 4.3 Locked-Rotor Current for Three-phase Motors at 230 V

Rated horsepower	Classes B, C, D, amperes	Rated horsepower	Classes B, C, D, amperes	Rated horsepower	Classes B, C, D, amperes	Rated horsepower	Classes B, C, D, amperes	Rated horsepower	Class B amperes
1	30	7½	127	30	435	100	1450	250	3650
1½	40	10	162	40	580	125	1815	300	4400
2	50	15	232	50	725	150	2170	350	5100
3	64	20	290	60	870	200	2900	400	5800
5	92	25	365	75	1085			450	6500
								500	7250

starting current for motors designed for other voltages is inversely proportional to the voltage.

4.2 Design Evaluation of Induction Motors

4.2.1 Introduction

The objective of this section is an understanding of induction motors from first principles. The analysis developed here results in the equivalent circuit models discussed in section 4.1, along with more complex models which give a fuller explanation of some of the loss mechanisms in induction machines.

The analysis starts with a circuit theoretical point of view of wound-rotor induction machines, which is used to explain the basics of induction machine operation. Following that, a model for squirrel-cage machines is developed. Finally, analysis of solid-rotor machines and mixed solid rotor plus squirrel-cage machines is developed.

The view taken in this chapter is relentlessly classical. All of the elements used here are calculated from first principles, with only limited use of numerical or empirical methods. While this may seem to be seriously limiting, it serves the basic objective, which is to achieve an understanding of how these machines work. Once that understanding exists, it is possible to employ further sophisticated methods of analysis to achieve more results for those elements of the machines which do not lend themselves to simple analysis.

An elementary picture of the induction machine is shown in Fig. 4.14. The rotor and stator are coaxial. The stator has a polyphase winding in slots. The rotor has either a winding or a cage, also in

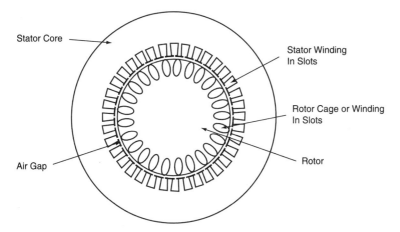

Figure 4.14 Axial view of an induction machine.

slots. Generally, this analysis is carried out assuming three phases. As with many systems, this generalizes to different numbers of phases with little difficulty.

4.2.2 Induction motor transformer model

This analysis is limited to polyphase motors. The induction machine has two electrically active elements: a rotor and a stator. In normal operation, the stator is excited by alternating voltage. The stator excitation creates a magnetic field in the form of a rotating, or traveling wave, which induces currents in the circuits of the rotor. Those currents, in turn, interact with the traveling wave to produce torque. To start the analysis of this machine, assume that both the rotor and the stator can be described by balanced, three-phase windings. The two sets are, of course, coupled by mutual inductances which are dependent on rotor position. Stator fluxes are $(\lambda_a, \lambda_b, \lambda_c)$ and rotor fluxes are $(\lambda_A, \lambda_B, \lambda_C)$. The flux vs. current relationship is given by

$$
\begin{bmatrix} \lambda_a \\ \lambda_b \\ \lambda_c \\ \lambda_A \\ \lambda_B \\ \lambda_C \end{bmatrix} = \begin{bmatrix} \underline{L}_S & \underline{M}_{SR} \\ \underline{M}_{SR}^T & \underline{L}_R \end{bmatrix} \begin{bmatrix} i_a \\ i_b \\ i_c \\ i_A \\ i_B \\ i_C \end{bmatrix}
\tag{4.15}
$$

where the component matrices are

$$
\underline{L}_S = \begin{bmatrix} L_a & L_{ab} & L_{ab} \\ L_{ab} & L_a & L_{ab} \\ L_{ab} & L_{ab} & L_a \end{bmatrix}
\tag{4.16}
$$

$$
\underline{L}_R = \begin{bmatrix} L_A & L_{AB} & L_{AB} \\ L_{AB} & L_A & L_{AB} \\ L_{AB} & L_{AB} & L_A \end{bmatrix}
\tag{4.17}
$$

The mutual inductance part of (1) is a circulant matrix

$$
\underline{M}_{SR} = \begin{bmatrix} M \cos(p\theta) & M \cos\left(p\theta + \dfrac{2\pi}{3}\right) & M \cos\left(p\theta - \dfrac{2\pi}{3}\right) \\ M \cos\left(p\theta - \dfrac{2\pi}{3}\right) & M \cos(p\theta) & M \cos\left(p\theta + \dfrac{2\pi}{3}\right) \\ M \cos\left(p\theta + \dfrac{2\pi}{3}\right) & M \cos\left(p\theta - \dfrac{2\pi}{3}\right) & M \cos(p\theta) \end{bmatrix}
\tag{4.18}
$$

To carry the analysis further, it is necessary to make some assumptions regarding operation. To start, assume balanced currents in both the stator and rotor

$$i_a = I_S \cos(\omega t)$$

$$i_b = I_S \cos\left(\omega t - \frac{2\pi}{3}\right)$$

$$i_c = I_S \cos\left(\omega t + \frac{2\pi}{3}\right) \qquad (4.19)$$

$$i_A = I_R \cos(\omega_R t + \xi_R)$$

$$i_B = I_R \cos\left(\omega_R t + \xi_R - \frac{2\pi}{3}\right)$$

$$i_C = I_R \cos\left(\omega_R t + \xi_R + \frac{2\pi}{3}\right) \qquad (4.20)$$

The rotor position θ can be described by

$$\theta = \omega_m t + \theta_0 \qquad (4.21)$$

Under these assumptions, the form of stator fluxes may be calculated. As it turns out, one need write out only the expressions for λ_a and λ_A to see what is going on

$$\lambda_a = (L_a - L_{ab})I_s \cos(\omega t) + MI_R \left[\cos(\omega_R t + \xi_R) \cos p(\omega_m + \theta_0) \right.$$

$$+ \cos\left(\omega_R t + \xi_R + \frac{2\pi}{3}\right) \cos\left(p(\omega_m t + \theta_0) - \frac{2\pi}{3}\right)$$

$$\left. + \cos\left(\omega_R t + \xi_R - \frac{2\pi}{3}\right) \cos\left(p(\omega_m t + \theta_0) + \frac{2\pi}{3}\right) \right] \qquad (4.22)$$

which, after reducing some of the trig expressions, becomes

$$\lambda_a = (L_a - L_{ab})I_s \cos(\omega t)$$

$$+ \frac{3}{2} MI_R \cos((p\omega_m + \omega_R)t + \xi_R + p\theta_0) \qquad (4.23)$$

Doing the same thing for the rotor phase, A yields

$$\lambda_A = MI_s \left[\cos p(\omega_m t + \theta_0) \cos(\omega t)) \right.$$

$$+ \cos \left(p(\omega_m t + \theta_0) - \frac{2\pi}{3} \right) \cos \left(\omega t - \frac{2\pi}{3} \right)$$

$$+ \cos \left(p(\omega_m t + \theta_0) + \frac{2\pi}{3} \right) \cos \left(\omega t + \frac{2\pi}{3} \right) \right]$$

$$+ (L_A - L_{AB})I_R \cos(\omega_R t + \xi_R) \tag{4.24}$$

This last expression is, after manipulating

$$\lambda_A = \frac{3}{2} MI_s \cos((\omega - p\omega_m)t - p\theta_0)$$

$$+ (L_A - L_{AB})I_R \cos(\omega_R t + \xi_R) \tag{4.25}$$

The two expressions give expressions for fluxes in the armature and rotor windings in terms of currents in the same two windings, assuming that both current distributions are sinusoidal in time and space and represent balanced distributions. The next step is to make another assumption—that the stator and rotor frequencies match through rotor rotation. That is

$$\omega - p\omega_m = \omega_R \tag{4.26}$$

It is important to keep straight the different frequencies here

ω is stator electrical frequency

ω_R is rotor electrical frequency

ω_m is mechanical rotation speed

so that $p\omega_m$ is *electrical* rotation speed.

To refer rotor quantities to the stator frame (i.e. non-rotating), and to work in complex amplitudes, the following definitions are made

$$\lambda_a = Re(\underline{\Lambda}_a e^{j\omega t}) \tag{4.27}$$

$$\lambda_A = Re(\underline{\Lambda}_A e^{j\omega_R t}) \tag{4.28}$$

$$i_a = Re(\underline{I}_a e^{j\omega t}) \tag{4.29}$$

$$i_A = Re(\underline{L}_A e^{j\omega_R t}) \tag{4.30}$$

With these definitions, the complex amplitudes embodied in Eq. (4.22) and (4.23) become

$$\underline{\Lambda}_a = L_S\underline{I}_a + \frac{3}{2}M\underline{I}_A e^{j(\xi_R + p\theta_0)} \tag{4.31}$$

$$\underline{\Lambda}_A = \frac{3}{2}M\underline{I}_a e^{-jp\theta_0} + L_R\underline{I}_A e^{j\xi_R} \tag{4.32}$$

There are two-phase angles embedded in these Equations: θ_0 which describes the rotor physical-phase angle with respect to stator current; and ξ_R which describes phase angle of rotor currents with respect to stator currents. Two new rotor variables are

$$\underline{\Lambda}_{AR} = \underline{\Lambda}_A e^{jp\theta_0} \tag{4.33}$$

$$\underline{I}_{AR} = \underline{I}_A e^{j(p\theta_0 + \xi_R)} \tag{4.34}$$

These are rotor flux and current referred to armature-phase angle. Note that $\underline{\Lambda}_{AR}$ and $\underline{I}_{AR}$ have the same phase relationship to each other as do $\underline{\Lambda}_A$ and $\underline{I}_A$. Using Eq. (4.33) and (4.34) in Eq. (4.31) and (4.32), the basic flux/current relationship for the induction machine becomes

$$\begin{bmatrix} \underline{\Lambda}_a \\ \underline{\Lambda}_{AR} \end{bmatrix} = \begin{bmatrix} L_S & \frac{3}{2}M \\ \frac{3}{2}M & L_R \end{bmatrix} \begin{bmatrix} \underline{I}_a \\ \underline{I}_{AR} \end{bmatrix} \tag{4.35}$$

This is an equivalent single-phase statement, describing the flux/current relationship in phase a, assuming balanced operation. The same expression will describe phases b and c.

Voltage at the terminals of the stator and rotor (possibly equivalent) windings is then

$$\underline{V}_a = j\omega\underline{\Lambda}_a + R_a\underline{I}_a \tag{4.36}$$

$$\underline{V}_{AR} = j\omega_R\underline{\Lambda}_{AR} + R_A\underline{I}_{AR} \tag{4.37}$$

or

$$\underline{V}_a = j\omega L_S\underline{I}_a + j\omega\frac{3}{2}M\underline{I}_{AR} + R_a\underline{I}_a \tag{4.38}$$

$$\underline{V}_{AR} = j\omega_R\frac{3}{2}M\underline{I}_a + j\omega_R L_R\underline{I}_{AR} + R_A\underline{I}_{AR} \tag{4.39}$$

To carry this further, it is necessary to go a little deeper into the machine's parameters. Note that L_S and L_R are synchronous inductances for the stator and rotor. These may be separated into space fundamental and "leakage" components as follows

$$L_S = L_a - L_{ab} = \frac{3}{2} \frac{4}{\pi} \frac{\mu_0 R l N_S^2 k_S^2}{p^2 g} + L_{Sl} \tag{4.40}$$

$$L_R = L_A - L_{AB} = \frac{3}{2} \frac{4}{\pi} \frac{\mu_0 R l N_R^2 k_R^2}{p^2 g} + L_{Rl} \tag{4.41}$$

Where the normal set of machine parameters holds

R is rotor radius

l is active length

g is the effective air-gap

p is the number of pole-pairs

N represents number of turns

k represents the winding factor

S as a subscript refers to the stator

R as a subscript refers to the rotor

L_l is "leakage" inductance

The two leakage terms L_{Sl} and L_{Rl} contain higher order harmonic stator and rotor inductances, slot inductances, end-winding inductances and, if necessary, a provision for rotor skew. Essentially, they are used to represent all flux in the rotor and stator that is not mutually coupled.

In the same terms, the stator-to-rotor mutual inductance, which is taken to comprise *only* a space fundamental term, is

$$M = \frac{4}{\pi} \frac{\mu_0 R l N_S N_R k_S k_R}{p^2 g} \tag{4.42}$$

Note that there are, of course, space harmonic mutual flux linkages. If they were to be included, they would complicate the analysis substantially. Ignore them here and note that they do have an effect on machine behavior, but that effect is second-order.

Air-gap permeance is defined as

$$\wp_{ag} = \frac{4}{\pi} \frac{\mu_0 R l}{p^2 g} \tag{4.43}$$

so that the inductances are

$$L_S = \frac{3}{2} \wp_{ag} k_S^2 N_S^2 + L_{Sl} \tag{4.44}$$

$$L_R = \frac{3}{2} \wp_{ag} k_R^2 N_R^2 + L_{Rl} \qquad (4.45)$$

$$M = \wp_{ag} N_S N_R k_S k_R \qquad (4.46)$$

Define "slip" s by

$$\omega_R = s\omega \qquad (4.47)$$

so that

$$s = 1 - \frac{p\omega_m}{\omega} \qquad (4.48)$$

Then, the voltage balance equations become

$$\underline{V}_a = j\omega \left(\frac{3}{2} \wp_{ag} k_S^2 N_S^2 + L_{Sl} \right) \underline{I}_a$$

$$+ j\omega \frac{3}{2} \wp_{ag} N_S N_R k_S k_R \underline{I}_{AR} + R_a \underline{I}_a \qquad (4.49)$$

$$\underline{V}_{AR} = js\omega \frac{3}{2} \wp_{ag} N_S N_R k_S k_R \underline{I}_a$$

$$+ js\omega \left(\frac{3}{2} \wp_{ag} k_R^2 N_R^2 + L_{Rl} \right) \underline{I}_{AR} + R_A \underline{I}_{AR} \qquad (4.50)$$

At this point, rotor current may be referred to the stator. This is done by assuming an effective turns ratio which, in turn, defines an equivalent stator current to produce the same fundamental MMF as a given rotor current

$$\underline{I}_2 = \frac{N_R k_R}{N_S k_S} \underline{I}_{AR} \qquad (4.51)$$

Now, if the rotor of the machine is shorted so that $\underline{V}_{AR} = 0$, some manipulation produces

$$\underline{V}_a = j(X_M + X_1)\underline{I}_a + jX_M \underline{I}_2 + R_a \underline{I}_a \qquad (4.52)$$

$$0 = jX_M \underline{I}_a + j(X_M + X_2)\underline{I}_2 + \frac{R_2}{s} \underline{I}_2 \qquad (4.53)$$

where the following definitions have been made

$$X_M = \frac{3}{2}\, \omega \wp_{ag} N_S^2 k_S^2 \qquad (4.54)$$

$$X_1 = \omega L_{Sl} \qquad (4.55)$$

$$X_2 = \omega L_{Rl} \left(\frac{N_S k_S}{N_R k_R}\right)^2 \qquad (4.56)$$

$$R_2 = R_A \left(\frac{N_S k_S}{N_R k_R}\right)^2 \qquad (4.57)$$

These expressions describe a simple equivalent circuit for the induction motor shown in Fig. 4.15.

4.2.3   Effective air-gap: Carter's coefficient

In induction motors, where the air-gap is usually quite small, it is necessary to correct the air-gap permeance for the effect of slot openings. These make the permeance of the air-gap slightly smaller than calculated from the physical gap, effectively making the gap a bit bigger. The ratio of effective to physical gap is

$$g_{\text{eff}} = g\,\frac{t + s}{t + s - g f(\alpha)} \qquad (4.58)$$

where

$$f(\alpha) = f\left(\frac{s}{2g}\right) = \alpha \tan(\alpha) - \log \sec \alpha \qquad (4.59)$$

4.2.4   Operation: energy balance

Now see how the induction machine actually works. Assume for the moment that Fig. 4.15 represents one phase of a polyphase system, that the machine is operated under balanced conditions, and that speed is constant or varying only slowly. "Balanced conditions" means that each phase has the same terminal voltage magnitude and that the phase difference between phases is a uniform. Under those con-

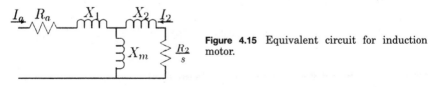

Figure 4.15 Equivalent circuit for induction motor.

ditions, it is possible to analyze each phase separately (as if it were a single-phase system). Assume an RMS voltage magnitude of V_t across each phase.

The "gap impedance," or the impedance looking to the right from the right-most terminal of X_1 is

$$Z_g = jX_m \left\| \left(jX_2 + \frac{R_2}{s} \right) \right. \tag{4.60}$$

A total, or terminal impedance is then

$$Z_t = jX_1 + R_a + Z_g \tag{4.61}$$

and terminal current is

$$I_t = \frac{V_t}{Z_t} \tag{4.62}$$

Rotor current is found by using a current divider

$$I_2 = I_t \frac{jX_m}{jX_2 + \dfrac{R_2}{s}} \tag{4.63}$$

"Air-gap" power is then calculated (assuming a three-phase machine)

$$P_{ag} = 3|I_2|^2 \frac{R_2}{s} \tag{4.64}$$

This is real (time-average) power crossing the air-gap of the machine. Positive slip implies rotor speed less than synchronous and positive air-gap power (motor operation). Negative slip means rotor speed is higher than synchronous, negative air-gap power (from the rotor to the stator) and generator operation.

This equivalent circuit represents a real physical structure, so it should be possible to calculate power dissipated in the physical rotor resistance, and that is

$$P_s = P_{ag}s \tag{4.65}$$

Note that since both P_{ag} and s will always have the same sign, dissipated power is positive. The rest of this discussion is framed in terms of *motor* operation, but the conversion to *generator* operation is simple. The difference between power crossing the air-gap and power dissipated in the rotor resistance must be converted from mechanical form

$$P_m = P_{ag} - P_s \tag{4.66}$$

and *electrical input* power is

$$P_{in} = P_{ag} + P_a \tag{4.67}$$

where armature dissipation is

$$P_a = 3|I_t|^2 R_a \tag{4.68}$$

Output (mechanical) power is

$$P_{out} = P_{ag} - P_w \tag{4.69}$$

where P_w describes friction, windage and certain stray losses, which will be discussed later.

And, finally, efficiency and power factor are

$$\eta = \frac{P_{out}}{P_{in}} \tag{4.70}$$

$$\cos \psi = \frac{P_{in}}{3V_t I_t} \tag{4.71}$$

EXAMPLE Torque and power vs. speed are shown in Figure 4.16 for the motor whose parameters are detailed in Table 4.4.

4.2.5 Squirrel-cage machine model

Now a circuit model for the squirrel-cage motor is derived using field analytical techniques. The model consists of two major parts. The first is a description of stator flux in terms of stator and rotor currents. The second is a description of rotor current in terms of air-gap flux. The result is a set of expressions for the elements of the circuit model for the induction machine.

To start, assume that the rotor is symmetrical enough to carry a surface current, the fundamental of which is

$$\overline{K}_r = \bar{i}_z Re(\underline{K}_r e^{j(s\omega t - p\phi')})$$

$$= \bar{i}_z Re(\underline{K}_r e^{j(\omega t - p\phi)}) \tag{4.72}$$

Note that in equation, use is made of the simple transformation between rotor and stator coordinates

$$\phi' = \phi - \omega_m t \tag{4.73}$$

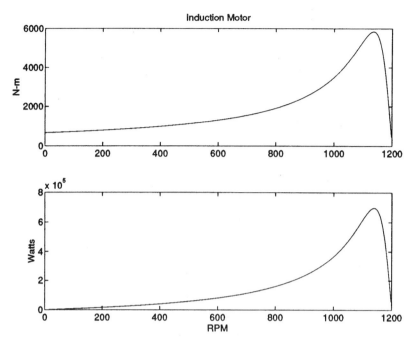

Figure 4.16   Torque and power vs. speed.

TABLE 4.4   Example, Standard Motor

Rating	300	kw
Voltage	440	VRMS, 1-1
	254	VRMS, 1-n
Stator Resistance R1	.0073	Ω
Rotor Resistance R2	.0064	Ω
Stator Reactance X1	.06	Ω
Rotor Reactance X2	.06	Ω
Magnetizing Reactance Xm	2.5	Ω
Synchronous Speed Ns	1200	RPM

and that

$$p\omega_m = \omega - \omega_r = \omega(1 - s) \tag{4.74}$$

Here, the following symbols are used

$\underline{K}_r$ is complex amplitude of rotor surface current

s is per-unit "slip"

ω is stator electrical frequency

ω_r is rotor electrical frequency

ω_m is rotational speed

The rotor current will produce an air-gap flux density of the form

$$B_r = Re(\underline{B}_r e^{j(\omega t - p\phi)}) \tag{4.75}$$

where

$$\underline{B}_r = -j\mu_0 \frac{R}{pg} \underline{K}_r \tag{4.76}$$

Note that this describes only radial magnetic flux density produced by the space fundamental of *rotor* current. Flux linked by the armature winding due to this flux density is

$$\lambda_{AR} = lN_S k_S \int_{-\pi/p}^{0} B_r(\phi) R d\phi \tag{4.77}$$

This yields a complex amplitude for λ_{AR}:

$$\lambda_{AR} = Re(\underline{\Lambda}_{AR} e^{j\omega t}) \tag{4.78}$$

where

$$\underline{\Lambda}_{AR} = \frac{2l\mu_0 R^2 N_S k_S}{p^2 g} \underline{K}_r \tag{4.79}$$

Adding this to flux produced by the stator currents, an expression for total stator flux is found

$$\underline{\Lambda}_a = \left(\frac{3}{2} \frac{4}{\pi} \frac{\mu_0 N_S^2 R l k_S^2}{p^2 g} + L_{Sl} \right) \underline{I}_a + \frac{2l\mu_0 R^2 N_S k_S}{p^2 g} \underline{K}_r \tag{4.80}$$

The expression produces a definition of an equivalent rotor current I_2 in terms of the space fundamental of rotor surface-current density

$$\underline{I}_2 = \frac{\pi}{3} \frac{R}{N_S k_S} \underline{K}_z \tag{4.81}$$

Then the simple expression for stator flux is

$$\underline{\Lambda}_a = (L_{ad} + L_{Sl})\underline{I}_a + L_{ad}\underline{I}_2 \tag{4.82}$$

where L_{ad} is the fundamental space harmonic component of stator inductance

$$L_{ad} = \frac{3}{2} \frac{4}{\pi} \frac{\mu_0 N_S^2 k_S^2 Rl}{p^2 g} \tag{4.83}$$

The second part of this derivation is the equivalent of finding a relationship between rotor flux and I_2. However, since this machine has no discrete windings, it is necessary to focus on the individual rotor bars.

Assume that there are N_R slots in the rotor. Each of these slots is carrying some current. If the machine is symmetrical and operating with balanced currents, we may write an expression for current in the kth slot as

$$i_k = Re(\underline{I}_k e^{js\omega t}) \tag{4.84}$$

where

$$\underline{I}_k = \underline{I}e^{-j(2\pi p/N_R)} \tag{4.85}$$

and $\underline{I}$ is the complex amplitude of current in slot number zero. Expression shows a uniform progression of rotor current *phase* about the rotor. All rotor slots carry the same current, but that current is phase-retarded (delayed) from slot-to-slot because of relative rotation of the current wave at slip frequency.

The rotor current density can then be expressed as a sum of impulses

$$K_z = Re \left[\sum_{k=0}^{N_R-1} \frac{1}{R} \underline{I}e^{j[\omega_r t - k(2\pi p/N_R)]} \, \delta\left(\phi' - \frac{2\pi k}{N_R} \right) \right] \tag{4.86}$$

The unit impulse function $\delta(\ )$ is a way of approximating the rotor current as a series of impulsive currents around the rotor.

This rotor surface current may be expressed as a fourier series of traveling waves

$$K_z = Re \left(\sum_{n=-\infty}^{\infty} \underline{K}_n e^{j(\omega_r t - np\phi')} \right) \tag{4.87}$$

Note that in (4.87), negative values of the space harmonic index n allow for reverse-rotating waves. This is really part of an expansion in both time and space, although we consider here only the fundamental time part. The nth space harmonic component is recovered by employing the following formula

$$\underline{K}_n = \left\langle \frac{1}{\pi} \int_0^{2\pi} K_r(\phi, t)e^{-j(\omega_r t - np\phi)} \, d\phi \right\rangle \tag{4.88}$$

Here the brackets $\langle \rangle$ denote time average and are here because of the two-dimensional nature of the expansion. To carry out (4.88), first expand it into its complex conjugate parts

$$K_r = \frac{1}{2} \sum_{k=0}^{N_R-1}$$

$$\times \left\{ \frac{I}{R} e^{j[\omega_r t - k(2\pi p/N_R)]} + \frac{I^*}{R} e^{-j[\omega_r t - k(2\pi p/N_R)]} \right\} \delta \left(\phi' - \frac{2\pi k}{N_R} \right) \quad (4.89)$$

If (4.89) is used in (4.88), the second half results in a sum of terms which time average to zero. The first half of the expression results in

$$\underline{K}_n = \frac{I}{2\pi R} \int_0^{2\pi} \sum_{k=0}^{N_R-1} e^{-j(2\pi pk/N_R)} e^{jnp\phi} \delta \left(\phi - \frac{2\pi k}{N_R} \right) d\phi \quad (4.90)$$

The impulse function turns the integral into an evaluation of the rest of the integrand at the impulses. What remains is the sum

$$\underline{K}_n = \frac{I}{2\pi R} \sum_{k=0}^{N_R-1} e^{j(n-1)(2\pi kp/N_R)} \quad (4.91)$$

The sum in is easily evaluated. It is

$$\sum_{k=0}^{N_R-1} e^{j[2\pi kp(n-1)/N_R]} = \begin{cases} N_R & \text{if } (n-1)\dfrac{P}{N_R} = \text{integer} \\ 0 & \text{otherwise} \end{cases} \quad (4.92)$$

The integer in (4.92) may be positive, negative or zero. As it turns out, only the first three of these (zero, plus and minus one) are important, because they produce the largest magnetic fields and therefore fluxes. These are

$$(n-1)\frac{p}{N_R} = -1 \quad \text{or } n = -\frac{N_{R-p}}{p}$$

$$= 0 \quad \text{or } n = 1$$

$$= 1 \quad \text{or } n = \frac{N_{R+p}}{p} \quad (4.93)$$

Note: that (4.93) appears to produce space harmonic orders that may be of non-integer order. This is not really true: it is necessary that np be an integer, and will always satisfy that condition.

So, the harmonic orders of interest are one and

$$n_+ = \frac{N_R}{p} + 1 \tag{4.94}$$

$$n_- = -\left(\frac{N_R}{p} - 1\right) \tag{4.95}$$

Each of the space harmonics of the squirrel-cage current will produce radial flux density. A surface current of the form

$$K_n = Re\left(\frac{N_R \underline{I}}{2\pi R} e^{j(\omega_r t - np\phi')}\right) \tag{4.96}$$

produces radial magnetic flux density

$$B_{rn} = Re(\underline{B}_{rn} e^{j(\omega_r t - np\phi')}) \tag{4.97}$$

where

$$\underline{B}_{rn} = -j\frac{\mu_0 N_R \underline{I}}{2\pi npg} \tag{4.98}$$

In turn, each of the components of radial flux density will produce a component of induced voltage. To calculate that, one must invoke Faraday's law

$$\nabla \times \overline{E} = -\frac{\partial \overline{B}}{\partial t} \tag{4.99}$$

Assuming that the fields do not vary with z, the radial component of flux density $\overline{B}$ is

$$\frac{1}{R}\frac{\partial}{\partial \phi} E_z = -\frac{\partial B_r}{\partial t} \tag{4.100}$$

Or, assuming an electric field component of the form

$$E_{zn} = Re(\underline{E}_n e^{j(\omega_r t - np\phi)}) \tag{4.101}$$

Using (4.98) and (4.101) in (4.100), we obtain an expression for electric field induced by components of air-gap flux

$$\underline{E}_n = \frac{\omega_r R}{np} \underline{B}_n \tag{4.102}$$

$$\underline{E}_n = -j\frac{\mu_0 N_R \omega_r R}{2\pi g(np)^2} \underline{I} \tag{4.103}$$

Now, the total voltage induced in a slot pushes current through the conductors in that slot. This is

$$\underline{E}_1 + \underline{E}_{n-} + \underline{E}_{n+} = \underline{Z}_{\text{slot}}\underline{I} \tag{4.104}$$

Now: in, there are three components of air-gap field. E_1 is the space fundamental field, produced by the space fundamental of rotor current as well as by the space fundamental of stator current. The other two components on the left of (4.104) are produced only by rotor currents and actually represent additional reactive impedance to the rotor. This is often called *zigzag* leakage inductance. The parameter Z_{slot} represents impedance of the slot itself: resistance and reactance associated with cross-slot magnetic fields. Then it can be re-written as

$$\underline{E}_1 = \underline{Z}_{\text{slot}}\underline{I} + j\,\frac{\mu_0 N_R \omega_r R}{2\pi g}\left(\frac{1}{(n_+p)^2} + \frac{1}{(n_-p)^2}\right)\underline{I} \tag{4.105}$$

To finish this model, it is necessary to translate back to the stator. See that and make the link between $\underline{I}$ and $\underline{I}_2$

$$\underline{I}_2 = \frac{N_R}{6N_S k_S}\,\underline{I} \tag{4.106}$$

Then the electric field at the surface of the rotor is

$$\underline{E}_1 = \left[\frac{6N_S k_S}{N_R}\,\underline{Z}_{\text{slot}} + j\omega_r\,\frac{3}{\pi}\,\frac{\mu_0 N_S k_S R}{g}\left(\frac{1}{(n_+p)^2} + \frac{1}{(n_-p)^2}\right)\right]\underline{I}_2 \tag{4.107}$$

This must be translated into an equivalent stator voltage. To do so, use (4.102) to translate (4.107) into a statement of radial magnetic field, then find the flux liked and hence stator voltage from that. Magnetic flux density is

$$\underline{B}_r = \frac{p\underline{E}_1}{\omega_r R}$$

$$= \left[\frac{6N_S k_S p}{N_R R}\left(\frac{R_{\text{slot}}}{\omega_r} + jL_{\text{slot}}\right)\right.$$

$$\left. + j\,\frac{3}{\pi}\,\frac{\mu_0 N_S k_S p}{g}\left(\frac{1}{(n_+p)^2} + \frac{1}{(n_-p)^2}\right)\right]\underline{I}_2 \tag{4.108}$$

where the slot impedance has been expressed by its real and imaginary parts

$$\underline{Z}_{\text{slot}} = R_{\text{slot}} + j\omega_r L_{\text{slot}} \tag{4.109}$$

Flux linking the armature winding is

$$\lambda_{ag} = N_S k_S lR \int_{-\pi/2p}^{0} Re\ (\underline{B}_r e^{j(\omega t - p\phi)})\ d\phi \qquad (4.110)$$

which becomes

$$\lambda_{ag} = Re(\underline{\Lambda}_{ag} e^{j\omega t}) \qquad (4.111)$$

where

$$\underline{\Lambda}_{ag} = j\ \frac{2N_S k_S lR}{p}\ \underline{B}_r \qquad (4.112)$$

Then, "air-gap" voltage is

$$\underline{V}_{ag} = j\omega\underline{\Lambda}_{ag} = -\frac{2\omega N_S k_S lR}{p}\ \underline{B}_r$$

$$= -\underline{I}_2 \left[\frac{12lN_S^2 k_S^2}{N_R} \left(j\omega L_{\text{slot}} + \frac{R_2}{s} \right) \right.$$

$$\left. + j\omega\ \frac{6}{\pi}\ \frac{\mu_0 RlN_S^2 k_S^2}{g} \left(\frac{1}{(n_+ p)^2} + \frac{1}{(n_- p)^2} \right) \right] \qquad (4.113)$$

This expression describes the relationship between the space fundamental air-gap voltage $\underline{V}_{ag}$ and rotor current $\underline{I}_2$. The expression also fits the equivalent circuit of Fig. 4.17 if the definitions made below hold

$$X_2 = \omega\ \frac{12lN_S^2 k_S^2}{N_R}\ L_{\text{slot}} + \omega\ \frac{6}{\pi}\ \frac{\mu_0 RlN_S^2 k_S^2}{g}$$

$$\times \left(\frac{1}{(N_R + p)^2} + \frac{1}{(N_R - p)^2} \right) \qquad (4.114)$$

$$R_2 = \frac{12lN_S^2 k_S^2}{N_R}\ R_{\text{slot}} \qquad (4.115)$$

Figure 4.17 Rotor equivalent circuit.

The first term in (4.114) expresses slot leakage inductance for the rotor. Similarly, (4.115) expresses rotor resistance in terms of slot resistance. Note that L_{slot} and R_{slot} are both expressed per unit length. The second term in (4.114) expresses the "zigzag" leakage inductance resulting from harmonics on the order of rotor slot pitch.

Next, see that armature flux is just equal to air-gap flux plus armature leakage inductance. It could be written as

$$\underline{\Lambda}_a = \underline{\Lambda}_{ag} + L_{al}\underline{I}_a \tag{4.116}$$

There are a number of components of stator slot leakage L_{al}, each representing flux paths that do not directly involve the rotor. Each of the components adds to the leakage inductance. The most prominent components of stator leakage are referred to as *slot, belt, zigzag, end winding, and skew.* Each of these will be discussed in the following paragraphs.

Belt and zigzag leakage components are due to air-gap space harmonics. As it turns out, these are relatively complicated to estimate, but we may get some notion from our first-order view of the machine. The trouble with estimating these leakage components is that they are not *really* independent of the rotor, even though they are called "leakage." *Belt* harmonics are of order $n = 5$ and $n = 7$. If there were no rotor coupling, the belt harmonic leakage terms would be

$$X_{ag5} = \frac{3}{2}\frac{4}{\pi}\frac{\mu_0 N_S^2 k_5^2 R l}{5^2 p^2 g} \tag{4.117}$$

$$X_{ag7} = \frac{3}{2}\frac{4}{\pi}\frac{\mu_0 N_S^2 k_7^2 R l}{7^2 p^2 g} \tag{4.118}$$

The belt harmonics link to the rotor, however, and actually appear to be parallel with components of rotor impedance appropriate to $5p$ and $7p$ pole-pair machines. At these harmonic orders, one can usually ignore rotor resistance so that rotor impedance is purely inductive. Those components are

$$X_{2,5} = \omega\frac{12 l N_S^2 k_5^2}{N_R}L_{\text{slot}}$$

$$+ \omega\frac{6}{\pi}\frac{\mu_0 R l N_S^2 k_5^2}{g}\left(\frac{1}{(N_R + 5p)^2} + \frac{1}{(N_R - 5p)^2}\right) \tag{4.119}$$

$$X_{2,7} = \omega\frac{12 l N_S^2 k_7^2}{N_R}L_{\text{slot}}$$

$$+ \omega\frac{6}{\pi}\frac{\mu_0 R l N_S^2 k_7^2}{g}\left(\frac{1}{(N_R + 7p)^2} + \frac{1}{(N_R - 7p)^2}\right) \tag{4.120}$$

In the simple model of the squirrel-cage machine, because the rotor resistances are relatively small and slip high, the effect of rotor resistance is usually ignored. Then the fifth and seventh harmonic components of belt leakage are

$$X_5 = X_{ag5}\|X_{2,5} \tag{4.121}$$

$$X_7 = X_{ag7}\|X_{2,7} \tag{4.122}$$

Stator zigzag leakage is from those harmonics of the orders $pn_s = N_{\text{slots}} \pm p$ where N_{slots} is the total number of slots around the periphery of the machine.

$$X_z = \frac{3}{2} \frac{4}{\pi} \frac{\mu_0 N_S^2 Rl}{g} \left(\frac{k_{n_s+}}{(N_{\text{slots}} + p)^2} + \frac{k_{n_s-}}{(N_{\text{slots}} - p)^2} \right) \tag{4.123}$$

Note that these harmonic orders do not tend to be shorted out by the rotor cage and so no direct interaction with the cage is ordinarily accounted for.

In order to reduce saliency effects that occur because the rotor teeth will tend to try to align with the stator teeth, induction motor designers always use a different number of slots in the rotor and stator. There still may be some tendency to become aligned, and this produces "cogging" torques which in turn produce vibration and noise and, in severe cases, can retard or even prevent starting. To reduce this tendency to "cog", rotors are often built with a little "skew", or twist of the slots from one end to the other. Thus, when one tooth is aligned at one end of the machine, it is un-aligned at the other end. A side effect of this is to reduce the stator and rotor coupling by just a little, and this produces leakage reactance. This is fairly easy to estimate. Consider, for example, a space-fundamental flux density $B_r = B_1 \cos p\theta$, linking a (possibly) skewed full-pitch current path

$$\lambda = \int_{-l/2}^{l/2} \int_{-(\pi/2p)+(\varsigma/p)(x/l)}^{(\pi/2p)+(\varsigma/p)(x/l)} B_1 \cos p\theta \, Rd\theta dx$$

Here, the skew in the rotor is ς *electrical* radians from one end of the machine to the other. Evaluation of this yields

$$\lambda = \frac{2B_1 Rl}{p} \frac{\sin \dfrac{\varsigma}{2}}{\dfrac{\varsigma}{2}}$$

Now, the difference between what would have been linked by a non-skewed rotor and what is linked by the skewed rotor is the skew leakage flux, now expressible as

$$X_k = X_{ag} \left(1 - \frac{\sin \frac{\varsigma}{2}}{\frac{\varsigma}{2}} \right)$$

The final component of leakage reactance is due to the end windings. This is perhaps the most difficult of the machine parameters to estimate, being essentially three-dimensional in nature. There are a number of ways of estimating this parameter, but a simplified, approximate parameter from Alger [1] is

$$X_e = \frac{14}{4\pi^2} \frac{q}{2} \frac{\mu_0 R N_a^2}{p^2} (p - 0.3)$$

As with all such formulae, extreme care is required here, since little guidance is available as to when this expression is correct or even close. Admittedly, a more complete treatment of this element of machine parameter construction would be an improvement.

4.2.6 Harmonic order-rotor resistance and stray load losses

It is important to recognize that the machine rotor "sees" each of the stator harmonics in essentially the same way, and it is quite straightforward to estimate rotor parameters for the harmonic orders, as we have done above. Now, particularly for the "belt" harmonic orders, there are rotor currents flowing in response to stator mmf's at the fifth and seventh space harmonic order. The resistances attributable to these harmonic orders are

$$R_{2,5} = \frac{12lN_s^2 k_5^2}{N_R} R_{\text{slot},5} \tag{4.124}$$

$$R_{2,7} = \frac{12lN_s^2 k_7^2}{N_R} R_{\text{slot},7} \tag{4.125}$$

The higher-order slot harmonics will have relative frequencies (slips) that are

$$s_n = 1 \mp (1 - s)n \begin{Bmatrix} n = 6k + 1 \\ n = 6k - 1 \end{Bmatrix} k \text{ an integer} \tag{4.126}$$

The induction-motor electromagnetic interaction can now be described by an augmented magnetic circuit as shown in Fig. 4.18. Note that the terminal flux of the machine is the sum of *all* of the harmonic fluxes, and each space harmonic is excited by the same current so the individual harmonic components are in series.

Each of the space harmonics will have an electromagnetic interaction similar to the fundamental: power transferred across the air-gap for each space harmonic is

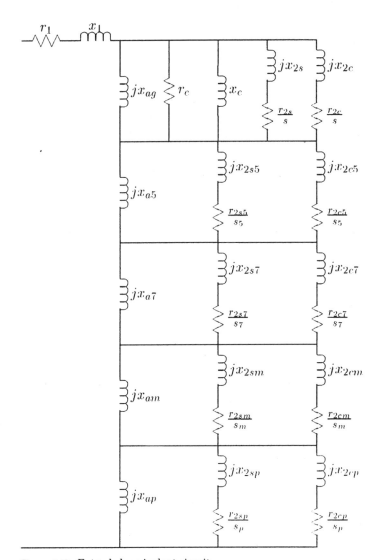

Figure 4.18 Extended equivalent circuit.

$$P_{em,n} = 3I_{2,n}^2 \frac{R_{2,n}}{s_n}$$

Of course dissipation in each circuit is:

$$P_{d,n} = 3I_{2,n}^2 R_{2,n}$$

leaving

$$P_{m,n} = 3I_{2,n}^2 \frac{R_{2,n}}{s_n} (1 - s_n)$$

Note that this equivalent circuit has provision for two sets of circuits which look like "cages." One of these sets is for the solid rotor body if that exists. There is also a provision (r_e) for loss in the stator core iron.

Power deposited in the rotor harmonic resistance elements is characterized as "stray load" loss because it is not easily computed from the simple machine equivalent circuit.

4.2.7 Slot models

Some of the more interesting things that can be done with induction motors have to do with the shaping of rotor slots to achieve particular frequency-dependent effects. Three cases are considered here, but there are many other possibilities.

First, suppose the rotor slots rectangular, as shown in Fig. 4.19 and assume that the slot dimensions are such that diffusion effects are not important so that the current in the slot conductor is approximately uniform. In that case, the slot resistance and inductance per unit length are

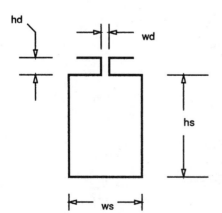

Figure 4.19 Single slot.

$$R_{\text{slot}} = \frac{1}{w_s h_s \sigma} \tag{4.127}$$

$$L_{\text{slot}} = \mu_0 \frac{h_s}{3 w_s} \tag{4.128}$$

The slot resistance is obvious. The slot inductance may be estimated by recognizing that if the current in the slot is uniform, the magnetic field crossing the slot must be

$$H_y = \frac{I}{w_s} \frac{x}{h_s}$$

Then energy stored in the field in the slot is simply

$$\frac{1}{2} L_{\text{slot}} I^2 = w_s \int_0^{h_s} \frac{\mu_0}{2} \left(\frac{Ix}{w_s h_s} \right)^2 dx = \frac{1}{6} \frac{\mu_0 h_s}{w_s} I^2$$

4.2.8 Deep slots

Now, suppose the slot is *not* small enough so that diffusion effects can be ignored. The slot becomes "deep" to the extent that its depth is less than (or even comparable to) the *skin depth* for conduction at slip frequency. Conduction in this case may be represented by using the Diffusion equation

$$\nabla^2 \overline{H} = \mu_0 \sigma \frac{\partial \overline{H}}{\partial t}$$

In the steady state, and assuming that only cross-slot flux (in the y direction) is important, and the only variation that is important is in the radial (x) direction

$$\frac{\partial^2 H_y}{\partial x^2} = j \omega_s \mu_0 \sigma H_y$$

This is solved by solutions of the form

$$H_y = H_{\pm} e^{\pm (1+j)(x/\delta)}$$

where the skin depth is

$$\delta = \sqrt{\frac{2}{\omega_s \mu_0 \sigma}}$$

Since H_y must vanish at the bottom of the slot, it must take the form

$$H_y = H_{\text{top}} \frac{\sinh(1 + j)\frac{x}{\delta}}{\sinh(1 + j)\frac{h_s}{\delta}}$$

Since current is the curl of magnetic field

$$J_z = \sigma E_z = \frac{\partial H_y}{\partial x} = H_{\text{top}} \frac{1 + j}{\delta} \cdot \frac{\cosh(1 + j)\frac{h_s}{\delta}}{\sinh(1 + j)\frac{h_s}{\delta}}$$

then slot impedance per unit length is

$$Z_{\text{slot}} = \frac{1}{w_s} \frac{1 + j}{\sigma\delta} \coth(1 + j)\frac{h_s}{\delta}$$

Of course the impedance (purely reactive) due to the slot depression must be added to this. It is possible to extract the real and imaginary parts of this impedance (the process is algebraically a bit messy) to yield

$$R_{\text{slot}} = \frac{1}{w_s\sigma\delta} \frac{\sinh 2\frac{h_s}{\delta} + \sin 2\frac{h_s}{\delta}}{\cosh 2\frac{h_s}{\delta} - \cos 2\frac{h_s}{\delta}}$$

$$L_{\text{slot}} = \mu_0 \frac{h_d}{w_d} + \frac{1}{w_s} \frac{1}{w_s\sigma\delta} \frac{\sinh 2\frac{h_s}{\delta} - \sin 2\frac{h_s}{\delta}}{\cosh 2\frac{h_s}{\delta} - \cos 2\frac{h_s}{\delta}}$$

4.2.9 Multiple cages

The purpose of a "deep" slot is to improve starting performance of a motor. When the rotor is stationary, the frequency seen by rotor conductors is relatively high, and current crowding due to the skin effect makes rotor resistance appear to be high. As the rotor accelerates, the frequency seen from the rotor drops, lessening the skin effect and making more use of the rotor conductor. This then gives the machine higher starting torque (requiring high resistance) without compromising running efficiency.

This effect can be carried even further by making use of *multiple cages,* such as is shown in Fig. 4.20. Here, there are two conductors in a fairly complex slot. Estimating the impedance of this slot is done in stages to build up an equivalent circuit.

Assume, for the purposes of this derivation, that each section of the multiple cage is small enough so that currents can be considered to be uniform in each conductor. Then the bottom section may be represented as a resistance in series with an inductance

$$R_a = \frac{1}{\sigma w_1 h_1}$$

$$L_a = \frac{\mu_0}{3} \frac{h_1}{w_1}$$

The narrow slot section with no conductor between the top and bottom conductors will contribute an inductive impedance

$$L_s = \mu_0 \frac{h_s}{w_s}$$

The top conductor will have a resistance:

$$R_b = \frac{1}{\sigma w_2 h_2}$$

Now, in the equivalent circuit, current flowing in the lower conductor

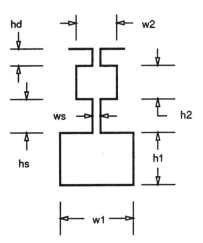

Figure 4.20 Double slot.

will produce a magnetic field across this section, yielding a series inductance of

$$L_b = \mu_0 \frac{h_2}{w_2}$$

By analogy with the bottom conductor, current in the top conductor flows through only one-third of the inductance of the top section, leading to the equivalent circuit of Fig. 4.21, once the inductance of the slot depression is added on

$$L_t = \mu_0 \frac{h_d}{w_d}$$

Now, this rotor bar circuit fits right into the framework of the induction-motor equivalent circuit, shown for the double cage case in Fig. 4.22, with

$$R_{2a} = \frac{12lN_S^2 k_S^2}{N_R} R_a$$

$$R_{2b} = \frac{12lN_S^2 k_S^2}{N_R} R_b$$

$$X_{2a} = \omega \frac{12lN_S^2 k_S^2}{N_R} \left(\frac{2}{3} L_b + L_s + L_a \right)$$

$$X_{2a} = \omega \frac{12lN_S^2 k_S^2}{N_R} \left(L_t + \frac{1}{3} L_b \right)$$

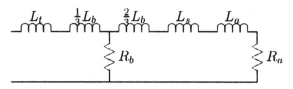

Figure 4.21   Equivalent circuit: Double bar.

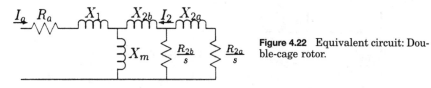

Figure 4.22   Equivalent circuit: Double-cage rotor.

4.2.10 Rotor end-ring effects

It is necessary to correct for "end-ring" resistance in the rotor. To do this, note that the magnitude of surface current density in the rotor is related to the magnitude of individual bar current by

$$I_z = K_z \frac{2\pi R}{N_R} \tag{4.129}$$

Current in the end-ring is

$$I_R = K_z \frac{R}{p} \tag{4.130}$$

It is then straightforward to calculate the ratio between power dissipated in the end-rings to power dissipated in the conductor bars themselves, considering the ratio of current densities and volumes. Assuming that the bars and end-rings have the same *radial* extent, the ratio of current densities is

$$\frac{J_R}{J_z} = \frac{N_R}{2\pi p} \frac{w_r}{l_r} \tag{4.131}$$

where w_r is the average width of a conductor bar and l_r is the axial end-ring length.

Now, the ratio of losses (and hence the ratio of resistances) is found by multiplying the square of current density ratio by the ratio of volumes. This is approximately

$$\frac{R_{\text{end}}}{R_{\text{slot}}} = \left(\frac{N_R}{2\pi p} \frac{w_r}{l_r} \right)^2 2 \frac{2\pi R}{N_R l} \frac{l_r}{w_r} = \frac{N_R R w_r}{\pi l l_r p^2} \tag{4.132}$$

4.2.11 Windage

Bearing friction, windage loss and fan input power are often regarded as elements of a "black art," of which motor manufacturers seem to take a highly empirical view. What follows is an attempt to build reasonable but simple models for two effects: loss in the air-gap due to windage; and input power to the fan for cooling. Some caution is required here, for these elements of calculation have *not* been properly tested, although they seem to give reasonable numbers.

The first element is gap windage loss. This is produced by shearing of the air in the relative rotation gap. It is likely to be a significant element only in machines with very narrow air-gaps or very high surface speeds. But these also include, of course, the high performance machines with which we are most interested. This may be approached with a simple "couette flow" model. Air-gap shear loss is approximately

$$P_w = 2\pi R^1 \Omega^3 l \rho_a f \tag{4.133}$$

where ρ_a is the density of the air-gap medium (possibly air) and f is the *friction factor*, estimated by

$$f = \frac{.0076}{R_n^{1/4}} \tag{4.134}$$

and the *Reynold's Number* R_n is

$$R_n = \frac{\Omega R g}{\nu_{\text{air}}} \tag{4.135}$$

and ν_{air} is the kinematic viscosity of the air-gap medium.

The second element is fan input power. An estimate of this rests on two hypotheses. The first is that the mass flow of air circulated by the fan can be calculated by the loss in the motor and an average temperature rise in the cooling air. The second hypothesis is that the pressure rise of the fan is established by the centrifugal pressure rise associated with the surface speed at the outside of the rotor. Taking these one at a time: if there is to be a temperature rise ΔT in the cooling air, then the mass flow volume is

$$\dot{m} = \frac{P_d}{C_p \Delta T}$$

and then volume flow is just

$$\dot{v} = \frac{\dot{m}}{\rho_{\text{air}}}$$

Pressure rise is estimated by centrifugal force

$$\Delta P = \rho_{\text{air}} \left(\frac{w}{p} r_{\text{fan}} \right)^2$$

then power is given by

$$P_{\text{fan}} = \Delta P \dot{v}$$

For reference, the properties of air are

Density	ρ_{air}	$1.18\ kg/m^2$
Kinematic Viscosity	v_{air}	$1.56 \times 10^{-5}\ m^2/sec$
Heat Capacity	C_p	1005.7 J/kg

4.2.12 Magnetic circuit loss and excitation

There will be some loss in the stator magnetic circuit due to eddy current and hysteresis effects in the core iron. In addition, particularly if the rotor and stator teeth are saturated, there will be MMF expended to push flux through those regions. These effects are very difficult to estimate from first principles, so resort to a simple model. Assume that the loss in saturated steel follows a law such as

$$P_d = P_B \left(\frac{\omega_e}{\omega_B}\right)^{\epsilon_f}\left(\frac{B}{B_B}\right)^{\epsilon_b} \tag{4.136}$$

This is not *too* bad an estimate for the behavior of core iron. Typically, ϵ_f is a bit less than 2.0 (between about 1.3 and 1.6) and ϵ_b is a bit more than two (between about 2.1 and 2.4). Of course, this model is only good for a fairly restricted range of flux density. Base dissipation is usually expressed in "watts per kilogram", so we first compute flux density and then the mass of the two principal components of the stator iron—the teeth and the back-iron.

In a similar way, one can model the exciting volt-amperes consumed by core iron by

$$Q_c = \left(Va_1 \left(\frac{B}{B_B}\right)^{\epsilon_{v1}} + Va_2 \left(\frac{B}{B_B}\right)^{\epsilon_{v2}}\right)\frac{\omega}{\omega_B} \tag{4.137}$$

This, too, is a form that appears to be valid for some steels. Quite obviously it may be necessary to develop different forms of curve "fits" for different materials.

Flux density (RMS) in the air-gap is

$$B_r = \frac{pV_a}{2RlN_a k_1 \omega_s} \tag{4.138}$$

Then flux density in the stator teeth is

$$B_t = B_r \frac{w_t + w_1}{w_t} \tag{4.139}$$

where w_t is tooth width and w_1 is slot-top width. Flux in the back-iron of the core is

$$B_e = B_r \frac{R}{pd_e} \tag{4.140}$$

where d_e is the radial depth of the core.

One way of handling this loss is to assume that the core handles flux corresponding to terminal voltage. Add up the losses and then compute an equivalent resistance and reactance

$$r_e = \frac{3|V_a|^2}{P_{\text{core}}}$$

$$x_e = \frac{3|V_a|^2}{Q_{\text{core}}}$$

then put this equivalent resistance in parallel with the air-gap reactance element in the equivalent circuit.

4.2.13 Solid iron rotor bodies

Solid steel rotor electric machines (SSRM) can be made to operate with very high surface speeds and are thus suitable for use in high RPM situations. They resemble hysteresis machines in form and function. However, asynchronous operation will produce higher power output because it takes advantage of higher flux density. Consider here the interactions to be expected from solid iron rotor bodies. The equivalent circuits can be placed in parallel (harmonic-by-harmonic) with the equivalent circuits for the squirrel cage, if there is a cage in the machine.

To estimate the rotor parameters R_{2s} and X_{2s}, assume that important field quantities in the machine are sinusoidally distributed in time and space, so that radial flux density is

$$B_r = Re(\underline{B}_r e^{j(\omega t - p\phi)}) \tag{4.141}$$

and, similarly, axially directed rotor surface current is

$$K_z = Re(\underline{K}_z e^{j(\omega t - p\phi)}) \tag{4.142}$$

Now, since by Faraday's law

$$\nabla \times \overline{E} = -\frac{\partial \overline{B}}{\partial t} \tag{4.143}$$

in this machine geometry

$$\frac{1}{R}\frac{\partial}{\partial \phi}E_z = -\frac{\partial B_r}{\partial t} \qquad (4.144)$$

The transformation between rotor and stator coordinates is

$$\phi' = \phi - \omega_m t \qquad (4.145)$$

where ω_m is rotor speed. Then

$$p\omega_m = \omega - \omega_r = \omega(1 - s) \qquad (4.146)$$

and now in the frame of the rotor, axial electric field is just

$$E_z = Re(\underline{E}_z e^{j(\omega t - p\phi)}) \qquad (4.147)$$

$$= Re(\underline{E}_z e^{j(\omega_r t - p\phi')}) \qquad (4.148)$$

and

$$\underline{E}_z = \frac{\omega_r R}{p}\underline{B}_r \qquad (4.149)$$

Of course, electric field in the rotor frame is related to rotor surface current by

$$\underline{E}_z = \underline{Z}_s\underline{K}_z \qquad (4.150)$$

Now these quantities can be related to the stator by noting that *air-gap* voltage is related to radial flux density by

$$\underline{B}_r = \frac{p}{2lN_a k_1 R\omega}\underline{V}_{ag} \qquad (4.151)$$

The stator-equivalent rotor current is

$$\underline{I}_2 = \frac{\pi}{3}\frac{R}{N_a k_a}\underline{K}_z \qquad (4.152)$$

Stator referred, rotor-equivalent impedance is

$$\underline{Z}_2 = \frac{\underline{V}_{ag}}{\underline{I}_2} = \frac{3}{2}\frac{4}{\pi}\frac{l}{R}N_a^2 k_a^2 \frac{\omega}{\omega_r}\frac{\underline{E}_z}{\underline{K}_a} \qquad (4.153)$$

Now, if rotor surface impedance can be expressed as

$$\underline{Z}_s = R_s + j\omega_r L_s \qquad (4.154)$$

then

$$\underline{Z}_2 = \frac{R_2}{s} + jX_2 \qquad (4.155)$$

where

$$R_2 = \frac{3}{2}\frac{4}{\pi}\frac{l}{R}N_a^2 k_1^2 R_s \qquad (4.156)$$

$$X_2 = \frac{3}{2}\frac{4}{\pi}\frac{l}{R}N_a^2 k_1^2 X_s \qquad (4.157)$$

Now, to find the rotor surface impedance, make use of a nonlinear eddy-current model proposed by Agarwal. First, define an equivalent penetration depth (similar to a skin depth)

$$\delta = \sqrt{\frac{2H_m}{\omega_r \sigma B_0}} \qquad (4.158)$$

where σ is rotor surface material-volume conductivity, B_0, "saturation flux density" is taken to be 75% of actual saturation flux density and

$$H_m = |\underline{K}_z| = \frac{3}{\pi}\frac{N_a k_a}{R}|\underline{I}_2| \qquad (4.159)$$

Then rotor surface resistivity and surface reactance are

$$R_s = \frac{16}{3\pi}\frac{1}{\sigma\delta} \qquad (4.160)$$

$$X_s = .5R_s \qquad (4.161)$$

Note that the rotor elements X_2 and R_2 depend on rotor current I_2, so the problem is nonlinear. However, a simple iterative solution can be used. First, make a guess for R_2 and find currents. Then use those currents to calculate R_2 and solve again for current. This procedure is repeated until convergence, and the problem seems to converge within just a few steps.

Aside from the necessity to iterate to find rotor elements, standard network techniques can be used to find currents, power input to the motor and power output from the motor, torque, etc.

Solution Not all of the equivalent circuit elements are known at the start of the solution. Assume a value for R_2, possibly some fraction of X_m, but the value chosen doesn't seem to matter much. The rotor reactance X_2 is just a fraction of R_2. Then proceed to compute an "air-gap" impedance, just the impedance looking into the parallel combination of magnetizing and rotor branches

$$Z_g = jX_m \left\| \left(jX_2 + \frac{R_2}{s} \right) \right. \tag{4.162}$$

(Note that, for a generator, slip s is negative).

A total impedance is then

$$Z_t = jX_1 + R_1 + Z_g \tag{4.163}$$

and terminal current is

$$I_t = \frac{V_t}{Z_t} \tag{4.164}$$

Rotor current is just

$$I_2 = I_t \frac{jX_m}{jX_2 + \dfrac{R_2}{s}} \tag{4.165}$$

Now it is necessary to iteratively correct rotor impedance. This is done by estimating flux density at the surface of the rotor using (Eq. 4.159), then getting a rotor surface impedance using (Eq. 4.160) and using *that* and (Eq. 4.157) to estimate a new value for R_2. Then we start again with Eq. 4.161. The process "drops through" this point when the new and old estimates for R_2 agree to some criterion.

4.2.14 Harmonic losses in solid steel

If the rotor of the machine is constructed of solid steel, there will be eddy-currents induced on the rotor surface by the higher-order space harmonics of stator current. These will produce magnetic fields and losses. This calculation assumes the rotor surface is linear and smooth and can be characterized by a conductivity and relative permeability. In this discussion, two space harmonics (positive- and negative-going) are considered. In practice, it may be necessary to carry four (or even more) harmonics, including both "belt-" and "zigzag-" order harmonics.

Terminal current produces magnetic field in the air-gap for each of the space harmonic orders, and each of these magnetic fields induces rotor currents of the same harmonic order.

The "magnetizing" reactances for the two harmonic orders (really the two components of the zigzag leakage) are

$$X_{zp} = X_m \frac{k_p^2}{N_p^2 k_1^2} \tag{4.166}$$

$$X_{zn} = X_m \frac{k_n^2}{N_n^2 k_1^2} \tag{4.167}$$

where N_p and N_n are the positive- and negative-going harmonic orders: For "belt" harmonics, these orders are 7 and 5. For "zigzag," they are

$$N_p = \frac{N_s + p}{p} \tag{4.168}$$

$$N_n = \frac{N_s - p}{p} \tag{4.169}$$

Now, there will be a current on the surface of the rotor at each harmonic order, and following Eq. (4.152), the equivalent rotor element current is

$$\underline{I}_{2p} = \frac{\pi}{3} \frac{R}{N_a k_p} \underline{K}_p \tag{4.170}$$

$$\underline{I}_{2n} = \frac{\pi}{3} \frac{R}{N_a k_n} \underline{K}_n \tag{4.171}$$

These currents flow in response to the magnetic field in the air-gap which in turn produces an axial electric field. Viewed from the rotor, this electric field is

$$\underline{E}_p = s_p \omega R \underline{B}_p \tag{4.172}$$

$$\underline{E}_n = s_n \omega R \underline{B}_n \tag{4.173}$$

where the *slip* for each of the harmonic orders is

$$s_p = 1 - N_p(1 - s) \tag{4.174}$$

$$s_n = 1 + N_p(1 - s) \tag{4.175}$$

and then the surface currents that flow in the surface of the rotor are

$$\underline{K}_p = \frac{\underline{E}_p}{Z_{sp}} \tag{4.176}$$

$$\underline{K}_n = \frac{\underline{E}_n}{Z_{sn}} \tag{4.177}$$

where Z_{sp} and Z_{sn} are the *surface* impedances at positive and negative harmonic slip frequencies, respectively. Assuming a linear surface, these are, approximately

$$Z_s = \frac{1 + j}{\sigma\delta} \tag{4.178}$$

where σ is material resistivity and the skin depth is

$$\delta = \sqrt{\frac{2}{\omega_s\mu\sigma}} \tag{4.179}$$

and ω_s is the frequency of the given harmonic from the rotor surface. We can postulate that the appropriate value of μ to use is the same as that estimated in the *nonlinear* calculation of the space fundamental, but this requires empirical confirmation.

The voltage induced in the stator by each of these space harmonic magnetic fluxes is

$$V_p = \frac{2N_a k_p l R \omega}{N_p p} \underline{B}_p \tag{4.180}$$

$$V_n = \frac{2N_a k_n l R \omega}{N_n p} \underline{B}_n \tag{4.181}$$

Then the equivalent circuit impedance of the rotor is just

$$Z_{2p} = \frac{V_p}{I_p} = \frac{3}{2}\frac{4}{\pi}\frac{N_a^2 k_p^2 l}{N_p R}\frac{Z_{sp}}{s_p} \tag{4.182}$$

$$Z_{2n} = \frac{V_n}{I_n} = \frac{3}{2}\frac{4}{\pi}\frac{N_a^2 k_n^2 l}{N_n R}\frac{Z_{sn}}{s_n} \tag{4.183}$$

The equivalent rotor circuit elements are now

$$R_{2p} = \frac{3}{2}\frac{4}{\pi}\frac{N_a^2 k_p^2 l}{N_p R}\frac{1}{\sigma\delta_p} \tag{4.184}$$

$$R_{2n} = \frac{3}{2}\frac{4}{\pi}\frac{N_a^2 k_n^2 l}{N_n R}\frac{1}{\sigma\delta_n} \tag{4.185}$$

$$X_{2p} = \frac{1}{2}R_{2p} \tag{4.186}$$

$$X_{2n} = \frac{1}{2}R_{2n} \tag{4.187}$$

4.2.15   Stray losses

The major elements of torque production and consequently of machine performance have been discussed and outlined, including the major sources of loss in induction machines. Using what has been outlined in this document will give a reasonable impression of how an induction machine works. Also discussed are some of the *stray load* losses: those which can be (relatively) easily accounted for in an equivalent circuit description of the machine. But there are other losses which will occur and which are harder to estimate. No claim is made of a particularly accurate job of estimating these losses, and fortunately they do not normally turn out to be very large. To be accounted for here are:

1. No-load losses in rotor teeth because of stator slot opening modulation of fundamental flux density.

2. Load losses in the rotor teeth because of stator zigzag mmf, and

3. No-load losses in the solid rotor body (if it exists) due to stator slot opening modulation of fundamental flux density.

Note that these losses have a somewhat different character from the other miscellaneous losses we compute. They show up as *drag* on the rotor, so we subtract their power from the mechanical output of the machine. The first and third of these are, of course, very closely related so we take them first.

The stator slot openings "modulate" the space-fundamental magnetic flux density. One may estimate a slot opening angle (relative to the slot pitch)

$$\theta_D = \frac{2\pi w_d N_s}{2\pi r} = \frac{w_d N_s}{r}$$

Then the amplitude of the magnetic field disturbance is

$$B_H = B_{r1} \frac{2}{\pi} \sin \frac{\theta_D}{2}$$

In fact, this flux disturbance is really in the form of two traveling waves—one going forward and one backward with respect to the stator at a velocity of ω/N_s. Since operating slip is relatively small, the two variations will have just about the same frequency viewed from the rotor, so it seems reasonable to consider them together. The frequency is

$$\omega_H = \omega \frac{N_s}{p}$$

Now, for laminated rotors, this magnetic field modulation will affect the tips of rotor teeth. Assume that the loss due to this magnetic field modulation can be estimated from ordinary steel data (as we estimated core loss above) and that only the rotor teeth, not any of the rotor body, are affected. The method to be used is straightforward and follows almost exactly what was done for core loss, with modification only of the frequency and field amplitude.

For solid steel rotors, the story is only a little different. The magnetic field will produce an axial electric field

$$\underline{E}_z = R \, \frac{\omega}{p} \, B_H$$

and that, in turn, will drive a surface current

$$\underline{K}_z = \frac{\underline{E}_z}{\underline{Z}_s}$$

Now, what is important is the magnitude of the surface current, and since $|\underline{Z}_s| = \sqrt{1 + .5^2} R_s \approx 1.118 R_s$, we can simply use rotor resistance. The nonlinear, surface penetration depth is

$$\delta = \sqrt{\frac{2 B_0}{\omega_H \sigma |\underline{K}_z|}}$$

A brief iterative substitution, re-calculating δ and then $|\underline{K}_z|$ quickly yields consistent values for δ and R_s. Then the full-voltage dissipation is

$$P_{rs} = 2 \pi R l \, \frac{|K_z|^2}{\sigma \delta}$$

and an equivalent resistance is

$$R_{rs} = \frac{3 |V_a|^2}{P_{rs}}$$

Finally, the zigzag-order current harmonics in the stator will produce magnetic fields in the air-gap which will drive magnetic losses in the teeth of the rotor. Note that this is a bit different from the modulation of the space fundamental produced by the stator slot openings (although the harmonic order will be the same, the spatial orientation will be different and will vary with load current). The magnetic flux in the air-gap is most easily related to the equivalent circuit voltage on the nth harmonic

$$B_n = \frac{np v_n}{2lRN_a k_n \omega}$$

This magnetic field variation will be substantial only for the zigzag-order harmonics: the belt harmonics will be essentially shorted out by the rotor cage and those losses calculated within the equivalent circuit. The frequency seen by the rotor is that of the space harmonics, already calculated, and the loss can be estimated in the same way as core loss, although as we have pointed out it appears as a "drag" on the rotor.

4.3 Classification of Motor Types

NEMA offers many ways of classifying motors. Most important of these are classifications according to:

- Application
- Electrical type
- Environmental protection and method of cooling

Given below are brief descriptions of these classifications. For more complete information, see NEMA MG1-1993, Fourth Revision.

4.3.1 Classification according to application

Three motor classes are recognized:

- General-purpose motor
- Definite-purpose motor
- Special-purpose motor.

4.3.1.1 General-purpose motor. A general-purpose motor is a motor with an open or enclosed construction, and is rated for continuous duty. This motor is designed with standard ratings, standard operating characteristics, and mechanical construction for use in usual service conditions and without restriction to a particular application or type of application. Depending on whether the motor is a small, medium or large motor, and whether it is an AC motor or a DC motor, it may have some additional characteristics defined in sections applicable to the particular motor type. Usual service conditions include an altitude no higher than 1000 m and an ambient temperature of 40°C or lower.

4.3.1.2 Definite-purpose motor. A definite-purpose motor is any motor designed in standard ratings with standard operating characteristics or mechanical construction for use in service conditions other than usual, or for use in a particular type of application.

4.3.1.3 Special-purpose motor. A special-purpose motor is a motor with special operating characteristics or special mechanical construction, or both, designed for a particular application, and not falling within the definition of a general-purpose or definite-purpose motor.

4.3.2 Classification according to electrical type

NEMA recognizes two broad classes:

- AC motor
- DC motor.

4.3.2.1 AC motor Two classes of alternating-current (ac) motors are recognized—the induction motor and the synchronous motor. The synchronous motor is further classified as the direct-current excited synchronous motor (usually called the synchronous motor), the permanent-magnet synchronous motor, and the reluctance synchronous motor.

Induction motor. An induction machine is an asynchronous machine that comprises a magnetic circuit which interlinks with two electric circuits, rotating with respect to each other and in which power is transferred from one circuit to the other by electromagnetic induction. An induction motor is an induction machine which transforms electric power into mechanical power, and in which one member (usually the stator) is connected to the power source, and a secondary winding on the other member (usually the rotor) carries induced current. In a squirrel-cage induction motor, the rotor circuit consists of a number of conducting bars shorted by rings at both ends. In a wound-rotor induction motor, the rotor circuit consists of an insulated winding whose terminals are either short-circuited or closed through external circuits.

Synchronous motor. A synchronous machine is an alternating-current machine in which the speed of operation is exactly proportional to the frequency of the system to which it is connected. A synchronous motor is a synchronous machine which transforms electric power into mechanical power. The stator winding is similar to the induction-motor winding. The rotor winding consists of an insulated-field winding, wound to produce the magnetic poles equal in number to the poles for which the stator is wound. The dc field excitation keeps the rotor speed in synchronism with that of the rotating field produced by the stator.

Permanent-magnet synchronous motor. A permanent magnet synchronous motor is a synchronous motor in which the field excitation is provided by permanent magnets.

Reluctance synchronous motor. A reluctance synchronous motor is a motor similar in construction to an induction motor, in which the rotor circuit has a cyclic variation of reluctance providing the effect of salient poles, without permanent magnets or direct current excitation. It has a squirrel-cage winding and starts as an induction motor but operates at synchronous speed.

4.3.2.2 DC motor

Direct-current (dc) motor. A direct-current machine is a machine consisting of a rotating armature winding connected to a commutator and stationary magnetic poles which are excited from a direct current source or permanent magnets. Direct-current motors are of four general types: shunt-wound; series-wound; compound-wound; and permanent-magnet.

Shunt-wound motor. A shunt-wound motor may be a straight shunt-wound or a stabilized shunt-wound motor. In a straight shunt-wound motor the field winding is connected in parallel with the armature circuit or to a source of separate excitation voltage. The shunt field is the only winding supplying field excitation. The stabilized shunt motor has two field windings, the shunt and the light series. The shunt-field winding is connected either in parallel with the armature circuit or to a separate source of excitation voltage. The light series winding is connected in series with the armature winding and is added to prevent a rise of speed or to obtain a slight reduction of speed with increase in load.

Series-wound motor. A series-wound motor is a dc motor in which the field winding is connected in series with the armature circuit.

Compound-wound motor. A compound-wound motor is a dc motor with two separate field windings: one connected as in a straight-shunt wound motor, and the other in series with the armature circuit.

Permanent magnet motor. A permanent magnet motor is a dc motor in which the field excitation is supplied by permanent magnets.

4.3.3 Classification according to environmental protection and method of cooling

Two main classifications are recognized: the open motor, and the totally enclosed motor. The open motor includes the following types:

- Dripproof
- Splashproof
- Semiguarded
- Guarded
- Dripproof-guarded
- Weather-protected I (WP I)
- Weather-protected II (WP II)

The **open** machine has ventilating openings which allow external cooling air passages over and around the windings, and does not restrict the flow of air other than that necessitated by mechanical construction. A **dripproof** motor is an open motor modified so that successful operation is not interfered with when drops of liquid or solid particles strike the enclosure at any angle between 0 to 15° downwards from the vertical. When this angle is up to 100°, the motor is a **splashproof** motor.

A **semiguarded** motor is an open motor whose openings in the top half are usually guarded, whereas a **guarded** motor is one in which all openings giving access to live or rotating parts are limited in size to less than 0.75″ in diameter. A **dripproof-guarded** motor is a dripproof motor whose ventilating openings are guarded as for a guarded motor. A **WP I** motor is an open motor with its ventilating passages so constructed as to minimize the ingress of rain, snow or airborne particles to the electric parts and having individual ventilating openings limited to 0.75″ in diameter.

A **WP II** motor has, in addition to the requirements of a WP I motor, its intake and exhaust ventilating passages arranged so that high velocity air and airborne particles blown into the motor can be discharged without entering the internal ventilating passages leading to the live parts of the machine. At entry three, abrupt right angle changes in direction for cooling air are also provided, and an area of low air velocity not exceeding 600 ft/min is also provided to minimize the entry of moisture or dirt entering the live parts of the motor.

The more common categories of **totally enclosed** motors include:

- Totally enclosed non-ventilated
- Totally enclosed fan-cooled
- Explosionproof
- Dust-ignition proof
- Totally enclosed air-to-water cooled
- Totally enclosed air-to-air cooled

A *totally enclosed* motor is constructed to prevent free exchange of air between the inside and the outside of the case. A **totally enclosed fan-cooled** motor is a totally enclosed motor equipped with external cooling by means of a fan or fans integral with the machine, but external to the enclosing parts.

An *explosionproof* motor is a totally enclosed motor whose enclosure is constructed to withstand an explosion of a specified gas or vapor which may occur within it.

A *dust-ignition proof* motor is a totally enclosed motor constructed to exclude ignitable material which may cause ignition by normal operation of the motor from entering the motor enclosure. The design is such that any arcs, sparks or heat generated inside the motor do not cause ignition of a specific dust in the vicinity of the motor.

A *totally enclosed water-to-air cooled* motor is a totally enclosed machine which is cooled by internally circulating air which, in turn, is cooled by circulating water. It is provided with a water-cooled heat exchanger. A *totally enclosed air-to-air cooled* motor, on the other hand, is a motor in which circulating air rather than water is used to cool the internal circulating air. An air-to-air heat exchanger is provided.

IEC classifies motors with the IP (protection) and IC (cooling) designations, which are much more extensive than NEMA classifications. An enclosure for IEC is completely defined by an IP code and an IC code. The IP designations are a suggested standard for future designs in NEMA.

4.4   Induction Motor Testing

All induction motors are tested before shipment from the factory. This testing can be subdivided in two groups:

- Routine tests
- Complete or prototype tests

IEEE Std 112-1996 applies to induction motor testing.

4.4.1   Routine tests

The primary purpose of the routine test is to insure freedom from electrical and mechanical defects, and to demonstrate by means of key tests the similarity of the motor to a "standard" motor of the same design. The "standard" motor is an imaginary motor whose performance characteristics would agree exactly with the expected performance predictions.

Depending on the size of the motor, some or all of the following tests could constitute routine tests:

- Winding resistance measurement
- No-load running current and power
- High-potential test
- Locked-rotor test
- Air-gap measurement
- Direction of rotation and phase sequence
- Current balance
- Insulation resistance measurement
- Bearing temperature rise
- Magnetic center at no-load
- Shaft voltages
- Noise
- Vibration

NEMA MG1 includes the first three tests for all motors, and the fourth test for medium motors only.

4.4.2 Prototype tests

The purpose of a prototype test is to evaluate all the performance characteristics of the motor. This test consists of the following tests in addition to the routine tests:

- No-load saturation characteristic
- Locked rotor saturation characteristic
- Locked rotor torque and current
- Loss measurement including stray load loss
- Determination or measurement of efficiency
- Temperature rise determination
- Surge withstand test

4.4.2.1 No-load running current and power. This is obtained by measuring volts, current and input power at rated voltage with motor unloaded. The no-load saturation curve is obtained by repeating this test at various voltages between 20% and 125% of rated voltage.

4.4.2.2 Current balance. With the motor running on no-load at rated voltage, the current in all three phases are measured and comparison can then be made between the highest and the lowest values for acceptability.

4.4.2.3 Winding resistance. This is measured usually using a digital bridge, or a calibrated ohmmeter if the resistance is greater than one ohm. The value is then corrected to 25°C for comparison with the expected value.

4.4.2.4 Insulation resistance measurement. Insulation resistance is useful as a long-term maintenance tool. Measured during the life of the motor, it provides an indication of the quality and relative cleanliness of the stator winding insulation. The test made in the factory before the motor is shipped is a good benchmark for this purpose. For this test, all accessories with leads located at the machine terminals are disconnected from the motor, and their leads are connected together and to the frame of the machine.

4.4.2.5 High potential test. This test entails applying a test voltage between the windings and ground for one minute, the test voltage being equal to twice the line voltage plus 1000 volts. The voltage is applied successively between each phase and the frame, with the windings not under test and the other metal parts connected to the frame. All motor accessories that have leads located in the main terminal box are disconnected during this test, with the leads connected together to the frame or core.

4.4.2.6 Vibration test. The normal test entails reading vibration at the bearing housing with the motor running uncoupled and on no-load at rated voltage and frequency. The limits are established in NEMA MG1. See Table 4.5. The unit of measure is peak velocity in in/sec, and the permissible magnitude is a function of speed.

4.4.2.7 Bearing temperature rise. This test is made by operating the motor unloaded for at least two hours while monitoring the bearing temperature. The test is continued until the bearing temperature stabilizes. A good indication of this is when there is less than 1°C rise between two consecutive readings taken half an hour apart.

4.4.2.8 Shaft voltage check. Any unbalances in the magnetic circuits can create flux linkages with the rotating systems which can produce a potential difference between the shaft ends. This is capable of driving circulating currents through the bearings resulting in premature bearing damage. See IEEE Std 112-1996 for details of this test.

4.4.2.9 Stray load loss. The stray load loss is that part of the total loss that does not lend itself to easy calculation. It consists of two parts, *viz.,* losses occurring at fundamental frequency, and losses occurring at high frequency.

The stray load loss can be determined by the indirect method or by the direct method. By the indirect method, the stray load loss is obtained by measuring the total losses using the input-output method and subtracting from them the sum of stator and rotor I^2R losses, the core lose and the friction and windage loss. The method thus entails subtracting two relatively large quantities from each other and is, therefore, not very accurate. For greater accuracy, and for the determination of efficiency by the loss segregation method, the direct measurement techniques must be used. In this, the fundamental frequency and high frequency components are measured separately and require two tests: the rotor removed test, and the reverse rotation test. The fundamental frequency losses can be measured by the rotor removed test, in which consists of measuring the power input with the rotor removed from the motor. The high frequency component is measured by the reverse rotation test, which entails measuring the power input to the motor, with the rotor being driven in the reverse direction to the stator revolving field, and at synchronous speed. For details of this test, see IEEE 112-1996.

4.4.3 Efficiency tests

Efficiency is the ratio of the motor output power and the motor input power.

Efficiency $=$ (output)/(input)

or $=$ (output)/(output $+$ losses)

or $=$ (input $-$ losses)/(input)

It can thus be calculated by a knowledge of power input and power output, or of power output and losses, or power input and losses.

The losses in the induction motor consist of the following:

- Stator I^2R loss
- Rotor I^2R loss
- Core loss
- Friction and windage loss
- Stray load loss

IEEE Std 112 gives 10 different methods for the measurement of efficiency. Only three of these methods will be described here, one each for fractional-horsepower, medium and larger induction motors. For a more complete description, see IEEE Std 112-1996.

4.4.3.1   Method A—input-output method. This method is suitable for fractional-horsepower motors. In this method, the motor is loaded by means of a brake or a dynamometer. Readings of electrical power input, voltage, current, frequency, slip, torque, ambient temperature and stator winding resistance are obtained at four load points, more-or-less equally spaced between 25% and 100% load, and two loads above the 100% point. Motor efficiency is then computed using the procedures laid out in Form A in IEEE Std 112.

4.4.3.2   Method B—input-output with loss segregation. This method is the only method suitable for testing motors designated energy efficient through 250 horsepower size range.

The method consists of several steps which need to be performed in a set order. By this method, the total loss (input minus output) is segregated into its various components with stray-load loss defined as the difference between the total loss and the sum of the conventional losses (stator and rotor I^2R losses, core loss, and friction and windage loss). Once the value of the stray load loss is determined, it is plotted against torque squared, and a linear regression is used to reduce the effect of random errors in the test measurements. The smoothed stray-load loss data are used to calculate the final value of the total loss and the efficiency.

The tests required to be performed to develop the loss information are described below.

- Stator I^2R loss is calculated from a knowledge of the rated stator current and the resistance of the stator winding corrected to the operating temperature.

- Rotor I^2R loss is calculated from a knowledge of the input power at rated load, the stator I^2R loss, the core loss and the per unit slip.

- Rotor I^2R loss = (measured input power − stator I^2R loss − core loss) × per unit slip.

- The core loss and friction and windage losses are determined from the no-load running current and power test. The motor is run with no load at rated voltage and frequency. The friction and windage loss is obtained by plotting the input power minus the stator I^2R loss vs. voltage, and extending this curve to zero voltage. The intercept with zero voltage axis is the friction and windage loss.

- The core loss is obtained by subtracting the sum of stator I^2R loss at no-load current and rated voltage, and the friction and windage loss from the no load power input at rated.

4.4.3.3 Method F (and variations)—equivalent-circuit method. This test is usually used for a motor whose size is greater than 250 hp, and its size is such that it is beyond the capabilities of the test equipment. This method uses the equivalent circuit of the induction motor to determine the performance from circuit parameters established from test measurements. The test provides acceptable accuracy for starting and running performance. It also yields the most accurate determination of the losses and hence the efficiency.

This method uses two locked rotor tests: one at line frequency, and the other at reduced frequency (a maximum of 25% of rated frequency). These tests, in conjunction with the running saturation test, delineate the classical equivalent circuit parameters of the motor. From the no-load saturation test, the magnetizing reactance, the stator leakage reactance and the magnetizing conductance can be determined. The rated-frequency locked-rotor test measures the stator and rotor reactance and the rotor resistance under initial starting conditions. The low-frequency locked-rotor test measures the stator and rotor leakage reactance and rotor resistance at close to the running frequency. The stator and rotor leakage reactances for equivalent circuit are separated using the ratio of these parameters provided by design. Also calculated value of full-load slip, and either tested value of stray load loss, or loss assumed according to Table 4.6 are used. The machine performance is then calculated using the parameters established from the test. Losses as determined from no-load tests are introduced at appropriate places in the calculation to obtain overall performance.

TABLE 4.5 Unfiltered Vibration Limits

Speed, r/min	Rotational frequency, Hz	Velocity in/sec, peak
3600	60	0.15
1800	30	0.15
1200	20	0.15
900	15	0.12
720	12	0.09
600	10	0.08

TABLE 4.6 Stray Load Loss Allowances per IEEE 112

Motor rating, hp	Stray-load Loss (percent of rated output)
1–125	1.8
126–500	1.5
501–2499	1.2
≥2500	0.9

4.4.4 Temperature test

The reason for doing temperature tests is to determine and verify the temperature rise of various parts and windings of the motor when operated at its design load, voltage and frequency, and to insure that unacceptably high temperatures do not exist in any part of the motor. Proper instrumentation of the motor by the installation of thermocouples, resistance temperature detectors, thermometers, together with prompt measurement of the winding temperature at shutdown is critical for this test. Also important is the need to ensure that the motor is shielded from conditions such as rapid change in ambient temperature and presence of air currents, since these will reduce the accuracy of the test. The motor can be loaded by coupling it to a dynamometer, or by the so-called dual frequency equivalent loading method. The test is continued until the motor is thermally stable.

In the dual frequency test method, the test machine is operated at no-load at rated voltage and frequency, and a low frequency power at a different frequency is superposed on the winding. For this test, the frequency of this auxiliary power is set at 10 Hz below the primary frequency, and the voltage is adjusted so that the stator winding current equals the rated load current. Since the motor is supplied with power at two different frequencies, it is subjected to oscillatory torques that will cause the motor vibration during the test to be higher than normal. The temperature rise determined by the dual frequency method is usually within a couple of degrees of that obtained by the direct loading method.

It is important that the hot resistance of the winding after shutdown be measured promptly to preserve accuracy in the calculation of losses and in the determination of temperature rise by the resistance method. IEEE 112 requires that this reading be taken within the time after shutdown shown in Table 4.7.

4.5   Data

4.5.1   Motor output rating categories

Induction motors are divided into three broad output power categories in NEMA MG1-1993, Revision 4: small motors; medium motors; and

TABLE 4.7   Permissible Time Lag

Rating	Time delay (seconds)
50 hp or less	30
51 hp to 200 hp	90
above 200 hp	120

large motors. Small motors are fractional-horsepower motors. Medium motors are integral-horsepower motors, 1 hp through 500 hp in size depending on the speed and number of poles and in maximum ratings given in Table 4.8. Large motors are motors in the horsepower range from 100 hp through 50,000 hp shown in Table 4.9. NEMA MG1 covers large motors for all the ratings in Table 4.9 at speeds of 450 r/min or slower. For higher speeds, only ratings greater than those given in Table 4.10 are included.

Since medium motors are used in very large quantities, much of the performance and dimensions are standardized in NEMA MG1 for ease in specification and application, and to make it possible for a motor

TABLE 4.8 Maximum Ratings for Medium-Class Motors

Synchronous Speed, r/min	Motor hp
1201–3600	500
901–1200	350
721–900	250
601–720	200
515–600	150
451–514	125

TABLE 4.9 Horsepower Ratings for Induction Motors

100	450	1500	4500	11000	19000	45000
125	500	1750	5000	12000	20000	50000
150	600	2000	5500	13000	22500	
200	700	2250	6000	14000	25000	
250	800	2500	7000	15000	27500	
300	900	3000	8000	16000	30000	
350	1000	3500	9000	17000	35000	
400	1250	4000	10000	18000	40000	

TABLE 4.10 NEMA Standard Ratings at ≥514 rev/min

Synchronous speed	HP
3600	500
1800	500
1200	350
900	250
720	200
600	150
514	125

Copyright by NEMA. Used by permission.

from one manufacturer to be interchanged with one from another. Large motors on the other hand are usually designed to particular user specifications, and therefore their performance is specified in a more general manner.

4.5.2 Performance of medium motors

4.5.2.1 Design letters. NEMA designates the design and performance of a motor by a code letter which represents the locked rotor kVA per horsepower for the motor. For selected design codes, NEMA also gives additional performance. These letter codes and the associated locked-rotor kVA are given in Table 4.11.

4.5.2.2 Efficiencies and energy efficient motors. The Energy Policy Act of 1992 (EPACT 1992) requires that all electric motors manufactured after October 24, 1997 be of the energy-efficient type, having an efficiency not less than that required by this statute. The **electric motor** defined here has the following attributes:

- General purpose
- T frame
- Single speed
- Polyphase squirrel-cage induction
- Horizontal, foot-mounted
- Open or enclosed
- 2-, 4- or 6-pole speeds (3600, 1800 and 1200 r/min)
- Operating on 230/460 volt 60 Hz line power
- Continuous rating
- Output in the range of 1 to 200 hp
- NEMA design A or B

TABLE 4.11 Letter Code Designations for Locked Rotor kVA

Letter Designation	kVA per Horsepower	Letter Designation	kVA per Horsepower	Letter Designation	kVA Horsepower
A	0–3.15	G	5.6–6.3	N	11.2–12.5
B	3.15–3.55	H	6.3–7.1	P	12.5–14.0
C	3.55–4.0	J	7.1–8.0	R	14.0–16.0
D	4.0–4.5	K	8.0–9.0	S	16.0–18.0
E	4.5–5.0	L	9.0–10.0	T	18.0–20.0
F	5.0–5.6	M	10.0–11.2	U	20.0–22.4
				V	22.4 AND UP

Copyright by NEMA. Used by permission.

Excluded from EPACT are:

- Definite- and special-purpose motors
- Inverter duty motors
- Footless designs
- U frame motors, and design C and D motors
- Horsepower sizes less than 1 and greater than 200
- 900 r/min and slower speeds
- 50 Hz motors
- Motors operating on 200/400 and 575 volts
- Single-phase motors

NEMA defines a polyphase squirrel-cage induction motor suitable for operation at less than or equal to 600 volts as energy efficient if it meets the efficiency requirements of NEMA MG1-Table 12.10 (Table 4.12), which includes 2-, 4-, 6- and 8-pole motors in outputs of 1 to 500 hp. It also includes the statutory efficiencies of 1 to 200 hp, 2-, 4- and 6-pole motors from EPACT 1992.

4.5.2.3 Starting performance (torque characteristics). Tables 4.13 through 4.15 give NEMA starting performance of design A and B energy efficient motors and include the starting torque, the breakdown torque and the pull-up torque. For others designs, see NEMA MG1-Part 12.

4.5.2.4 Dimensions. NEMA MG1 gives dimensions of medium motors in many horizontal and vertical configurations. For horizontal foot-mounted motors with a single shaft extension, the major dimensions are given in Fig. 4.23 and Table 4.16.

4.5.3 Performance of large induction motors

There are no energy efficiency requirements for large induction motors in NEMA MG1. The performance requirements are also not specified in as much detail as is the case with medium motors.

4.5.3.1 Starting performance. The locked rotor kVA per hp and the design letters are the same as for medium motors (see Table 4.11).

Two starting performance levels are specified—one for standard torque and the other for high-torque applications. Table 4.17 gives the torque requirements for these designs. NEMA requires that the de-

TABLE 4.12   Full-Load Efficiencies of Energy Efficient Motors (NEMA MG1—Table 12-10)

HP	2 Pole Nominal	2 Pole Minimum	4 Pole Nominal	4 Pole Minimum	6 Pole Nominal	6 Pole Minimum	8 Pole Nominal	8 Pole Minimum
				OPEN MOTORS				
1.0	—	—	82.5	80.0	80.0	77.0	74.0	70.0
1.5	82.5	80.0	84.0	81.5	84.0	81.5	75.5	72.0
2.0	84.0	81.5	84.0	81.5	85.5	82.5	85.5	82.5
3.0	84.0	81.5	86.5	84.0	86.5	84.0	86.5	84.0
5.0	85.5	82.5	87.5	85.5	87.5	85.5	87.5	85.5
7.5	87.5	85.5	88.5	86.5	88.5	86.5	88.5	86.5
10.0	88.5	86.5	89.5	87.5	90.2	88.5	89.5	87.5
15.0	89.5	87.5	91.0	89.5	90.2	88.5	89.5	87.5
20.0	90.2	88.5	91.0	89.5	91.0	89.5	90.2	88.5
25.0	91.0	89.5	91.7	90.2	91.7	90.2	90.2	88.5
30.0	91.0	89.5	92.4	91.0	92.4	91.0	91.0	89.5
40.0	91.7	90.2	93.0	91.7	93.0	91.7	91.0	89.5
50.0	92.4	91.0	93.0	91.7	93.0	91.7	91.7	90.2
60.0	93.0	91.7	93.6	92.4	93.6	92.4	92.4	91.0
75.0	93.0	91.7	94.1	93.0	93.6	92.4	93.6	92.4
100.0	93.0	91.7	94.1	93.0	94.1	93.0	93.6	92.4
125.0	93.6	92.4	94.5	93.6	94.1	93.0	93.6	92.4
150.0	93.6	92.4	95.0	94.1	94.5	93.6	93.6	92.4
200.0	94.5	93.6	95.0	94.1	94.5	93.6	93.6	92.4
250.0	94.5	93.6	95.4	94.3	95.4	94.5	94.5	93.6
300.0	95.0	94.1	95.4	94.5	95.4	94.5	—	—
350.0	95.0	94.1	95.4	94.5	95.4	94.5	—	—
400.0	95.4	94.5	95.4	94.5	—	—	—	—
450.0	95.8	95.0	95.8	95.0	—	—	—	—
500.0	95.8	95.0	95.8	95.0	—	—	—	—
				ENCLOSED MOTORS				
1.0	75.5	72.0	82.5	80.0	80.0	77.0	74.0	70.0
1.5	82.5	80.0	84.0	81.5	85.5	82.5	77.0	74.0
2.0	84.0	81.5	84.0	81.5	86.5	84.0	82.5	80.0
3.0	85.5	82.5	87.5	85.5	87.5	85.5	84.0	81.5
5.0	87.5	85.5	87.5	85.5	87.5	85.5	85.5	82.5
7.5	88.5	86.5	89.5	87.5	89.5	87.5	85.5	82.5
10.0	89.5	87.5	89.5	87.5	89.5	87.5	88.5	86.5
15.0	90.2	88.5	91.0	89.5	90.2	88.5	88.5	86.5
20.0	90.2	88.5	91.0	89.5	90.2	88.5	89.5	87.5
25.0	91.0	89.5	92.4	91.0	91.7	90.2	89.5	87.5
30.0	91.0	89.5	92.4	91.0	91.7	90.2	91.0	89.5
40.0	91.7	90.2	93.0	91.7	93.0	91.7	91.0	89.5
50.0	92.4	91.0	93.0	91.7	93.0	91.7	91.7	90.2
60.0	93.0	91.7	93.6	92.4	93.6	92.4	91.7	90.2
75.0	93.0	91.7	94.1	93.0	93.6	92.4	93.0	91.7
100.0	93.6	92.4	94.5	93.6	94.1	93.0	93.0	91.7
125.0	94.5	93.6	94.5	93.6	94.1	93.0	93.6	92.4
150.0	94.5	93.6	95.0	94.1	95.0	94.1	93.6	92.4
200.0	95.0	94.1	95.0	94.1	95.0	94.1	94.1	93.0
250.0	95.4	94.5	95.0	94.1	95.0	94.1	94.5	93.6
300.0	95.4	94.5	95.4	94.5	95.0	94.1	—	—
350.0	95.4	94.5	95.4	94.5	95.0	94.1	—	—
400.0	95.4	94.5	95.4	94.5	—	—	—	—
450.0	95.4	94.5	95.4	94.5	—	—	—	—
500.0	95.4	94.5	95.4	94.5	—	—	—	—

TABLE 4.13   Locked-Rotor Torque of Design A and B Single-Speed Polyphase Squirrel-Cage Medium Motors as Percentage of Full-Load Torque (NEMA MG1 Table 12-2)

		Synchronous Speed, R/min						
60 Hertz		3600	1800	1200	900	720	600	514
Hp	50 Hertz	3600	1500	1000	750	—	—	—
1		—	275	170	135	135	115	110
1½		175	250	165	130	130	115	110
2		170	235	160	130	125	115	110
3		160	215	155	130	125	115	110
5		150	185	150	130	125	115	110
7½		140	175	150	125	120	115	110
10		135	165	150	125	120	115	110
15		130	160	140	125	120	115	110
20		130	150	135	125	120	115	110
25		130	150	135	125	120	115	110
30		130	150	135	125	120	115	110
40		125	140	135	125	120	115	110
50		120	140	135	125	120	115	110
60		120	140	135	125	120	115	110
75		105	140	135	125	120	115	110
100		105	125	125	125	120	115	110
125		100	110	125	120	115	115	110
150		100	110	120	120	115	115	—
200		100	100	120	120	115	—	—
250		70	80	100	100	—	—	—
300		70	80	100	—	—	—	—
350		70	80	100	—	—	—	—
400		70	80	—	—	—	—	—
450		70	80	—	—	—	—	—
500		70	80	—	—	—	—	—

Copyright by NEMA. Used by permission.

veloped torque from locked rotor point to the breakdown torque point on the speed-torque curve exceed by 10% the torque obtained from a curve that varies as the square of the speed and is equal to 100% of rated full-load torque at 100% of rated speed.

4.5.3.2 Application information. Significant application-related information is included in MG1, such as sealed windings, surge withstand capabilities for insulation systems, bus transfer and terminal box sizes. For this and for a more complete description of performance standards for large induction motors, see the references below:

- NEMA MG1-1993, Rev 4, Part 20, "Motors and Generators"

- Ghai, Nirmal K., "IEC and NEMA Standards for Large Squirrel-cage Induction Motors—A Comparison," Paper submitted to IEEE

TABLE 4.14 Breakdown Torque of Single-Speed Design A and B Polyphase Squirrel-Cage Medium Motors with Continuous Ratings as Percentage of Full-Load Torque (NEMA MG1—12.39.1)

		Synchronous Speed, Rpm						
	60 Hertz	3600	1800	1200	900	720	600	514
Hp	50 Hertz	3600	1500	1000	750	–	–	–
1		–	300	265	215	200	200	200
1½		250	280	250	210	200	200	200
2		240	270	240	210	200	200	200
3		230	250	230	205	200	200	200
5		215	225	215	205	200	200	200
7½		200	215	205	200	200	200	200
10–125, INCLUSIVE		200	200	200	200	200	200	200
150		200	200	200	200	200	200	–
200		200	200	200	200	200	–	–
250		175	175	175	175	–	–	–
300–350		175	175	175	–	–	–	–
400–500, INCLUSIVE		175	175	–	–	–	–	–

Power Engineering Society for publication in IEEE Transactions on Energy Conversion.

4.6   Induction Motor Simulation and Control

4.6.1   Introduction

Developed in this section are models useful for simulating operation of induction machines. These models are similar to those used for synchronous machines, and in fact are somewhat easier to handle analytically.

Among the more useful impacts of modern power electronics and control, technology has made it possible to turn induction machines into high performance servomotors. A picture of how this is done is shown in this section. The objective is to emulate the performance of a dc machine, in which torque is a simple function of applied current. For a machine with one field winding, this is simply

$$T = GI_f I_a$$

This makes control of such a machine quite easy, for once the desired torque is known, it is easy to translate that torque command into a current and the motor does the rest.

Dc (commutator) machines are, at least in large sizes, expensive, not particularly efficient, have relatively high maintenance require-

TABLE 4.15 Pull-Up Torque of Single-Speed Design A and B Polyphase Squirrel-Cage Medium Motors with Continuous Ratings as Percentage of Full-Load Torque (NEMA MG1—12-40.1)

		Synchronous Speed, Rpm						
	60 Hertz	3600	1800	1200	900	720	600	514
Hp	50 Hertz	3600	1500	1000	750	—	—	—
1		—	190	120	100	100	100	100
1½		120	175	115	100	100	100	100
2		120	165	110	100	100	100	100
3		110	150	110	100	100	100	100
5		105	130	105	100	100	100	100
7½		100	120	105	100	100	100	100
10		100	115	105	100	100	100	100
15		100	110	100	100	100	100	100
20		100	105	100	100	100	100	100
25		100	105	100	100	100	100	100
30		100	105	100	100	100	100	100
40		100	100	100	100	100	100	100
50		100	100	100	100	100	100	100
60		100	100	100	100	100	100	100
75		95	100	100	100	100	100	100
100		95	100	100	100	100	100	100
125		90	100	100	100	100	100	100
150		90	100	100	100	100	—	
200		90	90	100	100	100	—	—
250		65	75	90	—	—	—	—
300		65	75	90	—	—	—	—
350		65	75	90	—	—	—	—
400		65	75	—	—	—	—	—
450		65	75	—	—	—	—	—
500		65	75	—	—	—	—	—

ments because of the sliding brush/commutator interface, provide environmental problems because of sparking and carbon dust and are environmentally sensitive. The induction motor is simpler and more rugged. Until fairly recently, the induction motor has not been widely used in servo applications because it was though to be "hard to control." It takes a little effort and even some computation to correctly do the controls, but this is becoming increasingly affordable.

4.6.2 Elementary model

In an elementary model of the induction motor, in ordinary variables, referred to the stator, the machine is described by flux-current relationships (in the d-q reference frame)

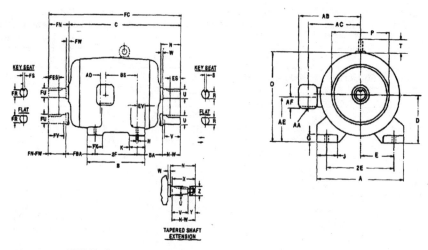

Figure 4.23 NEMA dimensions of medium motors for horizontal foot-mounted motors with a single shaft.

$$\begin{bmatrix} \lambda_{dS} \\ \lambda_{dR} \end{bmatrix} = \begin{bmatrix} L_S & M \\ M & L_R \end{bmatrix} \begin{bmatrix} i_{dS} \\ i_{dR} \end{bmatrix}$$

$$\begin{bmatrix} \lambda_{qS} \\ \lambda_{qR} \end{bmatrix} = \begin{bmatrix} L_S & M \\ M & L_R \end{bmatrix} \begin{bmatrix} i_{qS} \\ i_{qR} \end{bmatrix}$$

Note that the machine is symmetric (there is no saliency). The stator and rotor self-inductances include leakage terms

$$L_S = M + L_{S\ell}$$

$$L_R = M + L_{R\ell}$$

The voltage equations are

$$v_{dS} = \frac{d\lambda_{dS}}{dt} - \omega\lambda_{qS} + r_S i_{dS}$$

$$v_{qS} = \frac{d\lambda_{qS}}{dt} + \omega\lambda_{dS} + r_S i_{qS}$$

$$0 = \frac{d\lambda_{dR}}{dt} - \omega_s\lambda_{qR} + r_R i_{dR}$$

$$0 = \frac{d\lambda_{qR}}{dt} + \omega_s\lambda_{dR} + r_R i_{qR}$$

Note that both rotor and stator have "speed" voltage terms since

TABLE 4.16 Dimensions for Alternating-Current Foot-Mounted Machines with Single Straight-Shaft Extension (NEMA MG1—11.31)

Frame Designation	A Max	D	E	2F	BA	H	U	N-W	V Min	R	ES Min	S	AA Min
42	—	2.62	1.75	1.69	2.06	0.28 slot	0.3750	1.12	—	0.328	—	flat	—
48	—	3.00	2.12	2.75	2.50	0.34 slot	0.5000	1.50	—	0.453	—	flat	—
48H	—	3.00	2.12	4.75	2.50	0.34 slot	0.5000	1.50	—	0.453	—	flat	—
56	—	3.50	2.44	3.00	2.75	0.34 slot	0.6250	1.88	—	0.517	1.41	0.188	—
56H	—	3.50	2.44	5.00	2.75	0.34 slot	0.6250	1.88	—	0.517	1.41	0.188	—
143T	7.0	3.50	2.75	4.00	2.25	0.34 hole	0.8750	2.25	2.00	0.771	1.41	0.188	¾
145T	7.0	3.50	2.75	5.00	2.25	0.34 hole	0.8750	2.25	2.00	0.771	1.41	0.188	¾
182T	9.0	4.50	3.75	4.50	2.75	0.41 hole	1.1250	2.75	2.50	0.986	1.78	0.250	¾
184T	9.0	4.50	3.75	5.50	2.75	0.41 hole	1.1250	2.75	2.50	0.986	1.78	0.250	¾
213T	10.5	5.25	4.25	5.50	3.50	0.41 hole	1.3750	3.38	3.12	1.201	2.41	0.312	1
215T	10.5	5.25	4.25	7.00	3.50	0.41 hole	1.3750	3.38	3.12	1.201	2.41	0.312	1
254T	12.5	6.25	5.00	8.25	4.25	0.53 hole	1.625	4.00	3.75	1.416	2.91	0.375	1¼
256T	12.5	6.25	5.00	10.00	4.25	0.53 hole	1.625	4.00	3.75	1.416	2.91	0.375	1¼
284T	14.0	7.00	5.50	9.50	4.75	0.53 hole	1.875	4.62	4.38	1.591	3.28	0.500	1½
284TS	14.0	7.00	5.50	9.50	4.75	0.53 hole	1.625	3.25	3.00	1.416	1.91	0.375	1½
286T	14.0	7.00	5.50	11.00	4.75	0.53 hole	1.875	4.62	4.38	1.591	3.28	0.500	1½
286TS	14.0	7.00	5.50	11.00	4.75	0.53 hole	1.625	3.25	3.00	1.416	1.91	0.375	1½
324T	16.0	8.00	6.25	10.50	5.25	0.66 hole	2.125	5.25	5.00	1.845	3.91	0.500	2
324TS	16.0	8.00	6.25	10.50	5.25	0.66 hole	1.875	3.75	3.50	1.591	2.03	0.500	2
326T	16.0	8.00	6.25	12.00	5.25	0.66 hole	2.125	5.25	5.00	1.845	3.91	0.500	2

TABLE 4.16 (*Continued*)

Frame Designation	A Max	D	E	2F	BA	H	U	N-W	V Min	R	ES Min	S	AA Min
326TS	16.0	8.00	6.25	12.00	5.25	0.66 hole	1.875	3.75	3.50	1.591	2.03	0.500	2
364T	18.0	9.00	7.00	11.25	5.88	0.66 hole	2.375	5.88	5.62	2.021	4.28	0.625	3
364TS	18.0	9.00	7.00	11.25	5.88	0.66 hole	1.875	3.75	3.50	1.591	2.03	0.500	3
365T	18.0	9.00	7.00	12.25	5.88	0.66 hole	2.375	5.88	5.62	2.021	4.28	0.625	3
365TS	20.0	9.00	7.00	12.25	5.88	0.66 hole	1.875	3.75	3.50	1.591	2.03	0.500	3
404T	20.0	10.00	8.00	12.25	6.62	0.81 hole	2.875	7.25	7.00	2.450	5.65	0.750	3
404TS	20.0	10.00	8.00	12.25	6.62	0.81 hole	2.125	4.25	4.00	1.845	2.78	0.500	3
405T	20.0	10.00	8.00	13.75	6.62	0.81 hole	2.875	7.25	7.00	2.450	5.65	0.750	3
405TS	22.0	11.00	8.00	13.75	6.62	0.81 hole	2.125	4.25	4.00	1.845	2.78	0.500	3
444T	22.0	11.00	9.00	14.50	7.50	0.81 hole	3.375	8.50	8.25	2.880	6.91	0.875	3
444TS	22.0	11.00	9.00	14.50	7.50	0.81 hole	2.375	4.75	4.50	2.021	3.03	0.625	3
445T	22.0	11.00	9.00	16.50	7.50	0.81 hole	3.375	8.50	8.25	2.880	6.91	0.875	3
445TS	22.0	11.00	9.00	16.50	7.50	0.81 hole	2.375	4.75	4.50	2.021	3.03	0.625	3
447T	22.0	11.00	9.00	20.00	7.50	0.81 hole	3.375	8.50	8.25	2.880	6.91	0.875	3
447TS	22.0	11.00	9.00	20.00	7.50	0.81 hole	2.375	4.75	4.50	2.021	3.03	0.625	3
449T	22.0	11.00	9.00	25.00	7.50	0.81 hole	3.375	8.50	8.25	2.880	6.91	0.875	3
449TS	22.0	11.00	9.00	25.00	7.50	0.81 hole	2.375	4.75	4.50	2.021	3.03	0.625	3
440	—	11.00	9.00	—	7.50	—	—	—	—	—	—	—	—
500	—	12.50	10.00	—	8.50	—	—	—	—	—	—	—	—

Copyright by NEMA. Used by permission.

TABLE 4.17   NEMA Percent Starting Torques

Design	Locked rotor	Pull-up	Breakdown
Standard	60	60	175
High Torque	200	150	190

they are both rotating with respect to the rotating coordinate system. The speed of the rotating coordinate system is w with respect to the stator. Concerning the rotor, that speed is $\omega_s = \omega - p\omega_m$, where ω_m is the rotor mechanical speed. Note that this analysis does not require that the reference frame coordinate system speed ω be constant.

Torque is given by

$$T^e = \frac{3}{2} p(\lambda_{dS} i_{qS} - \lambda_{qS} i_{dS})$$

4.6.3   Simulation model

As a first step in developing a simulation model, see that the inversion of the flux-current relationship is, for the d-axis (the q-axis is identical)

$$i_{dS} = \frac{L_R}{L_S L_R - M^2} \lambda_{dS} - \frac{M}{L_S L_R - M^2} \lambda_{dR}$$

$$i_{dR} = \frac{M}{L_S L_R - M^2} \lambda_{dS} - \frac{L_S}{L_S L_R - M^2} \lambda_{dR}$$

Now, making the following definitions

$$X_d = \omega_0 L_S$$

$$X_{kd} = \omega_0 L_R$$

$$X_{ad} = \omega_0 M$$

$$X'_d = \omega_0 \left(L_S - \frac{M^2}{L_R} \right)$$

the currents become

$$i_{dS} = \frac{\omega_0}{X_d'} \lambda_{dS} - \frac{X_{ad}}{X_{kd}} \frac{\omega_0}{X_d'} \lambda_{dR}$$

$$i_{dR} = \frac{X_{ad}}{X_{kd}} \frac{\omega_0}{X_d'} \lambda_{dS} - \frac{X_d}{X_d'} \frac{\omega_0}{X_{kd}} \lambda_{dR}$$

The q-axis is the same.

With these calculations for the current, a torque may be written as

$$T_e = \frac{3}{2} p(\lambda_{dS} i_{qS} - \lambda_{qS} i_{dS}) = -\frac{3}{2} p \frac{\omega_0 X_{ad}}{X_{kd} X_d'} (\lambda_{dS} \lambda_{qR} - \lambda_{qS} \lambda_{dR})$$

Note that the foregong expression was written assuming that the variables are expressed as *peak* quantities.

With these, the simulation model is quite straightforward. The state equations are

$$\frac{d_{sd} \lambda_{dS}}{dt} = V_{dS} + \omega \lambda_{qS} - R_S i_{dS}$$

$$\frac{d\lambda_{qS}}{dt} = V_{qS} - \omega \lambda_{dS} - R_S i_{qS}$$

$$\frac{d\lambda_{dR}}{dt} = \omega_s \lambda_{qR} - R_R i_{dR}$$

$$\frac{d\lambda_{qR}}{dt} = -\omega_s \lambda_{dR} - R_S i_{qR}$$

$$\frac{d\Omega_m}{dt} = \frac{1}{J}(T_e + T_m)$$

where the rotor frequency (slip frequency) is

$$\omega_s = \omega - p\Omega_m$$

For simple simulations and constant excitation frequency, the choice of coordinate systems is arbitrary. For example, one might choose to fix the coordinate system to a synchronously rotating frame, so that stator frequency $\omega = \omega_0$. In this case, the stator voltage could lie on one axis or another. A common choice is $V_d = 0$ and $V_q = V$.

4.6.4 Control model

To turn the machine into a servomotor, it is necessary to be a bit more sophisticated about the coordinate system. In general, the principle of field-oriented control is much like emulating the function of a dc (commutator) machine. First, determine the location of the flux, then inject current to interact most directly with that flux.

As a first step, note that because the two stator flux linkages are the sum of air-gap and leakage flux

$$\lambda_{dS} = \lambda_{agd} + L_{S\ell} i_{dS}$$

$$\lambda_{qS} = \lambda_{agq} + L_{S\ell} i_{qS}$$

This means that torque may be expressed as

$$T^e = \frac{3}{2} p (\lambda_{agd} i_{qS} - \lambda_{agq} i_{dS})$$

Next, note that the rotor flux is, similarly, related to air-gap flux

$$\lambda_{agd} = \lambda_{dR} - L_{R\ell} i_{dR}$$

$$\lambda_{agq} = \lambda_{qR} - L_{R\ell} i_{qR}$$

Torque now becomes

$$T^e = \frac{3}{2} p (\lambda_{dR} i_{qS} - \lambda_{qR} i_{dS}) - \frac{3}{2} p L_{R\ell} (i_{dR} i_{qS} - i_{qR} i_{dS})$$

Now, since the rotor currents could be written as

$$i_{dR} = \frac{\lambda_{dR}}{M + L_{R\ell}} - \frac{M}{M + L_{R\ell}} i_{dS}$$

$$i_{qR} = \frac{\lambda_{qR}}{M + L_{R\ell}} - \frac{M}{M + L_{R\ell}} i_{qS}$$

That second term can be written as:

$$i_{dR} i_{qS} - i_{qR} i_{dS} = \frac{1}{M + L_{R\ell}} (\lambda_{dR} i_{qS} - \lambda_{qR} i_{dS})$$

so that torque is now

$$T^e = \frac{3}{2} p \left(1 - \frac{L_{R\ell}}{M + L_{R\ell}} \right) (\lambda_{dR} i_{qS} - \lambda_{qR} i_{dS})$$

$$= \frac{3}{2} p \frac{M}{M + L_{R\ell}} (\lambda_{dR} i_{qS} - \lambda_{qR} i_{dS})$$

4.6.5   Field-oriented strategy

Field-oriented control establishes a rotor flux in a known position (usually this position is the d-axis of the transformation). Put a current on the orthogonal axis (where it will be most effective in producing torque). Thus, attempt to set

$$\lambda_{dR} = \Lambda_0$$

$$\lambda_{qR} = 0$$

Torque is then produced by applying quadrature-axis current

$$T^e = \frac{3}{2} p \, \frac{M}{M + L_{R\ell}} \, \Lambda_0 i_{qS}$$

The process is almost that simple. There are a few details involved in determining the location of the quadrature axis and how hard to drive the direct axis (magnetizing) current.

Now, suppose one can succeed in putting flux on the right axis, so that $\lambda_{qR} = 0$, then the two rotor voltage equations are

$$0 = \frac{d\lambda_{dR}}{dt} - \omega_s \lambda_{qR} + r_R I_{dR}$$

$$0 = \frac{d\lambda_{qR}}{dt} + \omega_s \lambda_{dR} + r_R I_{qR}$$

Now, since the rotor currents are

$$i_{dR} = \frac{\lambda_{dR}}{M + L_{R\ell}} - \frac{M}{M + L_{R\ell}} i_{dS}$$

$$i_{qR} = \frac{\lambda_{qR}}{M + L_{R\ell}} - \frac{M}{M + L_{R\ell}} i_{qS}$$

The voltage expressions become, accounting for the fact that there is no rotor quadrature-axis flux

$$0 = \frac{d\lambda_{dR}}{dt} + r_R \left(\frac{\lambda_{dR}}{M + L_{R\ell}} - \frac{M}{M + L_{R\ell}} i_{dS} \right)$$

$$0 = \omega_s \lambda_{dR} - r_R \frac{M}{M + L_{R\ell}} i_{qS}$$

Noting that the rotor time constant is

$$T_R = \frac{M + L_{R\ell}}{r_R}$$

the result is

$$T_R \frac{d\lambda_{dR}}{dt} + \lambda_{dR} = Mi_{dS}$$

$$\omega_s = \frac{M}{T_R} \frac{i_{qS}}{\lambda_{dR}}$$

The first of these two expressions describes the behavior of the direct-axis flux: as one would think, it has a simple first-order relationship with direct-axis stator current. The second expression, which describes slip as a function of quadrature-axis current and direct-axis flux, actually describes how fast to turn the rotating coordinate system to hold flux on the direct axis.

Now, a real machine application involves phase currents i_a, i_b and i_c, and these must be derived from the model currents i_{dS} and i_{qS}. This is done, of course, with a mathematical operation which uses a transformation angle θ. That angle is derived from the rotor mechanical speed and computed slip

$$\theta = \int (p\omega_m + \omega_s)\, dt$$

A generally good strategy to make this sort of system work is to measure the three-phase currents and derive the direct- and quadrature-axis currents from them. A good estimate of direct-axis flux is made by running direct-axis flux through a first-order filter. The tricky operation involves dividing quadrature-axis current by direct-axis flux to get slip, but this is now easily done numerically (as are the trigonometric operations required for the rotating coordinate system transformation). An elementary block diagram of a (possibly) plausible scheme for this is shown in Fig. 4.24.

Start with commanded values of direct- and quadrature-axis currents, corresponding to flux and torque respectively, then translate by a rotating coordinate transformation into commanded phase currents. That transformation (simply the inverse Park's transform) uses the angle q derived as part of the scheme. In some (cheap) implementations of this scheme, the commanded currents are used rather than the measured currents to establish the flux and slip.

The commanded currents i_a^*, etc., are shown as inputs to an "Amplifier." This might be implemented as a PWM current-course, for example, and a tight loop here results in a rather high performance servo system.

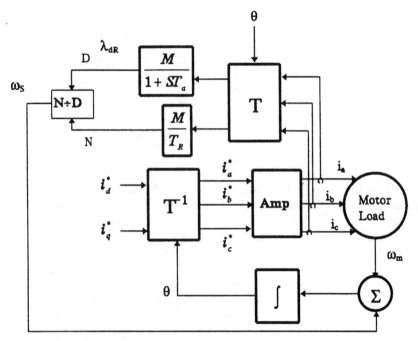

Figure 4.24 Field-oriented controller.

4.7 Induction Motor Speed Control

4.7.1 Introduction

The inherent attributes of induction machines make them very attractive for drive applications. They are rugged, economical to build and have no sliding contacts to wear. The difficulty with using induction machines in servomechanisms and variable speed drives is that they are "hard to control," since their torque-speed relationship is complex and nonlinear. However, with modern power electronics to serve as frequency changers and digital electronics to do the required arithmetic, induction machines are seeing increased use in drive applications.

4.7.2 Volts/Hz control

Induction machines generally tend to operate at relatively low *per unit* slip, so one very effective way of building an adjustable speed drive is to supply an induction motor with adjustable stator frequency. And this is, indeed, possible. One thing to remember is that flux is inversely proportional to frequency, so that to maintain constant flux, one must make stator voltage proportional to frequency (hence the name "constant volts/Hz"). However, voltage supplies are always limited, so that at some frequencies it is necessary to switch to constant voltage control. The analogy to dc machines is fairly direct here: below some "base" speed, the machine is controlled in constant flux ("volts/Hz") mode, while above the base speed, flux is inversely proportional to speed. It is easy to see that the maximum torque is then inversely proportional to the square of flux, or therefore to the square of frequency.

To get a first-order picture of how an induction machine works at adjustable speed, start with the simplified equivalent network that describes the machine, as shown in Fig. 4.25.

Torque can be calculated by finding the power dissipated in the virtual resistance R_2/s and dividing by electrical speed. For a three-phase machine, and assuming RMS magnitudes

$$T_e = 3 \frac{p}{\omega} |I_2|^2 \frac{R_2}{s}$$

where ω is the electrical frequency and p is the number of pole pairs. It is straightforward to find I_2 using network techniques. As an example, Fig. 4.26 shows a series of torque/speed curves for an induction machine operated with a wide range of input frequencies, both below and above its "base" frequency. The parameters of this machine are

Number of Phases	3
Number of Pole Pairs	3
RMS Terminal Voltage (line-line)	230
Frequency (Hz)	60
Stator Resistance R_1	.06 Ω
Rotor Resistance R_2	.055 Ω
Stator Leakage X_1	.34 Ω
Rotor Leakage X_2	.33 Ω
Magnetizing Reactance X_m	10.6 Ω

Strategy for operating the machine is to make terminal voltage magnitude proportional to frequency for input frequencies less than the "Base Frequency," in this case 60 Hz, and to hold voltage constant for frequencies above the "Base Frequency."

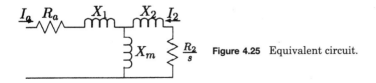

Figure 4.25   Equivalent circuit.

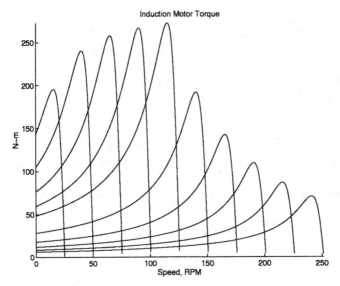

Figure 4.26 Induction machine torque-speed curves.

For high frequencies, the torque production falls fairly rapidly with frequency (as it turns out, it is roughly proportional to the inverse of the square of frequency). It also falls with very low frequency because of the effects of terminal resistance.

4.7.2.1 Idealized model: no stator resistance. Ignoring for the moment R_1, an equivalent circuit is shown in Fig. 4.27. From the rotor, the combination of source, armature leakage and magnetizing branch can be replaced by its equivalent circuit, as shown in in Fig. 4.28.

In the circuit of Fig. 4.28, the parameters are

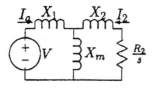

Figure 4.27 Idealized circuit (ignoring armature resistance).

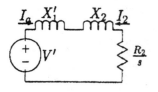

Figure 4.28 Idealized equivalent circuit.

$$V' = V \frac{X_m}{X_m + X_1}$$

$$X' = X_m \| X_1$$

If the machine is operated at variable frequency ω, but the reactance is established at frequency ω_B, current is

$$I = \frac{V'}{j(X_1' + X_2) \dfrac{\omega}{\omega_B} + \dfrac{R_2}{s}}$$

and then torque is

$$T_e = 3|I_2|^2 \frac{R_2}{s} = \frac{3p}{\omega} \frac{|V'|^2 \dfrac{R_2}{s}}{(X_1' + X_2)^2 + \left(\dfrac{R_2}{s}\right)^2}$$

What counts is the *absolute* slip of the rotor, defined with respect to base frequency

$$s = \frac{\omega_r}{\omega} = \frac{\omega_r}{\omega_B} \frac{\omega_B}{\omega} = s_B \frac{\omega_B}{\omega}$$

Then, assuming that voltage is applied proportional to frequency

$$V' = V_0' \frac{\omega}{\omega_B}$$

and with a little manipulation, this is

$$T_e = \frac{3p}{\omega_B} \frac{|V_0'|^2 \dfrac{R_2}{s_B}}{(X_1' + X_2)^2 + \left(\dfrac{R_2}{s_B}\right)^2}$$

If voltage is proportional to frequency, this would imply that torque means constantly applied flux, dependent only on absolute slip. The torque-speed curve is a constant, dependent only on the difference between synchronous and actual rotor speed.

This is fine, but eventually the notion of "volts per Hz" is invalid because at some number of Hz, there are no more volts available. This is generally taken to be the "base" speed for the drive. Above that speed, voltage is held constant, and torque is given by

$$T_e = \frac{3p}{\omega_B} \frac{|V'|^2 \frac{R_2}{s_B}}{(X_1' + X_2)^2 + \left(\frac{R_2}{s_B}\right)^2}$$

The peak of this torque has a square-inverse dependence on frequency, as can be seen from Fig. 4.29.

4.7.2.2 Peak torque capability. Assuming a sufficiently "smart" controller, the actual torque capability of the machine, at some voltage and frequency, is given by

$$T_e = 3|I_2|^2 \frac{R_2}{s} = \frac{3 \frac{p}{\omega} |V'|^2 \frac{R_2}{s}}{\left((X_1' + X_2)\left(\frac{\omega}{\omega_B}\right)\right)^2 + \left(R_1' + \frac{R_2}{s}\right)^2}$$

The *peak* value of that torque is given by the value of $R_2 s$, which maximizes power transfer to the virtual resistance. This is given by the matching condition

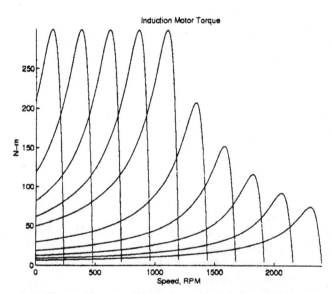

Figure 4.29 Idealized torque-speed curves with zero stator resistance.

$$\frac{R_2}{s} = \sqrt{R_1'^2 + \left((X_1' + X_2) \left(\frac{\omega}{\omega_B} \right) \right)^2}$$

Then maximum (breakdown) torque is given by

$$T_{max} = \frac{\frac{3p}{\omega} |V'|^2 \sqrt{R_1'^2 + \left((X_1' + X_2) \left(\frac{\omega}{\omega_B} \right) \right)^2}}{\left((X_1' + X_2) \left(\frac{\omega}{\omega_B} \right) \right)^2 + \left(R_1' + \sqrt{\left(R_1'^2 + \left((X_1' + X_2) \left(\frac{\omega}{\omega_B} \right) \right)^2 \right)} \right)^2}$$

This is plotted in Fig. 4.30. As a check, this was calculated assuming $R_1 = 0$, and the results are plotted in Fig. 4.31. This plot shows, as one would expect, a constant torque-limit region to zero speed.

4.8 Single Phase Induction Motors

4.8.1 Introduction

Single-phase induction motors are widely used in small power applications where the provision of polyphase power is impossible (as in domestic applications) or uneconomical. Single-phase induction motors share most advantages of polyphase motors: they are rugged and

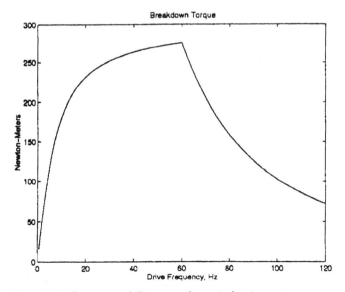

Figure 4.30 Torque-capability curve for an induction motor.

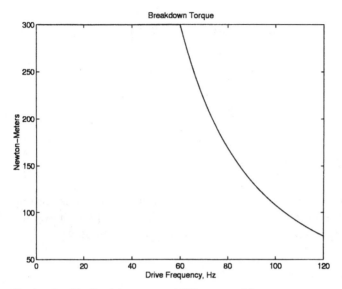

Figure 4.31 Idealized torque capability curve with zero stator resistance.

economical to build. Typically, single-phase motors are larger and less efficient than comparable polyphase motors and require special means for starting. They are therefore somewhat more expensive. In some cases, the starting mechanism involves a switch and this is an extra point of vulnerability which can lead to lower reliability.

This section includes the expressions required to analyze the basic operation of single-phase induction motors. It starts with the running operation of the motor on a single winding, using much of the notation developed for polyphase motors. It is then extended to include a second, *auxiliary* winding to show how these motors can be started.

Figure 4.32 shows a "cartoon-style" drawing of a single-phase motor. The "main" winding is contained in slots in the stator structure. Shown also is an "auxiliary" winding, usually situated in quadrature with the main winding, leading it by 90 electrical degrees. The auxiliary winding is, generally, *not* identical to the main winding.

Once a single-phase machine is turning, it will develop torque. In fact, many applications of single-phase motors use a starting mechanism which is removed once the motor is started, so that it actually runs as a true single-phase machine. For this reason, this section first develops the single-phase motor with a single-stator winding and shows how it works. Following that, it investigates, in a bit less depth, the mechanisms for starting these motors.

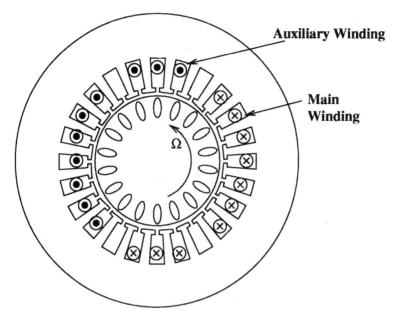

Figure 4.32   Single-phase induction motor windings.

4.8.2   Single-phase MMF

Single-phase induction motors come in a wide range of topologies, but in most cases, they are very similar to polyphase motors. The difference is that they have only one active winding on the stator. Recalling the development of winding inductances, express the space fundamental of MMF produced by a single-phase, p pole-pair winding carrying current i in N turns as

$$F_1 = \frac{4}{\pi} \frac{N i k_w}{2p} \sin p\theta \tag{4.188}$$

Now, if that winding is driven by a sinusoidal current $i = I \sin \omega t$, the MMF becomes

$$F_1 = \frac{4}{\pi} \frac{N I k_w}{2p} \sin p\theta \sin \omega t \tag{4.189}$$

$$= \frac{4}{\pi} \frac{N I k_w}{2p} \frac{1}{2} (\cos(p\theta - \omega t) - \cos(p\theta + \omega t)) \tag{4.190}$$

$$= \frac{4}{\pi} \frac{N I k_w}{2p} \frac{1}{2} \operatorname{Re}\{e^{j(p\theta - \omega t)} - e^{j(p\theta + \omega t)}\} \tag{4.191}$$

The result of Eq. (4.191) is illustrated in Fig. 4.33. The single-phase winding actually produces *two* traveling waves of MMF, each with half of the amplitude of the winding MMF. One rotates forward at synchronous speed and the other rotates backward at the same speed. The rotor of the machine interacts with both of these MMF waves. If the rotor is "round," or uniform in its azimuthal direction, and if the rotor is not heavily saturated so that the flux-MMF interaction is linear, the interactions of the two MMF waves with the rotor are independent of each other.

Assume that the rotor supports two independent MMF waves—one rotating forward and the other backward—and that the currents, referred to the stator, are $\underline{I}_{2f}$ and $\underline{I}_{2r}$, respectively. Assuming a common reference, the two rotor currents will induce a flux in the armature winding

$$\lambda_{ar} = M\mathrm{Re}\{\underline{I}_{2f} + \underline{I}_{2r}\}$$

At the same time, the armature current will induce flux equal to

$$\lambda_{ra} = \lambda_{fa} = \frac{M}{2}\underline{I}_1$$

Before writing a set of expressions that will permit calculation of these various currents, it is necessary to establish the frequencies of the two traveling waves on the rotor. Keeping in mind that the angular coordinate θ' with respect to the rotor is

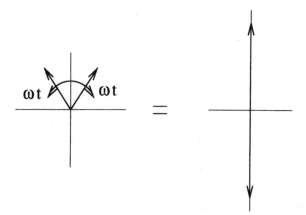

Figure 4.33 Single-phase induction phasor diagram.

$$\theta' = \theta - \frac{\omega_m}{p} t$$

where $\omega_m = \omega(1 - s)$ is the electrical equivalent of mechanical rotor speed and s is conventionally defined *slip*, the arguments of the two traveling waves may now be re-written in the rotor coordinate system

$$\omega t - p\theta = \omega t - (p\theta' + \omega(1 - s)t) \tag{4.192}$$

$$= s\omega t - p\theta' \tag{4.193}$$

$$\omega t + p\theta = \omega t + (p\theta' + \omega(1 - s)t) \tag{4.194}$$

$$= (2 - s)\omega t + p\theta' \tag{4.195}$$

Thus, if s is the conventionally defined *slip*, it is also the slip for the forward-going traveling wave, while $2 - s$ is the slip for the reverse-going wave.

Now it is possible to cast the single-phase motor interaction in terms which are already familiar. Currents are induced in the rotor in exactly the same fashion for both forward and reverse traveling waves as they are in the polyphase machine, taking note of the fact that the slips are different. The voltage equation is

$$\underline{V}_a = j(X_m + X_1)\underline{I}_1 + R_1\underline{I}_1 + jX_m (\underline{I}_{2f} + \underline{I}_{2r}) \tag{4.196}$$

$$0 = j\frac{X_m}{2} \underline{I}_1 + j(X_m + X_2)\underline{I}_{2f} + \frac{R_2}{s} \underline{I}_{2f} \tag{4.197}$$

$$0 = j\frac{X_m}{2} \underline{I}_1 + j(X_m + X_2)\underline{I}_{2r} + \frac{R_2}{2 - s} \underline{I}_{2r} \tag{4.198}$$

This set of expressions can be cast in the form of a convenient equivalent circuit by using $\underline{I}'_{2f} = 2\underline{I}_{2f}$ and $\underline{I}'_{2r} = 2\underline{I}_{2r}$. Then we have

$$\underline{V}_a = j(X_m + X_1)\underline{I}_1 + R_1\underline{I}_1 + j\frac{X_m}{2} (\underline{I}'_{2f} + \underline{I}'_{2r}) \tag{4.199}$$

$$0 = j\frac{X_m}{2} \underline{I}_1 + j \left(\frac{X_m}{2} + \frac{X_2}{2}\right) \underline{I}'_{2f} + \frac{R_2}{2s} \underline{I}'_{2f} \tag{4.200}$$

$$0 = j\frac{X_m}{2} \underline{I}_1 + j \left(\frac{X_m}{2} + \frac{X_2}{2}\right) \underline{I}'_{2r} + \frac{R_2}{2(2 - s)} \underline{I}'_{2r} \tag{4.201}$$

These expressions correspond with the equivalent circuit shown in Fig. 4.34.

4.8.3 Motor performance

The equivalent circuit of Fig. 4.34 may be used in a straightforward fashion to find energy transfer through the machine. Time average power across the air-gap is

$$P_{ag} = \frac{R_2}{2s} |\underline{I}'_{2f}|^2 + \frac{R_2}{2(2 - s)} |\underline{I}'_{2r}|^2$$

Time average power dissipated in the rotor is

$$P_d = \frac{R_2}{2} |\underline{I}'_{2f}|^2 + \frac{R_2}{2} |\underline{I}'_{2r}|^2$$

so that mechanical power is

$$P_m = \frac{R_2}{2s} |\underline{I}'_{2f}|^2 (1 - s) - \frac{R_2}{2} |\underline{I}'_{2r}|^2 (1 - s)$$

and then torque is

$$T = \frac{p}{\omega(1 - s)} P_m$$

$$= \frac{p}{\omega} R_2 s \left(|\underline{I}'_{2f}|^2 + \frac{|\underline{I}'_{2r}|^2}{2 - s} \right)$$

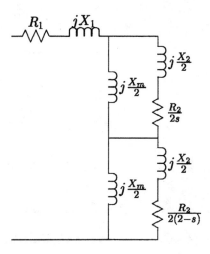

Figure 4.34 Single-phase motor equivalent circuit.

4.8.3.1   Example. From the equivalent circuit, it is straightforward to estimate elements of performance of the machine. In the example shown below, one more electrical element, a fixed element representing core loss, is added. Then, adding the power into that element, efficiency is estimated as power out divided by power in. Power factor is simply the phase angle of net current into the motor.

The example machine is a small motor used in a food processor. The parameters of the motor, looking from the "main" or "running" winding are

Stator Resistance	R_1	2	Ω
Rotor Resistance	R_2	4	Ω
Stator Leakage Reactance	X_1	1.3	Ω
Rotor Leakage Reactance	X_2	1.1	Ω
Magnetizing Reactance	X_m	40	Ω
Core Parallel Resistance	R_c	1540	Ω
Number of Pole Pairs	p	2	
Terminal Voltage	V	120	V, RMS
Electrical Frequency	f	60	Hz

Shown in Fig. 4.35 is a torque vs. speed curve for this motor, oper-

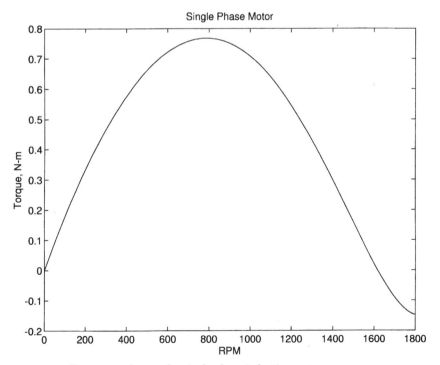

Figure 4.35   Torque-speed curve for single-phase induction motor.

ating against a nominal 60 Hz, 120 V RMS power supply, running on the "main" winding only. Note that in this type of motor, the torque does not go to zero-at-zero slip: at very low slips, the torque is actually slightly negative. The torque *does* go to zero-at-zero speed because of course, at that speed, the motor can make no distinction between the "forward" and "reverse" components of MMF.

Figure 4.36 shows efficiency and power factor for this motor as a function of output power. Efficiency of single-phase motors tends to be lower than that of polyphase motors because the "reverse" MMF produces torque in the wrong direction.

4.8.4 Starting

Single-phase motors create no net torque-at-zero speed, which means that they cannot "start" without some additional influence. To make single-phase motors work, a number of methods have been used, including building a repulsion motor into the machine, using a second-phase winding driven in such a way as to provide currents that are both out of phase in time and space, and using parasitic "shaded"

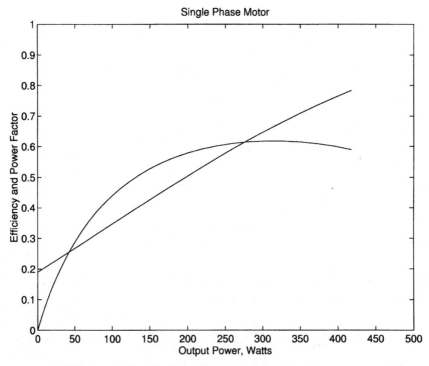

Figure 4.36 Efficiency and power factor for single-phase induction motor.

poles. The first of these methods is now obsolete because it requires an additional commutator. Capacitive and resistive split-phase motors are actually quite common, and we will describe how they work. Shaded-pole motors are also used in small machines.

4.8.4.1 Split-phase windings. Split-phase motors utilize a second armature winding, connected in such a way that its current is temporarily out of phase with the current in the main winding, at least when the rotor is stalled. This phase shift may be produced by a series capacitor or, if large starting torque is not required, by a resistor.

Figure 4.37 shows a *capacitor-start* motor, one of the most common of split-phase types. In such motors there are two windings, typically arranged to be *spatially* in quadrature, but not necessarily identical. The strategy in starting a motor is to try to get the current phasors to appear as they do in Fig. 4.38. During starting, the current in the main winding lags terminal voltage because the winding is inductive. If a capacitor is inserted in series with the auxiliary winding, and if

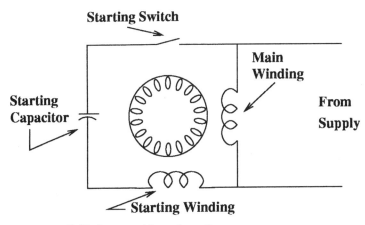

Figure 4.37   Split-phase machine schematic.

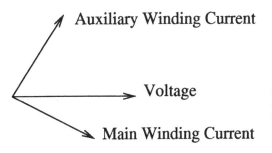

Figure 4.38 Idealized phasor diagram at start for single-phase induction motor.

its impedance dominates, its current will tend to *lead* terminal voltage. If everything is sized correctly, the current in the auxiliary winding can be made to lead that of the main winding by approximately 90°.

Note that since the terminal impedance of the induction motor will increase and become more resistive as its speed increases, a capacitor sized to produce optimal currents at start will not produce optimal currents when the machine is running. Further, the large capacitance (economical capacitors often used for starting duty) are relatively lossy and would overheat if left connected to the motor. For this reason, in most capacitor-start motors a switch is used to disconnect the starting capacitor once the machine is turning.

The starting switch is usually operated by a speed-sensitive mechanism (a variation of flyballs) which holds the switch closed when the motor is stationary, but which withdraws and allows the switch to open as the rotor turns at a high enough speed. Such a mechanism is simple and cheap, but has moving parts and always employs a pair of contacts. Such switches are therefore subject to wearout failures and environmental hazards.

An alternative to the mechanical, centrifugal switch is an electronic equivalent, usually employing a simple timer and a Triac, which connects the auxiliary winding and capacitor for a fixed period of time after energization.

It is not necessary to employ a capacitor in a split-phase motor. If the auxiliary winding is made to have relatively high resistance, compared with its inductance at start, the current in that winding will lead the current in the main winding by some amount (although it is not possible to get the ideal 90° phase shift). This will tend to produce a rotating flux wave. Such motors have smaller starting torque than capacitive split-phase motors, and cannot be used in applications such as pumps where the motor must start against a substantial "head". For low starting duty applications (for example driving most fans) however, resistive split-phase windings are satisfactory for starting.

"Permanent" split-phase motors leave the auxiliary winding connected all of the time. For example, permanent capacitive split-phase motors employ a capacitor selected to be able to operate all of the time. The sizing of that capacitor is a compromise between running and starting performance. Where running efficiency and noise are important and starting duty is light, a permanent split-phase capacitor motor may be the appropriate choice. Permanent split-phase motors do not have the reliability problems associated with the starting switch.

In some applications, it is appropriate to employ *both* a starting and running capacitor, as shown in Fig. 4.39. While this type of motor employs a starting switch, it can achieve both good starting and running performance.

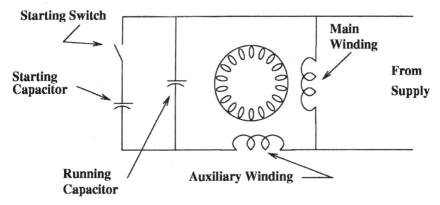

Figure 4.39 Split-phase machine schematic with running capacitor.

4.8.4.2 Analytical model. Assume a single-phase motor with two windings (geometrically, like a two-phase motor in which the two phases are not necessarily "balanced"). If the number of turns and winding factor of the second phase together produce a flux per ampere that is, say α times that of the primary winding, and if that winding is in space quadrature with the main winding as shown in Fig. 4.40, note that

The *forward* component of stator MMF is

$$\underline{F}_f = \frac{4}{\pi}\frac{N}{2p}(\underline{I_a} - j\alpha\underline{I_b}) \qquad (4.202)$$

The *reverse* component of stator MMF is

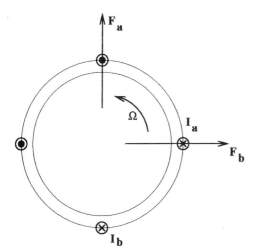

Figure 4.40 Split phasors for single-phase induction motor.

$$\underline{F}_f = \frac{4}{\pi}\frac{N}{2p}(\underline{I}_a + j\alpha\underline{I}_b) \tag{4.203}$$

Then the voltage and current equations which govern the machine are

$$\underline{V}_a = j(X_m + X_1)\underline{I}_a + j\frac{X_m}{2}(\underline{I}'_{2f} + \underline{I}'_{2r}) \tag{4.204}$$

$$\underline{V}_b = j(\alpha^2 X_m + X_b)\underline{I}_b + j\alpha\frac{X_m}{2}(j\underline{I}'_{2f} - j\underline{I}'_{2r}) \tag{4.205}$$

$$0 = j\frac{X_m}{2}(\underline{I}_a - j\alpha\underline{I}_b) + j\left(\frac{X_m}{2} + \frac{X_2}{s}\right)\underline{I}'_{2f} + \frac{R_2}{2s}\underline{I}'_{2f} \tag{4.206}$$

$$0 = j\frac{X_m}{2}(\underline{I}_a + j\alpha\underline{I}_b) + j\left(\frac{X_m}{2} + \frac{X_2}{s}\right)\underline{I}'_{2r} + \frac{R_2}{2(2 - s)}\underline{I}'_{2r} \tag{4.207}$$

This set of expressions does not lend itself to a nice equivalent circuit, but may be solved in a straightforward way. Of course for a split-phase motor, some auxiliary impedance is placed in series with winding "b," and the terminal voltage is the same as the main winding. The set is therefore

$$\underline{V} = j(X_m + X_1)\underline{I}_a + j\frac{X_m}{2}(\underline{I}'_{2f} + \underline{I}'_{2r}) \tag{4.208}$$

$$\underline{V} = j(\alpha^2 X_m + X_b + \underline{Z}_e)\underline{I}_b + j\alpha\frac{X_m}{2}(j\underline{I}'_{2f} - j\underline{I}'_{2r}) \tag{4.209}$$

$$0 = j\frac{X_m}{2}(\underline{I}_a - j\alpha\underline{I}_b) + j\left(\frac{X_m}{2} + \frac{X_2}{s}\right)\underline{I}'_{2f} + \frac{R_2}{2s}\underline{I}'_{2f} \tag{4.210}$$

$$0 = j\frac{X_m}{2}(\underline{I}_a + j\alpha\underline{I}_b) + j\left(\frac{X_m}{2} + \frac{X_2}{s}\right)\underline{I}'_{2r} + \frac{R_2}{2(2 - s)}\underline{I}'_{2r} \tag{4.211}$$

Torque and mechanical power are estimated in the same way as for the motor operating with a single winding. Time average power across the air-gap is

$$P_{ag} = \frac{R_2}{2s}|\underline{I}'_{2f}|^2 + \frac{R_2}{2(2 - s)}|\underline{I}'_{2r}|^2 \tag{4.212}$$

Time average power dissipated in the rotor is

$$P_d = \frac{R_2}{2} |\underline{I}'_{2f}|^2 + \frac{R_2}{2} |\underline{I}'_{2r}|^2 \tag{4.213}$$

so that mechanical power is

$$P_m = \frac{R_2}{2s} |\underline{I}'_{2f}|^2 (1 - s) - \frac{R_2}{2} |\underline{I}'_{2r}|^2 \frac{(1 - s)}{(2 - s)} \tag{4.214}$$

and then torque is

$$T = \frac{p}{\omega(1 - s)} P_m \tag{4.215}$$

$$= \frac{p}{\omega} R_2 s \left(|\underline{I}'_{2f}|^2 + \frac{|\underline{I}'_{2r}|^2}{2 - s} \right) \tag{4.216}$$

4.8.4.3 Example. The motor presented in an earlier section is actually a capacitor-start motor, with these parameters

Stator resistance	R_1	2	Ω
Rotor resistance	R_2	4	Ω
Stator leakage reactance	X_1	1.3	Ω
Rotor leakage reactance	X_2	1.1	Ω
Magnetizing reactance	X_m	40	Ω
Core parallel resistance	R_c	1540	Ω
Number of pole pairs	p	2	
Terminal voltage	V	120	V, RMS
Electrical frequency	f	60	Hz
Auxiliary winding factor	α	1.05	
Auxiliary winding resistance	R_b	4.6	Ω
Auxiliary winding leakage reactance	X_b	1.2	Ω
Starting capacitance	C	150	μF

Shown in Fig. 4.41 are starting and running torques for this motor as a function of shaft speed. Note that starting torque is substantially larger than running torque for this particular motor.

4.8.4.4 Shaded poles. There is yet another starting scheme that is widely used, particularly for low-power motors with small starting requirements such as fan motors. This is often referred to as the "shaded-pole" motor. An illustration of this motor type is in Fig. 4.42.

In the shaded-pole motor, part of each pole is surrounded by a "shading coil," usually a single short-circuited turn of copper. This shorted turn links the main flux, but has some inductance itself. Therefore, it tends to reduce the flux through that part of the pole and to retard it in phase. Thus the flux pattern across the pole has a component which

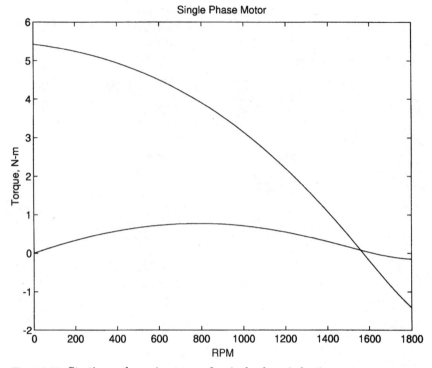

Figure 4.41 Starting and running torque for single-phase induction motor.

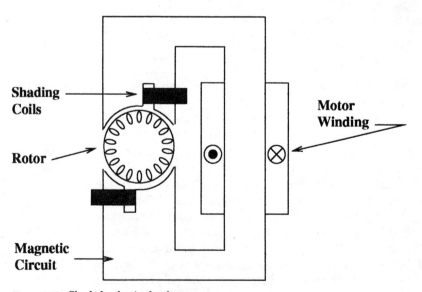

Figure 4.42 Shaded-pole single-phase motor.

tends to move from the unshaded part of the pole to the shaded part of the pole.

Shaded-pole motors tend to have low efficiency and low power density because part of the active pole is permanently short-circuited. They are used primarily for small rating applications in which starting torque is not important, such as blowers. A major application for this motor type is now largely obsolete—the synchronous motors used to drive electric clocks. These always started as shaded-pole induction motors, but ran as (weakly) salient-pole synchronous machines.

4.9 Reference

1. Alger, Philip L. "Induction Machines." Gordon and Breach, New York, 1970.

Synchronous Motors

J. Kirtley and N. Ghai

5.1 Definition

A synchronous motor is a machine that transforms electric power into mechanical power. The average speed of normal operation is exactly proportional to the frequency of the system to which it is connected. Unless otherwise stated, it is generally understood that a synchronous motor has field poles excited with direct current.

5.2 Types

The synchronous motor is built with one set of ac polyphase distributed windings, designated the *armature,* which is usually on the stator and is connected to the ac supply system. The configuration of the opposite member, usually the rotor, determines the type of synchronous motor. Motors with dc excited field windings on silent-pole or round rotors, rated 200 to 100,000 hp and larger, are the dominant industrial type. In the brushless synchronous motor, the excitation (field current) is supplied through shaft-mounted rectifiers from an ac exciter. In the slip-ring synchronous motor, the excitation is supplied from a shaft-mounted exciter or a separate dc power supply. Synchronous-induction motors rated below 5 hp, usually supplied from adjustable-speed drive inverters, are designed with a different reluctance across the air gap in the direct and quadrature axis to develop reluctance torque. The motors have no excitation source for synchronous operation. Synchronous motors employing a permanent-magnetic field excitation and driven by a transistor inverter from a dc source are termed *brushless dc motors.* These are described in Chapter 6.

5.3 Theory of Operation

The operation of the dc separately excited synchronous motor can be explained in terms of the air-gap magnetic-field model, the circuit model, or the phasor diagram model of Fig. 5.1.

In the magnetic-field model of Fig. 5.1a, the stator windings are assumed to be connected to a polyphase source, so that the winding currents produce a rotating wave of current density J_a and radial armature reaction field B_a. The rotor carrying the main field poles is rotating in synchronism with these waves. The excited field poles produce a rotating wave of field B_d. The net magnetic field B_t is the spatial sum of B_a and B_d; it induces an air-gap voltage V_{ag} in the stator windings, nearly equal to the source voltage V_t. The current-density distribution J_a is shown for the current I_a in phase with the voltage V_t, and (in this case) pf = 1. The electromagnetic torque acting between the rotor and the stator is produced by the interaction of the main field B_d and the stator current density J_a, as a $J \times B$ force on each unit volume of stator conductor. The force on the conductors is to the left $(-\phi)$; the reaction force on the rotor is to the right $(+\phi)$, and in the direction of rotation.

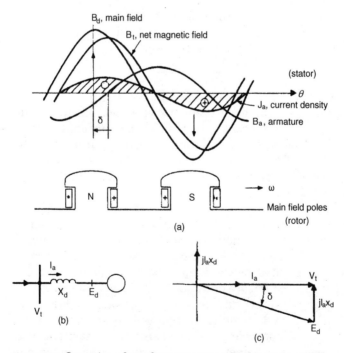

Figure 5.1 Operation of synchronous motor. (a) Air-gap magnetic-field model; (b) circuit model; (c) phasor-diagram model.

The operation of the synchronous motor can be represented by the circuit model of Fig. 5.1b. The motor is characterized by its synchronous reactance x_d and the excitation voltage E_d behind x_d. The model neglects saliency (poles), saturation, and armature resistance, and is suitable for first-order analysis, but not for calculation of specific operating points, losses, field current, and starting.

The phasor diagram of 5.1c is drawn for the field model and circuit model previously described. The phasor diagram neglects saliency and armature resistance. The phasors correspond to the waves in the field model. The terminal voltage V_t is generated by the field B_t; the excitation voltage E_d is generated by the main field B_d; the voltage drop $jI_a x_d$ is generated by the armature reaction field B_a; and the current I_a is the aggregate of the current-density wave J_a. The power angle δ is that between V_t and E_d, or between B_i and B_d. The excitation voltage E_d, in pu, is equal to the field current I_{fd}, in pu, on the air-gap line of the no-load (open-circuit) saturation curve of the machine.

5.4 Power-Factor Correction

Synchronous motors were first used because they were capable of raising the power factor of systems having large induction-motor loads. Now they are also used because they can maintain the terminal voltage on a weak system (high source impedance) and they are more efficient than corresponding induction motors, particularly the low-speed motors. Synchronous motors are built for operation at pf = 1.0, or pf = 0.8 lead, the latter being higher in cost and slightly less efficient at full load.

The selection of a synchronous motor to correct an existing power factor is merely a matter of bookkeeping of active and reactive power. The synchronous motor can be selected to correct the overall power factor to a given value, in which case it must also be large enough to accomplish its motoring functions; or it can be selected for its motoring function and required to provide the maximum correction that it can when operating at pf = 0.8 lead. In Fig. 5.2, a power diagram shows how the active and reactive power components P_s and Q_s of the synchronous motor are added to the components P_i and Q_i of an induction motor to obtain the total P_t and Q_t components, the kVA$_t$, and the power factor. The Q_s of the synchronous motor is based on the rated kVA and pf = 0.8 lead, rather than the actual operating kVA.

The synchronous motor can support the voltage of a weak system, so that a larger-rating synchronous motor can be installed than an induction motor for the same source impedance. With an induction motor, both the P and Q components produce voltage drops in the source impedance. With a synchronous motor, operating at leading

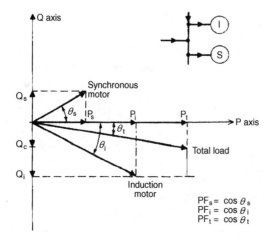

Figure 5.2 Power diagrams of induction motor and synchronus motor operating in parallel, showing component and net values of P and Q.

power factor, the P component produces a voltage drop in the source resistance, but the Q component produces a voltage rise in the source reactance that can offset the drop and allow the terminal voltage to be normal. If necessary, the field current of the synchronous motor can be controlled by a voltage regulator connected to the motor bus. The leading current of a synchronous motor is able to develop a sufficient voltage rise through the source reactance to overcome the voltage drop and maintain the motor voltage equal to the source voltage.

5.5 Starting

The interaction of the main field produced by the rotor and the armature current of the stator will produce a net average torque to drive the synchronous motor only when the rotor is revolving at speed n in synchronism with the line frequency f; $n = 120\ f/p$, p = poles. The motor must be started by developing other than synchronous torques. Practically, the motor is equipped with an induction-motor-type squirrel-cage winding on the rotor, in the form of a damper winding, in order to start the motor.

The motor is started on the damper windings with the field winding short-circuited, or terminated in a resistor, to attenuate the high "transformer"-induced voltages. When the motor reaches the lowest slip speed, nearly synchronous speed, the field current is applied to the field winding, and the rotor poles accelerate and pull into step with the synchronously rotating air-gap magnetic field. The damper windings see zero slip and carry no further current, unless the rotor oscillates with respect to the synchronous speed.

Starting curves for a synchronous motor are shown in Fig. 5.3. The damper winding is designed for high starting torque, as compared to

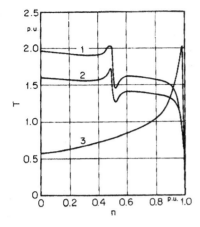

Figure 5.3 Characteristic torque curves for 5000 hp synchronous induction motor during runup at full voltage, (1) Synchronous motor for pf = 1; (2) synchronous motor for pf = 0.8; (3) squirrel-cage induction motor.

an induction motor of the same rating. The closed field winding contributes to the starting torque in the manner of a three-phase induction motor with a one-phase rotor. The field winding produces positive torque to half speed, then negative torque to full speed, accounting for the anomaly at half speed. The maximum and minimum torque excursion at the anomaly is reduced by the resistance in the closed field winding circuit during starting. The effect is increased by the design of the damper winding.

The velocity of the rotor during the synchronizing phase, after field current is applied, is shown in Fig. 5.4. The rotor is assumed running at 0.05 pu slip on the damper winding. The undulation in speed, curve 1, is the effect of the poles attempting to synchronize the rotor just by reluctance torque. The added effect of the field current is shown by curve 2, and the resultant by curve 3. The effect of the reluctance torque of curve 1 is not dependent on pole polarity. The synchronizing

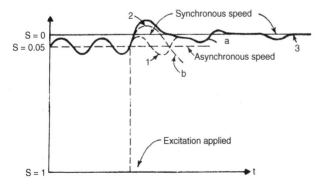

Figure 5.4 Relationship between slip and time for a synchronous motor pulling into synchronism. (a) Successful; (b) unsuccessful.

torque of curve 2, with the field current applied, is pole polarity dependent; the poles want to match the air-gap field in the forward torque direction. Curve a shows a successful synchronization. Curve b shows the condition of too much load or inertia to synchronize.

The method used to start a synchronous motor depends upon two factors: the required torque to start the load and the maximum starting current permitted from the line. Basically, the motor is started by using the damper windings to develop asynchronous torque or by using an auxiliary motor to bring the unloaded motor up to synchronous speed. Solid-state converters have also been used to bring up to speed large several-hundred-MVA synchronous motor/generators for pumped storage plants.

Techniques for asynchronous starting on the damper windings are the same as for squirrel-cage induction motors of equivalent rating. Across-the-line starting provides the maximum starting torque, but requires the maximum line current. The blocked-rotor kVA of synchronous motors as a function of pole number is shown in Fig. 5.5. If the ac line to the motor supplies other loads, the short-circuit kVA of the line must be at least 6 to 10 times the blocked rotor kVA of the motor to limit the line-voltage dip on starting. The starting and pull-in torques for three general classes of synchronous motors are shown in Fig. 5.6. The torques are shown for rated voltage; for across-the-line starting, the values will be reduced to V_t^2 (pu).

Reduced-voltage starting is used where the full starting torque of the motor is not required and/or the ac line cannot tolerate the full starting current. The starter includes a three-phase open-delta or three-winding autotransformer, which can be set to apply 50, 65, or 80% of line voltage to the motor on the first step. The corresponding torque is reduced to 25, 42, or 64%. the starter switches the motor to full voltage when it has reached nearly synchronous speed, and then applies the field excitation to synchronize the motor.

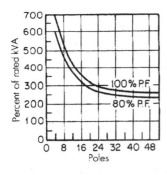

Figure 5.5 Approximate blocked-rotor kVA of synchronous motors.

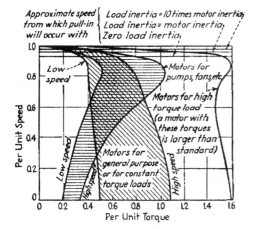

Figure 5.6   Approximate starting performance of synchronous motors.

ANSI C50.11 limits the number of starts for a synchronous motor, under its design conditions of Wk^2, load torque, nominal voltage, and starting method, to the following:

1. Two starts in succession, coasting to rest between starts, with the motor initially at ambient temperature, or
2. One start with the motor initially at a temperature not exceeding its rated load operating temperature.

If additional starts are required, it is recommended that none be made until all conditions affecting operation have been thoroughly investigated and the apparatus examined for evidence of excessive heating. It should be recognized that the number of starts should be kept to a minimum since the life of the motor is affected by the number of starts.

5.6   Torque Definitions

The torques described in the following paragraphs are listed in the Standards [8,11]. The minimum values are given in Table 5.1.

Locked-rotor torque is the minimum torque which the synchronous motor will develop at rest for all angular positions of the rotor, with rated voltage at rated frequency applied.

Pull-in torque is the maximum constant-load torque under which the motor will pull into synchronism, at rated voltage and frequency, when its rated field current is applied. Whether the motor can pull the load into step from the slip running on the damper windings de-

TABLE 5.1   Locked-Rotor, Pull-in, and Pull-out Torques for Synchronous Motors

			Percent of rated full-load torque*		
		Locked	Pull-in (based on normal Wk^2	Pull-out†	
r/min	hp	rotor	of load)†	1.0 pf	0.8 pf
514 to 1800	200 and below; 1.0 pf 150 and below; 0.8 pf	100	100	150	175
	250 to 1000; 1.0 pf 200 to 1000; 0.8 pf	60	60	150	175
	1250 and larger	40	60	150	175
450 and below	All ratings	40	30	150	200

*The torque values with other than rated voltage applied are approximately equal to the rated voltage values multiplied by the ratio of the actual voltage to rated voltage in the case of the pull-out torque, and multiplied by the square of this ratio in the case of the locked-rotor and pull-in torque.
†With rated excitation current applied.

pends on the speed-torque character of the load and the total inertia of the revolving parts. A typical relationship between maximum slip and percent of normal Wk^2 for pulling into step is shown in Fig. 5.7. Table 5.1 species minimum values of pull-in torque with the motor loaded with normal Wk^2. *Nominal pull-in torque* is the value at 95% of synchronous speed, with rated voltage at rated frequency applied, when the motor is running on the damper windings.

Pull-out torque is the maximum sustained torque which the motor will develop at synchronous speed for one min, with rated voltage at rated frequency applied, and with rated field current.

In addition, the *pull-up torque* is defined as the minimum torque developed between stand-still and the pull-in point. This torque must

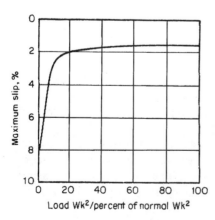

Figure 5.7 Typical relationship between load inertia and maximum slip for pulling synchronous motors into step.

exceed the load torque by a sufficient margin to assure satisfactory acceleration of the load during starting.

The *reluctance torque* is a component of the total torque when the motor is operating synchronously. It results from the saliency of the poles and is a manifestation of the poles attempting to align themselves with the air-gap magnetic field. It can account for up to 30% of the pull-out torque.

The *synchronous torque* is the total steady-state torque available, with field excitation applied, to drive the motor and the load at synchronous speed. The maximum value as the motor is loaded is the pull-out torque, developed as a power angle $\delta = 90°$.

5.7 Synchronization

Synchronization is the process by which the synchronous motor "pulls into step" during the starting process, when the field current is applied to the field winding. Initially, the rotor is revolving at a slip with respect to the synchronous speed of the air-gap magnetic-field waves. The rotor torque, produced by the damper windings, is in equilibrium with the load torque at that slip. The ability of the rotor to accelerate and synchronize depends upon the total inertia (Wk^2), the initial slip, and the closing angle of the poles with respect to the field wave at the instant field current is applied.

Figure 5.8 shows the torque versus angle δ locus for the rotor during a successful synchronization. The rotor is subjected to the synchronous torque T_s, which is a function of δ, and the damper torque T_d, which is a function of the slip velocity $(n_0 - n)$. The torque T_a available to accelerate the rotor is the residual of $T_a = T_s + T_d - T_l$. In the figure, the closing angle is assumed zero at point a. Furthermore, $T_d = T_l$,

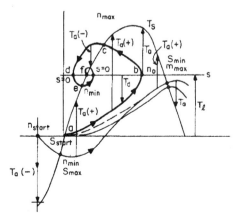

Figure 5.8 Locus of torque and speed vs. power angle δ for a synchronous motor during a successful and an unsuccessful attempt to synchronize.

so that the residual torque T_a is zero. The rotor has a finite slip, so that the power angle δ increases. As it does, the synchronous torque T_s increases, T_a increases and the rotor accelerates to point b, where $n = n_0$, $T_d = 0$. The slip goes negative, reverses the direction of the damper torque, but the rotor continues to accelerate to point c, where the speed is maximum and the accelerating torque is zero. The rotor falls back to points d and e at minimum speed, accelerates again, and finally synchronizes at point f.

If the initial slip is excessive, or if the inertia and/or load too great, the locus in Fig. 5.8 could follow the path ab'. The condition of $T_a = 0$ is reached below synchronous speed; the rotor never pulls into step, but oscillates around the initial slip velocity until the machine is tripped off.

5.8 Damper Windings

Damper windings are placed on the rotors of synchronous motors for two purposes: for starting and for reducing the amplitude of power-angle oscillation. The damper windings consist of copper or brass bars inserted through holes in the pole shoes and connected at the ends to rings to form the equivalent of a squirrel cage. The rings can extend between the poles to form a complete damper. Synchronous motors with solid pole shoes, or solid rotors, perform like motors with damper windings.

The design of the damper winding requires the selection of the bar and ring material to meet the torque and damping requirements. Figure 5.9 shows the effect on the starting curves for the damper winding of varying the material from a low-resistance copper in curve 1, to a higher-resistance brass or aluminum-bronze alloy in curve 2. Curve 1 gives a starting torque of about 0.25 pu, and a pull-in torque of 1.0 pu, of the nominal synchronous torque. Curve 2 gives a higher starting torque of about 0.5 to 1.0 pu, but a pull-in torque of about 0.4 pu of

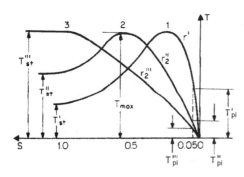

Figure 5.9 Effect of resistivity of damper material on the starting and pull-in torque of the synchronous motor. Damper winding 1, least resistance; damper winding 3, maximum resistance.

the nominal value. The additional starting torque of the field winding is super-imposed on the torque of the damper alone. The damper winding must be designed to meet the characteristics of the load.

To design the damper winding so that the amplitude of the natural-frequency oscillation is reduced, the bar currents during the low-frequency sweeping of the air-gap flux across the pole faces must be maximized. Since the slip frequency is low, the currents and damper effectiveness are maximized by making the dampers low-resistance, corresponding to curve 1 in Fig. 5.9. This design coincides with the requirement for low starting torque and high pull-in torque. In special cases, the equivalent of a deep-bar or double-bar damper can be used, if there is adequate space on the pole shoe.

5.9   Exciters

Exciters are classified into slip-ring types and brushless types. The slip-ring consists of a dc generator, whose output is fed into the motor field winding through slip rings and stationary brushes. The brushless type consists of an ac generator, with rotating armature and stationary field; the output is rectified by solid-state rectifier elements mounted on the rotating structure and fed directly to the motor field winding. In each type, the motor field current is controlled by the exciter field current. Typical kilowatt ratings for exciters for 60-Hz synchronous motors are given in MG1-21.16 as a function of hp rating, speed, and power factor. For a given hp rating, the excitor kW increases as the speed is reduced, and as the power factor is shifted from pf = 1.0 to pf = 0.8 lead.

During starting, the motor field winding must be disconnected from the exciter and loaded with a resistor, to limit the high induced voltage, to prevent damage to the rectifier elements of the brushless type, and to prevent the circulation of ac current through a slip-ring-type dc exciter. The switching is done with a contactor for the slip-ring type, and with thyristors on the rotating rectifier assembly for the brushless type. Except for the disconnection for starting, the synchronous-motor excitation system is practically the same as for an ac generator of the same rating.

Brushless-type exciters are now used on all new high-speed synchronous motors (two to eight poles) that formerly were built with direct-drive dc exciters and slip rings. The brushless-type exciters require minimum maintenance and can be used in explosive-atmospheres. The circuit of a typical brushless-type excitation system is shown in Fig. 5.10. The semicontrolled bridge with three diodes and three thyristors rectifies the output of the ac exciter generator and supplies the motor field winding. The thyristors act as a switch to open

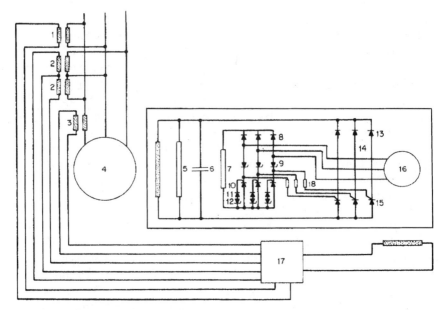

Figure 5.10   Brushless-type excitation system for a synchronous motor.

the rectifier during starting and to close it during running, whereas the ac exciter generator is excited with its own field current. The resistor is permanently connected across the motor field winding during starting and running. It improves the torque characteristics during starting, and protects the bridge elements against transient overvoltages during running. The capacitor protects the diodes and thyristors against commutation overvoltages.

The control system (Fig. 5.10) comprises a simple auxiliary rectifier arrangement connected in parallel with the main rectifier bridge and loaded with an auxiliary resistor 7. Each main thyristor has an auxiliary thyristor which provides the gate current and operates on the same phase of the ac excitor voltage. Consequently the trigger signal always occurs at the correct instant, that is, when the thyristors have a forward loading. No trigger signal is given during the blocking period. There is no excitation at the exciter during run-up, and therefore no trigger signal is applied to the gates of the thyristors and they remain blocking. The alternating current induced in the field winding flows in both directions through the protection resistor 5. When the machine has been run-up to normal speed, the field voltage is applied to the ac exciter. It then supplies the control current and the thyristors are fired. Control losses are only 0.1 to 0.2% of the exciter power and are therefore negligible. The auxiliary thyristor 10 together with the diode 11 and zener diode 12 prevents preignition of the thyristors during run-up due to high residual voltage in the exciter. On the other hand the gates of the other thyristors are protected against overload by zener diode 9 and resistor 18. If the voltage exceeds the zener voltage, the zener diode conducts the excess current.

5.10 Standard Ratings

Standard ratings for dc separately excited synchronous motors are given in NEMA MG1, Part 21. Standard horsepowers range from 20 to 100,000 hp. Speed ratings extend from 3600 r/min (2-pole) to 80 r/min (90-pole) for 60-Hz machines, and five-sixths of the values for 50-Hz machines. The power factor shall be unity or 0.8 leading. The voltage ratings for 60-Hz motors are 200, 230, 460, 575, 2300, 4000, 4600, 6600, and 13,200 V. It is not practical to build motors of all horsepower ratings at these speeds and voltages.

5.11 Efficiency

Efficiency and losses shall be determined in accordance with IEEE Test Procedures for Synchronous Machines, Publication No. 115. The efficiency shall be determined at rated output, voltage, frequency, and power factor. The following losses shall be included in determining the efficiency: (1) I^2R loss of armature and field; (2) core loss; (3) stray-load loss; (4) friction and windage loss; (5) exciter loss for shaft-driven exciter. The resistances should be corrected for temperature.

Typical synchronous motor efficiencies are shown in Fig. 5.11. The unity-power-factor synchronous motor is generally 1 to 3% more efficient than the NEMA Design B induction motors. The 0.8 of synchronous motor, because of the increased copper loss, is lower in efficiency; its efficiency is closer to that of the induction motor at high speed, but better at low speed.

5.12 Standard Tests

Tests on synchronous motors shall be made in accordance with IEEE Test Procedure for Synchronous Machines, Publication No. 115, and ANSI C50.10. The following tests shall be made on motors completely assembled in the factory and furnished with shaft and complete set of bearings: resistance test of armature and field windings; dielectric test of armature and field windings; mechanical balance; current balance at no load; direction of rotation. The following tests may be specified on the same or duplicate motors: locked-rotor current; temperature rise; locked-rotor torque; overspeed; harmonic analysis and TIF; seg-

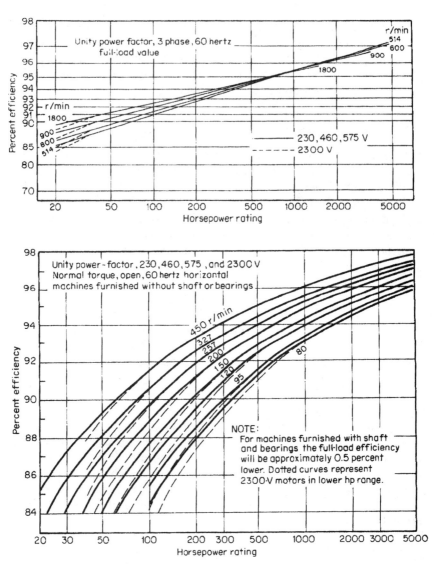

Figure 5.11   Full-load efficiencies of (a) high-speed general-purpose synchronous motors and (b) low-speed synchronous motors.

regated losses; short-circuit tests at reduced voltage to determine re-actances and time constants; field-winding impedance; and speed-torque curve.

The following tests shall be made on all motors not completely assembled in the factory: resistance and dielectric tests of armature and field windings. The following field tests are recommended after installation: resistance and dielectric tests of armature and field windings not completely assembled in the factory; mechanical balance; bearing insulation; current balance at no load; direction of rotation. The following field tests may be specified on the same or duplicate motors: temperature rise; short-circuit tests at reduced voltage to determine reactances and time constants; and field-winding impedance.

The dielectric test for the armature winding shall be conducted for one min, with an ac rms voltage of 1000 V plus twice the rated voltage. For machines rated 6 kV and above, the test may be conducted with a dc voltage of 1.7 times the ac rms test value. The dielectric test for the field winding depends upon the connection for starting. For a short-circuited field winding, the ac rms test voltage is 10 times the rated excitation voltage, but no less than 2500 V, nor more than 5000 V. For a field winding closed through a resistor, the ac rms test voltage is twice the rms value of the IR drop, but not less than 2500 V, where the current is the value that would circulate with a short-circuited winding. When a test is made on an assembled group of several pieces of new apparatus, each of which has passed a high-potential test, the test voltage shall not exceed 85% of the lowest test voltage for any part of the group. When a test is made after installation of a new machine which has passed its high-potential test at the factory and whose windings have not since been disturbed, the test voltages should be 75% of the original values.

5.13 Cycloconverter Drive

A unique application for large low-speed synchronous motors is for gearless ball-mill drives for the cement industry. For a recently installed drive, the motor is rated 8750 hp, 1.0 pf, 6850 kVA, 14.5 r/min 1900 V, 4.84 Hz, 40 poles, Class B. The power is provided by a cycloconverter over the range 0 to 4.84 Hz, as shown in Fig. 5.12. The cycloconverter consists of six thyristor rectifiers, each of which generates the polarity of the three-phase ac voltage wave applied to the motor. The cycloconverter can be used effectively up to about one-third of the line frequency. The motor can be controlled in speed by the cycloconverter frequency, or in torque by the angle between the armature voltage and the field-pole position, approximately the power angle δ.

5.14 Inverter-Synchronous Motor Drive

Synchronous motors over about 1000 hp are being driven by machine-commutated inverters for adjustable-speed drives for large fans, pumps, and other loads. The machine-commutated inverter drive con-

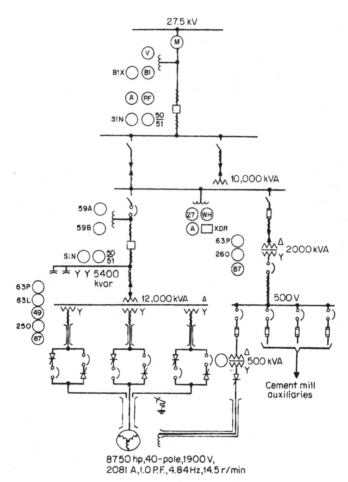

Figure 5.12 Cycloconverter-synchronous motor gearless-drive system for ball mill.

sists of two converters interconnected by a dc link as shown in Fig. 5.13*a*. The synchronous motor operates at constant volts per hertz, i.e., voltage proportional to frequency and speed. The converter characteristics are shown in Fig. 5.13*b* and *c*. The $\pm V_d$ values are 1.35 times the line-line voltage on the ac side of each converter. For a given motor speed, frequency, and voltage, the firing angle of the rectifier is set to α_r to yield the required dc voltage V_l for the link. The firing angle of the inverter is set at α_i in the inverting quadrant of the converter so that the link voltage V_l matches the internal ac voltage generated by the motor at the given speed. Power flows from the rectifier at $V_e I_d$ into the inverter and the motor. The inverter firing signals are

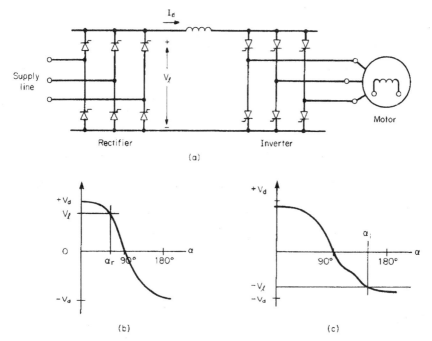

Figure 5.13   (a) Diagram of machine-commutated synchronous motor drive; (b) DC voltage vs. firing angle α_r, characteristic of rectifier; (c) DC voltage vs. firing angle α_i, characteristic of inverter.

synchronized to the motor voltage. For decelerating the motor, the rectifier and inverter functions are reversed by shifting the firing angles. Power flows from the motor into the dc link and to the supply line.

5.15   Synchronous Motor Applications

The motor choice for any application is determined by an evaluation of the load requirements and by economic considerations. In sizes greater than one horsepower per r/min, the synchronous motor has inherent advantages over other motor types (the induction motor and the direct-current motor) that often make it the logical choice for many industrial-drive applications. The synchronous motor is unique in that it runs at constant speed, at leading power factor and has low starting or inrush current. The efficiency is higher than for other motor types. The torque characteristics can be modified by design to match the driven load needs and the characteristics of the available power supply to which the motor is connected. The starting, pull-in, and pull-out torques can be varied over a wide range. The synchronous motor usually runs at unity power factor and can be designed for 80% leading power factor also, thereby providing system power factor improvement.

5.15.1 Motor selection by speed

Specific load and application considerations are critical in the selection of a motor for a given load. In general, the following also applies:

5.15.1.1 3600 to 3000 r/min. Synchronous motors are not economical for this speed range because of the high cost of their rotor construction. Although large two-pole, 3600 r/min synchronous motors from 2000 to 20,000 horsepower (hp) have been manufactured, slower speed motors with step-up gears or squirrel-cage induction motors are usually a more economic choice. The exception will be the adjustable speed motor for use at speeds higher than 4000 r/min. In this case, the higher cost of the rotor is more than compensated by system economics.

5.15.1.2 1800 to 900 r/min. Synchronous motors above 1000 hp are widely used for pumps and for centrifugal compressors (with speed increasers) as well as for fans, pulverizers, rubber mills, banbury mixers, refiners, and motor-generator sets. The need for power-factor correction, high efficiency, low inrush, or constant speed may sometimes also favor synchronous motors below 1000 hp.

For compressors, applications above 2500 hp requiring speed increasers, 1200 r/min unity-power-factor synchronous motors might provide a better fit than the 1800 r/min motors. For 1200 r/min loads such as pumps above 1250 hp and other 900 r/min loads above 1000 hp, the synchronous motor may be more economical than the induction motor. In this speed range however, induction motors are more economical below 1000 hp.

5.15.1.3 720 to 514 r/min. Synchronous motors are often selected in sizes above 1 hp per r/min, that is, 750 hp at 720 r/min, 600 hp at 600 r/min, and 500 hp at 514 r/min.

5.15.1.4 Below 514 r/min. Because of higher efficiency, improved power factor, and possibly lower cost, the synchronous motor should be considered for sizes down to 200 hp. At voltages above 4000 volts, the synchronous motor is usually more economical than other motor types at even lower horsepower sizes.

5.15.2 Load torque considerations

Motor torques are a consideration both during starting as well as during rated speed operation under load.

The maximum torque that a synchronous motor can carry at synchronous speed, rated voltage, and with rated field current, or the torque which if exceeded, will cause the motor to pull out of synchronism is the *pull-out* torque. Since the pull-out torque is directly proportional to voltage, possible line voltage variations must be taken into consideration when specifying a value for it. Adequate torque must be specified to provide ample margin over maximum torque required by the driven machine at the lowest motor-line voltage that will be encountered. However, it is not desirable to specify a pull-out torque greatly above actual requirements because the size and therefore the cost of the machine increases as the pull-out torque increases. High pull-out torque also results in a higher starting current.

Since pull-out torque is also proportional to the motor field excitation, most synchronous motors should be operated at rated field excitation, even at light loads, if the load is subject to fluctuations. Decreasing field excitation to maintain rated power factor at light loads can cause the motor to pull out of synchronism if the load suddenly increases.

Motors rated at 0.8 power factor (leading) usually have higher pull-out torque than unity-power-factor motors because of their higher field excitation. Therefore, for high pull-out torque applications, motors designed for 0.8 power factor may be more economical than unity power factor motors.

Motors with high pull-out torque are capable of handling high momentary overloads. However, they do not necessarily have greater capability for handling continuous overloads. The size of motors with high pull-out torque is bigger in order to increase the thermal capacity of the field winding, but the thermal capacity of the stator winding is not increased in the same proportion. Therefore, unless the stator has been designed for low temperature rise operation, it will reach its limiting operating temperature under rated load conditions. A continuous overload will cause overheating of this winding.

Special consideration must be given to drives for loads with cyclic torque pulsations, such as reciprocating compressors. The torque pulsations produce a cyclic pulsation in the stator current drawn by the motor. These applications require close coordination between the motor designer and the compressor designer so that a limiting value for the stator line current pulsation can be achieved to avoid high voltage fluctuations and other system problems. This is done by providing adequate inertia in the rotating masses.

Current pulsation is defined as the difference between the maximum and minimum current peaks divided by the peak of the rated line current drawn. The NEMA standard current pulsation is set at 66%, because experience has shown that this will still give satisfactory operation in nearly all installations.

5.15.3 Motor starting considerations

The synchronous motor has two starting windings, the field winding and the amortisseur winding. The field discharge resistor used for fixed winding protection also acts as the starting winding. The torque developed by a motor at any instant during acceleration is the resultant of the torques developed by all of these windings. The starting characteristics of the synchronous motor can thus be adjusted by controlling the resistances and the reactances of these rotor circuits.

The starting performance characteristics of the synchronous motor are completely defined by the starting torque, the pull-in torque and the starting current. These characteristics are interdependent: any change in one affects the others.

The torques are expressed as a percentage of the torque delivered by the motor when it is carrying its rated load at synchronous speed. The starting current is expressed as a percentage of rated full-load current.

5.15.3.1 Starting torque. The starting torque is the torque developed by the motor at standstill or zero speed when voltage is applied to the stator winding. The torque must be sufficient to break the load loose and accelerate it to the speed from which the motor pulls into synchronism when field excitation is applied.

5.15.3.2 Pull-in torque. The pull-in torque is the torque developed by a motor when it is operating as an induction motor at the speed from which it can pull into synchronism when the field winding is energized. The speed at which this can happen successfully is generally a function of the total inertia of the rotating masses (motor and the driven equipment). Most synchronous motors pull in from approximately 5% slip, i.e., 95% speed.

The torque developed by a synchronous motor while it is operating as an induction motor varies as the square of the applied voltage. When motors are started at reduced voltage, the control usually is arranged so that transfer to full voltage will be made before the pull-in speed is reached. Field excitation is not applied before the motor is up to pull-in speed.

The pull-in torque capability depends on the inertia of the connected load. Nominal pull-in torque is the torque against which the motor can synchronize a normal inertia load (see load inertia, below) from approximately 95% speed with full voltage and rated excitation applied. Nominal pull-in torque is a function of the motor alone and is independent of the load characteristics.

5.15.3.3 Starting current. The starting current is the current in the synchronous motor stator winding when rated voltage is applied to it at standstill or zero speed. If the motor is being started at reduced voltage, this current will vary approximately with the voltage. The torque developed by the motor, however, varies approximately as the square of the voltage.

5.15.4 Specifying motor torques

Synchronous motors can be designed with any combination of torques required for practical applications. Although NEMA MG1 publishes recommended values for minimum torques for most synchronous motor applications, the torque specified for any given condition should be in line with the needs of that particular application which may be higher, lower or the same as NEMA values. If the torque is not correctly specified, either the motor will be inadequate for the application, or of inadequate design in regard to its cost and economic performance. For example, if the driven machine can be started unloaded with low starting and pull-in torques, specifying lower torque values than standard will permit the motor to be designed with a lower value of starting current and with higher efficiencies under load. If the driven machine cannot be started unloaded, and the torque requirements of the load are higher than standard, the necessary starting and pull-in torques must be specified. This will increase the cost of the motor. The starting current will also be higher. A motor suitable for pulverizer application, for instance, might require 225% starting torque, 150% pull-in torque and 220% pull-out torque. A motor suitable for this application may cost as much as 50% more than a low-torque motor. Further, the starting current is likely to be more than twice that of a low-torque motor.

Typical load torque requirements for standard NEMA motors are given in Table 5.2 and for special industrial applications in Table 5.3. These values should be used as a guide if more precise information is not available.

5.15.5 Load inertia

The torques given in Table 5.2 are expected to be adequate in starting a load whose inertia is no higher than the so-called NEMA nominal inertia which is calculated by the following equation

$$\text{Nominal inertia} = \frac{[0.375 \times (\text{Horsepower rating})^{1.15}]}{[\text{speed in (r/min)}/1000]^2}$$

TABLE 5.2   Torque Values (NEMA MG1-1993, Rev. 4)

Speed r/min	Horsepower	Power factor	Torques, percent of rated full-load torque		
			Locked-rotor	Pull-in (based on normal Wk^2 of load)	Pull-out
500 to 1800	200 and below	1.0	100	100	150
	150 and below	0.8	100	100	175
	250 to 1000	1.0	60	60	150
	200 to 1000	0.8	60	60	175
	1250 and larger	1.0	40	60	150
		0.8	40	60	175
450 and below	All ratings	1.0	40	30	150
		0.8	40	30	200

Copyright by NEMA. Used by permission.

Loads with inertia values lower than the nominal value are considered to be low inertia loads, and those with higher values will be the high inertia loads.

The load inertia is critical in selecting and specifying a synchronous motor because the motor must develop adequate torques not only to start the load, but also accelerate it to the pull-in speed fast enough to ensure that the motor windings do not attain unacceptably high temperatures. If load torque is small compared to available motor torque, the heat developed in the amortisseur winding of the motor during acceleration is equivalent to the kinetic energy of the rotating mass. If the load torque is high, additional rotor heating results. Unless the winding components are large enough, the temperature may become excessive and damage to the windings will be the result.

Loads such as blowers, chippers, compressors, centrifugal fans and refiners are high inertia loads, whereas loads such as pumps, refiners and rolling mills are low inertia loads. The developed torques for the motor at available voltage at the line terminals must be greater than the load requirements for successful acceleration and pulling into synchronism of the load.

In all applications, it is thus important that accurate load requirements (inertia, torques, and voltage during starting) be determined for an adequate motor to be selected. In this regard it is necessary, especially for high inertia loads, that the possibility of starting unloaded be considered, since in such cases the load torque requirements will be lower. A better matched design for the motor will be possible which will be of smaller size and will have higher efficiencies than may otherwise be the case.

5.15.6   Power supply

Limitations of the power supply must always be evaluated when applying large synchronous motors. Starting current is the most impor-

TABLE 5.3   Torques for Special Applications (NEMA MG1-1993, Rev. 4)

Application	Torque-percent of motor full-load torque			Ratio of load inertia to normal inertia*
	Locked-rotor	Pull-in	Pull-out	
Ball mills (for rock and coal)	140	110	175	2–4
Band mills	40	40	250	50–110
Blowers, centrifugal-starting with:				
a. Inlet or discharge valve closed	30	40–60	150	3–30
b. Inlet or discharge valve open	30	100	150	3–30
Banbury mixers	125	125	250	0.2–1
Blowers, positive displacement, rotary-bypassed for starting	50	50	150	3–8
Chipper-starting empty	60	50	250	10–100
Compressors, centrifugal-starting with: Inlet or discharge valve closed	30	40–60	150	3–30
Inlet and discharge valve open	30	100	150	3–30
Compressors, reciprocating-starting unloaded				
Air and gas	30	25	150	0.2–15
Ammonia (discharge 100–250 psi)	30	25	150	0.2–15
Freon	30	40	150	0.2–15
Crushers, cone-starting unloaded	100	100	250	1–2
Crushers-starting loaded (ball or rod mills)	160	120	175	2–4
Fans, centrifugal-starting with: Inlet or discharge valve closed	30	40–60*	150	5–60
Inlet and discharge valve open	30	100	150	5–60
Generators, alternating current	20	10	150	2–15
Generators, direct current (except electroplating)				
a. 150 kW and smaller	20	10	150	2–3
b. Over 150kW	20	10	200	2–3
Generators, dc (m-g sets)	40	10	200	2–15
Grinders, pulp-starting unloaded	50	40	150	2–5
Hammer mills-starting unloaded	100	80	250	30–60
Pulverizers	200	100	175	4–10
Pumps, axial flow, fixed blade-starting with:				
Casing dry	5–40	15	150	0.2–2
Casing filled, discharged closed	5–40	175–250	150	0.2–2
Casing filled, discharge open	5–40	100	150	0.2–2

TABLE 5.3 Continued

Application	Torque-percent of motor full-load torque			Ratio of load inertia to normal inertia*
	Locked-rotor	Pull-in	Pull-out	
Pumps centrifugal-starting with:				
Casing dry	5–40	15	150	0.2–2
Casing filled, discharge closed	5–40	60–80	150	0.2–2
Casing filled, discharge open	5–40	100	150	0.2–2
Pumps mixed flow-starting with:				
Casing dry	5–40	15	150	0.2–2
Casing filled, discharge closed	5–40	80–125	150	0.2–2
Casing filled, discharge open	5–40	100	150	0.2–2
Pumps, reciprocating-starting with:				
a. Cylinders dry	40	30	150	0.2–15
b. By-pass open	40	40	150	0.2–15
c. No by-pass (three cylinder)	150	100	150	0.2–15
Refiners, disc type-starting unloaded	50	50	150	1–20
Refiners, pulp	50	50	150	2–20
Rod mills (for ore grinding)	160	120	175	1.5–4
Rolling mill, roughing stands of hot-strip mills	50	40	250	0.5–1
Rolling mills				
a. structural and rail finishing mills	40	30	250	0.5–1
b. hot strip mills, continuous, individual drive rough stands	50	40	250	0.5–1
Rubber mills	125	125	250	0.5–1
Vacuum pumps, reciprocating-starting unloaded	40	60	150	0.2–15

tant characteristic affecting the power supply because it causes a system voltage dip. For weak systems, this dip may be significant.

Full-voltage starting is the most economical and simple form of motor starting because the motor is energized by simply connecting it across the line. But when motor starting current is too high for the power supply, other starting methods must be evaluated. Reduced-voltage starting can be used only if the motor is capable of developing adequate torques to accelerate the load to running speed without causing the windings to exceed their safe temperatures. Otherwise special motors with higher torque capabilities must be specified.

5.15.7 Power factor

Synchronous motors are usually rated at either unity power factor or 0.8 leading power factor. Substantial power factor improvement can be obtained with leading power factor machines by allowing the motor to operate lightly loaded with rated field current applied to the field winding. Unity power factor motors may provide a small amount of leading kVAR at reduced loads. A machine typically has about 25% reactive kVA at no load and zero reactive kVA at full load. If higher leading kVAR is desired at all reduced loads, field current should be set at rated value at all lower loads for these machines.

If reactive kVA is not required, the field excitation of both unity and leading power factor machines can be adjusted to maintain unity power factor at all loads. This results in reduced losses and higher operating efficiency. However, since pull-out torque of leading power factor motors is reduced under these conditions, synchronous motors operated with reduced field excitation must have automatic provision for increasing field excitation when the load returns to normal values.

5.15.8 Twice slip frequency torque pulsations

The average motor torque during the starting and acceleration of the silent pole synchronous motor is developed by the interaction of the rotating magnetic field produced by the stator winding currents and the induced currents in the amortisseur and field windings. The motor has two axes of symmetry—the direct axis which is the centerline of the poles, and the quadrature axis which is the centerline between the poles. This dissymmetry causes an oscillating torque to be developed in addition to the average motor torque. The frequency of oscillation of this torque equals twice the slip frequency and is twice the line frequency at zero speed and zero at the synchronous speed. The average value of this torque is zero and does not contribute to the accelerating torque. It is simply superimposed on the average torque developed by the motor.

The torque pulsation is defined as the single amplitude of the oscillating torque as a percentage of the rated motor torque.

The rotor systems, which are systems of inertias connected by shafting, generally have torsional natural frequencies in the range of zero to twice the line frequency. Since the torque pulsations also lie in the same frequency range, at some time during the acceleration of the motor and the driven equipment, these two frequencies will coincide and the various components of the rotor system will be subjected to high alternating stresses as the motor passes through the torsional natural frequency of the system. The phenomenon is more troublesome in high-speed motors accelerating high inertia loads.

Whereas the successful system design must take into consideration this phenomenon and design, the number of alternating stress cycles and the torque amplification at resonance can be controlled to some extent by design. To this end, a good motor design will limit torque pulsations to less than 60% of rated torque.

5.16 Synchronous Machine and Winding Models

5.16.1 Introduction

The objective of this section is to develop a simple but physically meaningful model of the synchronous machine as well as to provide an analysis and understanding of the machine, particularly in cases where one or another analytical picture is more appropriate than others. Both operation and sizing of the machine are also discussed.

Machine windings are approached from two points of view. On the one hand, they may be approximated as sinusoidal distributions of current and flux linkage. Alternately, one may take a concentrated coil point of view and generalize that into a more realistic and useful winding model.

5.16.2 Physical picture: current sheet description

Consider the simple picture shown in Fig. 5.14. The machine consists of a cylindrical rotor and a cylindrical stator which are coaxial and which have sinusoidal current distributions on their surfaces: the outer surface of the rotor and the inner surface of the stator.

The rotor and stator bodies are made of highly permeable material, approximated as being infinite for the time being, but this is something that needs to be looked at carefully later. Assume also that the

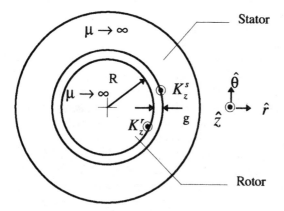

Figure 5.14 Axial view of elementary machine model.

rotor and stator have current distributions are axially (z) directed and sinusoidal

$$K_z^S = K_S \cos p\theta$$

$$K_z^R = K_R \cos p\,(\theta - \phi)$$

Here, the angle ϕ is the physical angle of the rotor. The current distribution on the rotor goes along. Now assume that the air-gap dimension g is much less than the radius: $g \ll R$. It is not difficult to show that with this assumption the radial flux density B_r is nearly uniform across the gap (i.e. not a function of radius) and obeys

$$\frac{\partial B_r}{\partial \theta} = -\mu_0 \frac{K_z^S + K_z^R}{g}$$

Then the radial magnetic flux density for this case is simply

$$B_r = -\frac{\mu_0 R}{pg} (K_S \sin p\theta + K_R \sin p\,(\theta - \phi))$$

Now it is possible to compute the traction on rotor and stator surfaces by recognizing that the surface current distributions are the azimuthal magnetic fields: at the surface of the stator $H_\theta = -K_z^S$, and at the surface of the rotor $H_\theta = K_z^R$. So at the surface of the rotor, traction is

$$\tau_\theta = T_{r\theta} = -\frac{\mu_0 R}{pg} (K_S \sin p\theta + K_R \sin p\,(\theta - \phi))\, K_R \cos p\,(\theta - \phi)$$

The average of that is simply

$$\langle \tau_\theta \rangle = -\frac{\mu_0 R}{2pg} K_S K_R \sin p\phi$$

The same exercise done at the surface of the stator yields the same results (with opposite sign). To find torque, use

$$T = 2\pi R^2 l \langle \tau_\theta \rangle = \frac{\mu_0 \pi R^3 l}{pg} K_S K_R \sin p\phi$$

Pause here to make a few observations:

1. For a given value of surface currents K_S and K_r, torque goes as the fourth power of linear dimension. The volume of the machine goes as the third power, so this implies that torque capability goes as

the ⅓ power of machine volume. Actually, this understates the situation since the assumed surface current densities are the products of volume current densities and winding depth, which one would expect to increase with machine size. Thus, machine torque (and power) densities tend to increase somewhat faster with size.

2. The current distributions want to align with each other. In actual practice, what is done is to generate a stator current distribution which is not static as implied here, but which rotates in space

$$K_z^S = K_S \cos (p\theta - \omega t)$$

and this pulls the rotor along.

3. For a given pair of current distributions, there is a maximum torque that can be sustained, but as long as the torque that is applied to the rotor is less than that value, the rotor will adjust to the correct angle.

5.16.3 Continuous approximation to winding patterns

Those surface current distributions cannot be produced *exactly* with physical windings, but they may be approximated by a turns distribution that looks like

$$n_S = \frac{N_S}{2R} \cos p\theta$$

$$n_R = \frac{N_R}{2R} \cos p (\theta - \phi)$$

Note that this implies that N_S and N_R are the total number of turns on the rotor and stator, i.e.

$$p \int_{-\pi/2}^{\pi/2} n_S R d\theta = N_S$$

Then the assumed surface current densities are as above, with

$$K_S = \frac{N_S I_S}{2R} \qquad K_R = \frac{N_R I_R}{2R}$$

So far nothing is different, but with an assumed number of turns, one can proceed to computing inductances. It is important to remember what these assumed winding distributions mean: they are the *density* of wires along the surface of the rotor and stator. A positive value implies a wire with sense in the $+z$ direction, a negative value implies a wire with sense in the $-z$ direction. That is, if terminal current for a winding is positive, current is in the $+z$ direction if n is positive, and in the $-z$ direction if n is negative. In fact, such a winding would be made of elementary coils with one-half (the negatively-going half) separated from the other half (the positively-going half) by a physical angle of π/p. So the flux linked by that elemental coil would be

$$\Phi_i(\theta) = \int_{\theta - \pi/p}^{\theta} \mu_0 H_r(\theta') \ell R d\theta'$$

So, if only the stator winding is excited, radial magnetic field is

$$H_r = -\frac{N_S I_S}{2gp} \sin p\theta$$

and thus the elementary coil flux is

$$\Phi_i(\theta) = \frac{\mu_0 N_S I_S \ell R}{p^2 g} \cos p\theta$$

Now, this is flux linked by an elementary coil. To get flux linked by a whole winding, we must "add up" the flux linkages of all the elementary coils. In our continuous approximation to the real coil, this is the same as integrating over the coil distribution

$$\lambda_S = p \int_{-\pi/2p}^{\pi/2p} \Phi_i(\theta) n_S(\theta) \, R d\theta$$

This evaluates fairly easily to

$$\lambda_S = \mu_0 \frac{\pi}{4} \frac{\ell R N_S^2}{gp^2} I_s$$

which implies a self-inductance for the stator winding of

$$L_S = \mu_0 \frac{\pi}{4} \frac{\ell R N_S^2}{gp^2}$$

The same process can be used to find self-inductance of the rotor winding (with appropriate changes of spatial variables), and the answer is

$$L_R = \mu_0 \frac{\pi}{4} \frac{\ell R N_R^2}{gp^2}$$

To find the mutual inductance between the two windings, excite one and compute flux linked by the other. All of the expressions here can be used, and the answer is

$$M(\phi) = \mu_0 \frac{\pi}{4} \frac{\ell R N_S N_R}{g p^2} \cos p\phi$$

Now it is fairly easy to compute torque using conventional methods. Assuming both windings are excited, magnetic co-energy is

$$W'_m = \frac{1}{2} L_S I_S^2 + \frac{1}{2} L_R I_R^2 + M(\phi) I_S I_R$$

and then torque is

$$T = \frac{\partial W'_m}{\partial \phi} = -\mu_0 \frac{\pi}{4} \frac{\ell R N_S N_R}{g p} I_S I_R \sin p\phi$$

and then substituting for $N_S I_S$ and $N_R I_R$

$$N_S I_S = 2 R K_S$$

$$N_R I_R = 2 R K_R$$

we get the same answer for torque as with the field approach

$$T = 2\pi R^2 \ell \langle \tau_\theta \rangle = \frac{\mu_0 \pi R^3 \ell}{p g} K_S K_R \sin p\phi$$

5.16.4 Classical, lumped-parameter synchronous machine

Examine the simplest model of a polyphase synchronous machine. Assume a machine in which the rotor is the same as the one previously considered, but in which the stator has three separate windings, identical but with spatial orientation separated by an electrical angle of $120° = 2\pi/3$. The three stator windings have the same self-inductance (L_a).

With a little bit of examination it can be seen that the three stator windings will have mutual inductance, and that inductance will be characterized by the cosine of $120°$. Since the physical angle between any pair of stator windings is the same

$$L_{ab} = L_{ac} = L_{bc} = -\frac{1}{2} L_a$$

There will also be a mutual inductance between the rotor and each phase of the stator. Using M to denote the magnitude of that inductance

$$M = \mu_0 \frac{\pi}{4} \frac{\ell R N_a N_f}{g p^2}$$

$$M_{af} = M \cos{(p\phi)}$$

$$M_{bf} = M \cos{\left(p\phi - \frac{2\pi}{3}\right)}$$

$$M_{cf} = M \cos{\left(p\phi + \frac{2\pi}{3}\right)}$$

Torque for this system is

$$T = -pMi_a i_f \sin{(p\phi)} - pMi_b i_f \sin{\left(p\phi - \frac{2\pi}{3}\right)}$$

$$- pMi_c i_f \sin{\left(p\phi + \frac{2\pi}{3}\right)}$$

5.16.5 Balanced operation

Now, suppose the machine is operated in this fashion: the rotor turns at a constant velocity, the field current is held constant, and the three stator currents are sinusoids in time, with the same amplitude and with phases that differ by 120°.

$$p\phi = \omega t + \delta_i$$

$$i_f = I_f$$

$$i_a = I \cos{(\omega t)}$$

$$i_b = I \cos{\left(\omega t - \frac{2\pi}{3}\right)}$$

$$i_c = I \cos{\left(\omega t + \frac{2\pi}{3}\right)}$$

Straightforward (but tedious) manipulation yields an expression for torque

$$T = -\frac{3}{2} pMII_f \sin{\delta_i}$$

Operated in this way, with balanced currents and with the mechanical speed consistent with the electrical frequency ($p\Omega = \omega$), the machine exhibits a *constant* torque. The phase angle δ_i is called the torque angle, but it is important to use some caution, as there is more than one torque angle.

Now, look at the machine from the electrical terminals. Flux linked by Phase A will be

$$\lambda_a = L_a i_a + L_{ab} i_b + L_{ac} i_c + MI_f \cos p\phi$$

Noting that, under balanced conditions, the sum of phase currents is zero and that the mutual phase-phase inductances are equal, this simplifies to

$$\lambda_a = (L_a - L_{ab}) i_a + MI_f \cos p\phi = L_d i_a + MI_f \cos p\phi$$

where the notation L_d denotes synchronous inductance.

Now, if the machine is turning at a speed consistent with the electrical frequency, it is said to be operating *synchronously*, and it is possible to employ complex notation in the sinusoidal steady state. Then, note

$$i_a = I \cos (\omega t + \theta_i) = \text{Re}\{Ie^{j\omega t + \theta_i}\}$$

The complex amplitude of flux is

$$\lambda_a = \text{Re}\{\underline{\Lambda}_a e^{j\omega t}\}$$

where

$$\underline{I} = Ie^{j\theta_i}$$

$$\underline{I}_f = I_f e^{j\theta_m}$$

Now, terminal voltage of this system is

$$v_a = \frac{d\lambda_a}{dt} = \text{Re}\{j\omega\underline{\Lambda}_a e^{j\theta t}\}$$

This system is described by the equivalent circuit shown in Fig. 5.15, where the internal voltage is

$$\underline{E}_{af} = j\omega MI_f e^{j\theta_m}$$

If that is connected to a voltage source (i.e. if is fixed), terminal current is

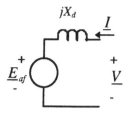

Figure 5.15 Round-rotor synchronous machine equivalent circuit.

$$I = \frac{V - E_{af}e^{j\delta}}{jX_d}$$

where $X_d = \omega L_d$ is the *synchronous reactance.*
Then real and reactive power (in Phase A) are:

$$P + jQ = \frac{1}{2} VI^*$$

$$= \frac{1}{2} V \left(\frac{V - E_{af}e^{j\delta}}{jX_d} \right)^*$$

$$= \frac{1}{2} \frac{|V|^2}{-jX_d} - \frac{1}{2} \frac{VE_{af}e^{j\delta}}{-jX_d}$$

This makes real and reactive power

$$P_a = -\frac{1}{2} \frac{VE_{af}}{X_d} \sin \delta$$

$$Q_a = \frac{1}{2} \frac{V^2}{X_d} - \frac{1}{2} \frac{VE_{af}X_d}{\cos} \delta$$

If we consider all three phases, real power is

$$P = -\frac{3}{2} \frac{VE_{af}}{X_d} \sin \delta$$

Now, look at actual operation of these machines, which can serve either as motors or as generators.

Vector diagrams that describe operation as a motor and as a generator are shown in Fig. 5.16 and 5.17, respectively.

Operation as a generator is not much different from operation as a motor, but it is common to make notations with the terminal current given the opposite ("generator") sign.

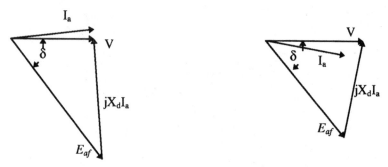

Figure 5.16   Under-excited (a) and over-excited (b) motor operation.

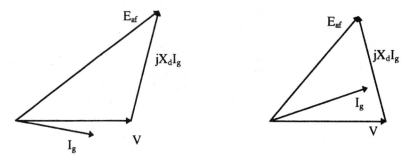

Figure 5.17   Under-excited (a) and over-excited (b) generator operation.

5.16.6 Reconciliation of models

Power and/or torque characteristics can be predicted from two points of view: first, by knowing currents in the rotor and stator, one may derive an expression for torque vs. a power angle

$$T = -\frac{3}{2} pMII_f \sin \delta_i$$

From a circuit point of view, it is possible to derive an expression for power

$$P = -\frac{3}{2} \frac{VE_{af}}{X_d} \sin \delta$$

and of course since power is torque times speed, this implies that

$$T = -\frac{3}{2} \frac{VE_{af}}{\Omega X_d} \sin \delta = -\frac{3}{2} \frac{pVE_{af}}{\omega X_d} \sin \delta$$

To reconcile these notions, look a bit more at what they mean. A generalization of the simple theory to salient pole machines follows as an introduction to two-axis theory of electric machines.

5.16.6.1 Torque angles. Figure 5.18 shows a vector diagram that shows operation of a synchronous motor. It represents the MMF's and fluxes from the rotor and stator in their respective positions in *space* during normal operation. Terminal flux is chosen to be "real," or occupy the horizontal position. In motor operation, the rotor lags by angle δ, so the rotor flux MI_f is shown in that position. Stator current is also shown, and the torque angle between it and the rotor, δ_i is also shown. Now, note that the dotted line OA, drawn perpendicular to a line drawn between the stator flux $L_d I$ and terminal flux Λ_t, has length

$$|OA| = L_d I \sin \delta_i = \Lambda_t \sin \delta$$

Then, noting that terminal voltage $V = \omega \Lambda_t$, $E_a = \omega MI_f$ and $X_d = \omega L_d$, straightforward substitution yields

$$\frac{3}{2} \frac{p V E_{af}}{\omega X_d} \sin \delta = \frac{3}{2} p MII_f \sin \delta_i$$

So the current- and voltage-based pictures *do* give the same result for torque.

5.16.7 Per-unit systems

The per-unit system is a notational device that, in addition to being convenient, is conceptually helpful. The basic notion is quite simple: for most variables, note a base quantity and then divide the *ordinary*

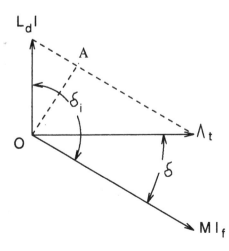

Figure 5.18 Synchronous machine phasor addition.

variable by the *base* to derive a *per-unit* version of that variable. Generally, the base quantity is tied to some aspect of normal operation. So, for example, the base voltage and current might correspond with machine rating. If that is the case, then power base becomes

$$P_B = 3V_B I_B$$

and one can define, in similar fashion, an impedance base

$$Z_B = \frac{V_B}{I_B}$$

Now, a little caution is required here. Voltage base is defined as line-neutral and current base as line-current (both RMS). That is not necessary. In a three-phase system, one could very well have defined base voltage to have been line-line and base current to be current in a delta connected element

$$V_{B\Delta} = \sqrt{3}V_B \qquad I_{B\Delta} = \frac{I_B}{\sqrt{3}}$$

In that case the base power would be unchanged, but base impedance would differ by a factor of three

$$P_B = V_{B\Delta} I_{B\Delta} \qquad Z_{B\Delta} = 3Z_B$$

However, to be consistent with actual impedances (note that a delta connection of elements of impedance $3Z$ is equivalent to a wye connection of Z), the per-unit impedances of a given system are not dependent on the particular connection. In fact one of the major advantages of using a per-unit system is that per-unit values are uniquely determined, while ordinary variables can be line-line, line-neutral, RMS, peak, etc., for a large number of variations.

Perhaps unfortunately, base quantities are usually given as line-line voltage and base power. So that

$$I_B = \frac{P_B}{\sqrt{3}V_{B\Delta}} \qquad Z_B = \frac{V_B}{I_B} = \frac{1}{3}\frac{V_{B\Delta}}{I_{B\Delta}} = \frac{V_{B\Delta}^2}{P_B}$$

Usually, per-unit variables are written as lower-case versions of the ordinary variables

$$v = \frac{V}{V_B} \qquad p = \frac{P}{P_B} \qquad \text{etc.}$$

Thus, written in per-unit notation, real and reactive power for a synchronous machine operating in steady state are

$$p = -\frac{ve_{af}}{x_d} \sin \delta \qquad q = \frac{v^2}{x_d} - \frac{ve_{af}}{x_d} \sin \delta$$

These are, of course, in motor reference coordinates, and represent real and reactive power into the terminals of the machine.

5.16.8 Normal operation

The synchronous machine is used, essentially interchangeably, as a motor and as a generator. Note that, as a motor, this type of machine produces torque only when it is running at synchronous speed. This is not, of course, a problem for a turbogenerator which is started by its prime mover (e.g. a steam turbine). Many synchronous motors are started as induction machines on their damper cages (sometimes called starting cages) and of course with power electronic drives, the machine can often be considered to be "in synchronism", even down to zero speed.

As either a motor or as a generator, the synchronous machine can either produce or consume reactive power. In normal operation, real power is dictated by the load (if a motor) or the prime mover (if a generator), and reactive power is determined by the real power and by field current.

Figure 5.19 shows one way of representing the capability of a synchronous machine. This picture represents operation as a generator, so the signs of p and q are reversed, but all of the other elements of operation are as one would ordinarily expect. Plot p and q (calculated in the normal way) against each other. If one starts at a location $q = -v^2/x_d$, (and remember that normally $v = 1$ per-unit), then the locus

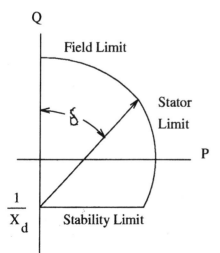

Figure 5.19 Synchronous generator capability diagram.

of p and q is what would be obtained by swinging a vector of length ve_{af}/x_d over an angle δ. This is called a *capability chart* because it is an easy way of visualizing what the synchronous machine (in this case generator) can do. There are three easily noted limits to capability. The upper limit is a circle (the one traced out by that vector) which is referred to as *field* capability. The second limit is a circle that describes constant $|p + jq|$. This is, of course, related to the magnitude of armature current and so this limit is called *armature* capability. The final limit is related to machine stability, since the torque angle cannot go beyond 90°. In actuality, there are often other limits that can be represented on this type of a chart. For example, large synchronous generators typically have a problem with heating of the stator iron when they attempt to operate in highly underexcited conditions (q strongly negative), so that one will often see another limit that prevents the operation of the machine near its stability limit. In very large machines with more than one cooling state (e.g. different values of cooling hydrogen pressure), there may be multiple curves for some or all of the limits.

Another way of describing the limitations of a synchronous machine is embodied in the *Vee Curve*. An example is shown in Fig. 5.20. This is a cross-plot of magnitude of armature current with field current. Note that the field and armature current limits are straightforward (and are the right-hand and upper boundaries, respectively, of the chart). The machine stability limit is what terminates each of the

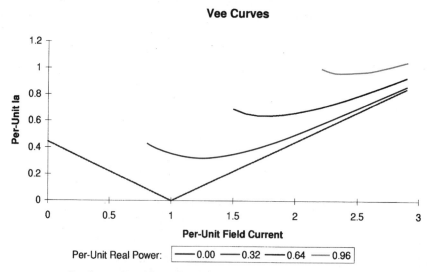

Figure 5.20 Synchronous machine Vee Curve.

curves at the upper left-hand edge. Note that each curve has a minimum at unity power factor. In fact, there is yet another cross-plot possible, called a *compounding curve*, in which field current is plotted against real power for fixed power factor.

5.16.9 Salient-pole machines: two-reaction theory

So far, the machines described are referred to as "round rotor" machines, in which stator reactance is not dependent on rotor position. This is a reasonable approximation for large turbine generators and many smaller two-pole machines, but it is not a fair approximation for many synchronous motors nor for slower speed generators. For many such applications, it is more cost-effective to wind the field conductors around steel bodies (called poles) which are then fastened onto the rotor body, with bolts or dovetail joints. These produce magnetic anisotropies into the machine which affect its operation. The theory which follows is an introduction to two-reaction theory and consequently for the rotating field transformations that form the basis for most modern dynamic analyses.

Figure 5.21 shows a very schematic picture of the salient-pole machine, intended primarily to show how to frame this analysis. As with the round rotor machine, the stator winding is located in slots in the surface of a highly permeable stator core annulus. The field winding is wound around steel pole pieces. Separate the stator current sheet into two components: one aligned with and one in quadrature to the field. Remember that these two current components are themselves (linear) combinations of the stator-phase currents. The transformation

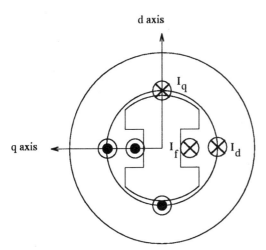

Figure 5.21 Cartoon of a salient-pole synchronous machine.

between phase currents and the d- and q-axis components is straight-forward.

The key here is to separate MMF and flux into two orthogonal components and to pretend that each can be treated as sinusoidal. The two components are aligned with the direct axis and with the quadrature axis of the machine. The direct axis is aligned with the field winding, while the quadrature axis leads the direct by 90°. Then, if ϕ is the angle between the direct axis and the axis of Phase a, we can write for flux linking Phase a

$$\lambda_a = \lambda_d \cos \phi - \lambda_q \sin \phi$$

Then, in steady state operation, if $V_a = d\lambda_a/dt$ and $\phi = \omega t + \delta$

$$V_a = -\omega\lambda_d \sin \phi - \omega\lambda_q \cos \phi$$

which allows these definitions

$$V_d = -\omega\lambda_q$$

$$V_q = \omega\lambda_d$$

one might think of the "voltage" vector as leading the "flux" vector by 90°.

Now, if the machine is linear, those fluxes are given by

$$\lambda_d = L_d I_d + M I_f$$

$$\lambda_q = L_q I_q$$

Note that, in general, $L_d \neq L_q$. In wound-field synchronous machines, usually $L_d > L_q$. The reverse is true for most salient (buried magnet) permanent magnet machines.

Referring to Fig. 5.22, one can resolve terminal voltage into these components

$$V_d = V \sin \delta$$

$$V_q = V \cos \delta$$

or

$$V_d = -\omega\lambda_q = -\omega L_q I_q = V \sin \delta$$

$$V_q = \omega\lambda_d = \omega L_d I_d + \omega M I_f = V \cos \delta$$

which is easily inverted to produce

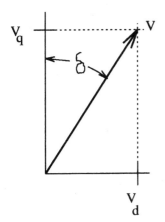

Figure 5.22 Resolution of terminal voltage.

$$I_d = \frac{V \cos \delta - E_{af}}{X_d}$$

$$I_q = -\frac{V \sin \delta}{X_q}$$

where

$$X_d = \omega L_d \qquad X_q = \omega L_q \qquad E_{af} = \omega M I_f$$

These variables are easily cast into a complex frame of reference

$$\underline{V} = V_d + jV_q$$

$$\underline{I} = I_d + jI_q$$

Complex power is

$$P + jQ = \frac{3}{2}\,\underline{V}\underline{I}^* = \frac{3}{2}\{(V_dI_d + V_qI_q) + j(V_qI_d - V_dI_q)\}$$

or

$$P = -\frac{3}{2}\left(\frac{VE_{af}}{X_d}\sin\delta + \frac{V^2}{2}\left(\frac{1}{X_q} - \frac{1}{X_d}\right)\sin 2\delta\right)$$

$$Q = \frac{3}{2}\left(\frac{V^2}{2}\left(\frac{1}{X_d} + \frac{1}{X_q}\right) - \frac{V^2}{2}\left(\frac{1}{X_q} - \frac{1}{X_d}\right)\cos 2\delta - \frac{VE_{af}}{X_d}\cos\delta\right)$$

A phasor diagram for a salient-pole machine is shown in Fig. 5.23.

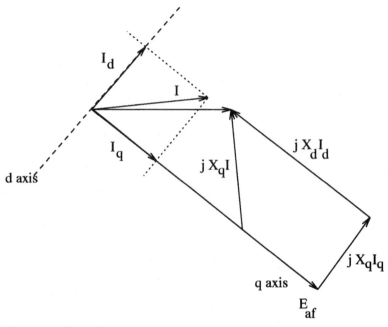

Figure 5.23 Phasor diagram of a salient-pole machine.

This is a little different from the equivalent picture for a round-rotor machine, in that stator current has been separated into its d- and q-axis components, and the voltage drops associated with those components have been drawn separately. It is interesting and helpful to recognize that the internal voltage E_{af} can be expressed as

$$E_{af} = E_1 + (X_d - X_q) I_d$$

where the voltage E_1 is on the quadrature axis. In fact, E_1 would be the internal voltage of a round rotor machine with reactance X_q and the same stator current and terminal voltage. Then the operating point is found fairly easily

$$\delta = -\tan^{-1} \left(\frac{X_q I \sin \psi}{V + X_q I \cos \psi} \right)$$

$$E_1 = \sqrt{(V + X_q I \sin \psi)^2 + (X_q I \cos \psi)^2}$$

A comparison of torque-angle curves for a pair of machines, one with a round, one with a salient rotor is shown in Fig. 5.24. It is not too difficult to see why power systems analysts often neglect saliency in doing things like transient stability calculations.

Power-Angle Curves

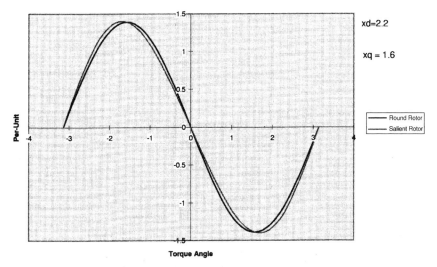

Figure 5.24   Torque-angle curves for round-rotor and salient-pole machines.

5.16.10   Relating rating to size

It is possible, even using the simple model developed so far, to establish a quantitative relationship between machine size and rating, depending (of course) on elements such as useful flux and surface current density. To start, note that the rating of a machine (motor or generator) is

$$|P + jQ| = qVI$$

where q is the number of phases, V is the RMS voltage in each phase and I is the RMS current. To establish machine rating it is necessary to establish voltage and current. These are done separately.

5.16.10.1   Voltage. Assume that the sinusoidal approximation for turns density is valid

$$n_a(\theta) = \frac{N_a}{2R} \cos p\theta$$

and suppose that working flux density is

$$B_r(\theta) = B_0 \sin p(\theta - \phi)$$

Now, to compute flux linked by the winding (and consequently to compute voltage), first compute flux linked by an incremental coil

$$\lambda_i(\theta) = \int_{\theta-\pi/p}^{\theta} \ell B_r(\theta')R d\theta'$$

Then flux linked by the whole coil is

$$\lambda_a = p \int_{-\pi/2p}^{\pi/2p} \lambda_i(\theta)n_a(\theta) \, R d\theta = \frac{\pi}{4} \frac{2\ell R N_a}{p} B_0 \cos p\phi$$

This is instantaneous flux linked when the rotor is at angle ϕ. If the machine is operating at some electrical frequency ω with a phase angle so that $p\phi = \omega t + \delta$, the RMS magnitude of terminal voltage is

$$V_a = \frac{\omega}{p} \frac{\pi}{4} 2\ell R N_a \frac{B_0}{\sqrt{2}}$$

Finally, note that the useful peak current density that can be used is limited by the fraction of machine periphery used for slots

$$B_0 = B_s(1 - \lambda_s)$$

where B_s is the flux density in the teeth, limited by saturation of the magnetic material.

5.16.10.2 Current. The (RMS) magnitude of the current sheet produced by a current of (RMS) magnitude I is

$$K_z = \frac{q}{2} \frac{N_a I}{2R}$$

and then the current is, in terms of the current sheet magnitude

$$I = 2RK_z \frac{2}{qN_a}$$

Note that the surface current density is, in terms of area current density J_s, slot space factor λ_s and slot depth h_s

$$K_z = \lambda_s J_s h_s$$

This gives terminal current in terms of dimensions and useful current density

$$I = \frac{4R}{qN_a} \lambda_s h_s J_s$$

5.16.10.3 Rating. Assembling these expressions, machine rating becomes

$$|P + jQ| = qVI = \frac{\omega}{p} 2\pi R^2 \ell \frac{B_s}{\sqrt{2}} \lambda_s(1 - \lambda_s)h_s J_s$$

This expression is actually fairly easily interpreted. The product of slot factor times one minus slot factor optimizes rather quickly to ¼ (when $\lambda_s = 1$). We could interpret this as

$$|P = jQ| = A_s u_s \tau^*$$

where the interaction *area* is

$$A_s = 2\pi R \ell$$

The surface velocity of interaction is

$$u_s = \frac{\omega}{p} R = \Omega R$$

and the fragment of expression which "looks like" traction is

$$\tau^* = h_s J_s \frac{B_s}{\sqrt{2}} \lambda_s(1 - \lambda_s)$$

Note that this is not quite traction since the current and magnetic flux may not be ideally aligned, and this is why the expression incorporates reactive as well as real power.

This is not yet the whole story. The limit of B_s is easily understood to be caused by saturation of magnetic material. The other important element on shear stress density, $h_s J_s$ is a little more involved.

Per-unit, or normalized synchronous reactance is

$$x_d = X_d \frac{I}{V} = \frac{\mu_0 R}{pg} \frac{\lambda_s}{1 - \lambda_s} \sqrt{2} \frac{h_s J_s}{B_s}$$

While this may be somewhat interesting by itself, it becomes useful if, when solved for $h_s J_a$

$$h_s J_a = x_d g \frac{p(1 - \lambda_s)B_s}{\mu_0 R \lambda_s \sqrt{2}}$$

That is, if x_d is fixed, $h_s J_a$ (and so power) are directly related to air-gap g. Now, to get a limit on g, we must consider the question of how far the field winding can "throw" effective air-gap flux. To understand this question, first calculate the field current to produce rated voltage, no-load, and then the excess of field current required to accommodate load current.

Under *rated* operation, per-unit field voltage is

$$e_{af}^2 = v^2 + (x_d i)^2 + 2x_d i \sin \psi$$

Or, if at rated conditions v and i are both unity (one per unit), then

$$e_{af} = \sqrt{1 + x_d^2 + 2x_d \sin \psi}$$

Thus, given a value for x_d and ψ, per-unit internal voltage e_{af} is also fixed. Then field current required can be calculated by first estimating field winding current for "no-load operation"

$$B_r = \frac{\mu_0 N_f I_{fnl}}{2gp}$$

and *rated* field current is

$$I_f = I_{fnl} e_{af}$$

or, required rated field current is

$$N_f I_f = \frac{2gp(1 - \lambda_p)B_s}{\mu_0} e_{af}$$

Next, I_f can be related to a field current *density*

$$N_f I_f = \frac{N_{RS}}{2} A_{RS} J_f$$

where N_{RS} is the number of rotor slots and the rotor slot area A_{RS} is

$$A_{RS} = w_R h_R$$

where h_R is rotor slot *height* and w_R is rotor slot width

$$w_R = \frac{2\pi R}{N_{RS}} \lambda_R$$

Then

$$N_f I_f = \pi R \lambda P h_R J_f$$

Now air-gap g is

$$g = \frac{2\mu_0 k_f R \lambda_R h_R J_f}{p(1 - \lambda_s) B_s e_{af}}$$

This in turn gives useful armature surface current density

$$h_s J_s = \sqrt{2} \, \frac{x_d}{e_{af}} \frac{\lambda_R}{\lambda_s} h_R J_f$$

Note that the ratio of x_d/e_{af} can be quite small (if the per-unit reactance is small). It will never be a very large number for any practical machine, and is generally less than one. As a practical matter, it is unusual for the per-unit synchronous reactance of a machine to be larger than about 2 or 2.25 per unit. This means that either the rotor or the stator of a machine can produce the dominant limitation on shear stress density (and so on rating). The best designs are "balanced," with both limits being reached at the same time.

5.16.11 Winding inductance calculation

This section shows how the inductances of windings in round-rotor machines with narrow air-gaps may be calculated. It deals only with the idealized air-gap magnetic fields, and does not consider slot, end winding, peripheral or skew reactances. It does consider the space harmonics of winding magneto-motive force (MMF).

To start, consider the MMF of a full-pitch, concentrated winding. Assuming that the winding has a total of N turns over p pole-pairs, the MMF is

$$F = \sum_{\substack{n=1 \\ nodd}}^{\infty} \frac{4}{n\pi} \frac{NI}{2p} \sin np\phi$$

This leads directly to magnetic flux density in the air-gap

$$B_r = \sum_{\substack{n=1 \\ nodd}}^{\infty} \frac{\mu_0}{g} \frac{4}{n\pi} \frac{NI}{2p} \sin np\phi$$

Note that a real winding, which will most likely not be full-pitched and concentrated, will have a *winding factor* which is the product of pitch and breadth factors. This will be discussed later.

Now, suppose that there is a polyphase winding, consisting of more than one phase (we will use three phases), driven with one of two types of current. The first of these is *balanced* current

$$I_a = I \cos(\omega t)$$

$$I_b = I \cos\left(\omega t - \frac{2\pi}{3}\right)$$

$$I_c = I \cos\left(\omega t + \frac{2\pi}{3}\right)$$

Conversely, consider *zero-sequence* currents

$$I_a = I_b = I_c = I \cos \omega t$$

Then it is possible to express magnetic flux density for the two distinct cases. For the *balanced* case

$$B_r = \sum_{n=1}^{\infty} B_{rn} \sin(np\phi \mp \omega t)$$

where

- The upper sign holds for $n = 1, 7, \ldots$
- The lower sign holds for $n = 5, 11, \ldots$
- all other terms are zero

and

$$B_{rn} = \frac{3}{2} \frac{\mu_0}{g} \frac{4}{n\pi} \frac{NI}{2p}$$

The zero-sequence case is simpler: it is non-zero only for the *triplen* harmonics

$$B_r = \sum_{n=3,9,\ldots}^{\infty} \frac{\mu_0}{g} \frac{4}{n\pi} \frac{NI}{2p} \frac{3}{2} (\sin(np\phi - \omega t) + \sin(np\phi + \omega t))$$

Next, consider the flux from a winding on the rotor: that will have the same form as the flux produced by a single armature winding, but will be referred to the rotor position

$$B_{rf} = \sum_{\substack{n=1 \\ nodd}}^{\infty} \frac{\mu_0}{g} \frac{4}{n\pi} \frac{NI}{2p} \sin np\phi'$$

which is, substituting $\phi' = \phi - \omega t/p$

$$B_{rf} = \sum_{\substack{n=1 \\ nodd}}^{\infty} \frac{\mu_0}{g} \frac{4}{n\pi} \frac{NI}{2p} \sin n(p\phi - \omega t)$$

The next step here is to find the flux linked by a winding if air-gap flux density is of the form

$$B_r = \sum_{n=1}^{\infty} B_{rn} \sin(np\phi \pm \omega t)$$

Now, it is possible to calculate flux linked by a single-turn, full-pitched winding by

$$\phi = \int_0^{\pi/p} B_r R l d\phi$$

and this is

$$\phi = 2Rl \sum_{n=1}^{\infty} \frac{B_{rn}}{np} \cos(\omega t)$$

This allows one to compute self and mutual inductances, since winding flux is

$$\lambda = N\phi$$

The end of this is a set of expressions for various inductances. It should be noted that, in the real world, most windings are not full-pitched nor concentrated. Fortunately, these shortcomings can be accommodated by the use of *winding factors*.

The simplest and perhaps best definition of a winding factor is the ratio of flux linked by an actual winding to flux that would have been linked by a full-pitch, concentrated winding with the same number of turns. That is

$$k_w = \frac{\lambda_{actual}}{\lambda_{full\text{-}pitch}}$$

It is relatively easy to show, using reciprocity arguments, that the winding factors are also the ratio of effective MMF produced by an actual winding to the MMF that would have been produced by the same winding were it to be full-pitched and concentrated. The argument goes as follows: mutual inductance between any pair of windings is reciprocal. That is, if the windings are designated *one* and *two,* the mutual inductance is flux induced in winding *one* by current in winding *two,* and it is also flux induced in winding *two* by current in winding *one.* Since each winding has a winding factor that influences its linking flux, and since the mutual inductance must be reciprocal, the same winding factor must influence the MMF produced by the winding.

The winding factors are often expressed for each space harmonic, although sometimes when a winding factor is referred to without reference to a harmonic number, what is meant is the space factor for the space fundamental.

Two winding factors are commonly specified for ordinary, regular windings. These are usually called *pitch* and *breadth* factors, reflecting the fact that often windings are not *full-pitched,* which means that individual turns do not span a full π electrical radians and that the windings occupy a range or breadth of slots within a phase belt. The breadth factors are ratios of flux linked by a given winding to the flux that would be linked by that winding were it full-pitched and concentrated. These two winding factors are discussed in a little more detail below. What is interesting to note, although we do not prove it here, is that the winding factor of any given winding is the *product* of the pitch and breadth factors

$$k_w = k_p k_b$$

With winding factors as defined here and in the sections below, it is possible to define winding inductances. For example, the *synchronous* inductance of a winding will be the apparent inductance of one phase when the polyphase winding is driven by a *balanced* set of currents. This is, approximately

$$L_d = \sum_{n=1,5,7,\ldots}^{\infty} \frac{3}{2} \frac{4}{\pi} \frac{\mu_0 N^2 R l k_{wn}^2}{p^2 g n^2}$$

This expression is approximate because it ignores the asynchronous interactions between higher order harmonics and the rotor of the machine. These are beyond the scope of this section.

Zero-sequence inductance is the ratio of flux to current if a winding is excited by zero-sequence currents

$$L_0 = \sum_{n=3,9,\ldots}^{\infty} 3 \frac{4}{\pi} \frac{\mu_0 N^2 R l k_{wn}^2}{p^2 g n^2}$$

and then mutual inductance, as between a *field* winding (f) and an *armature* winding (a), is

$$M(\theta) = \sum_{\substack{n=1 \\ nodd}}^{\infty} \frac{4}{\pi} \frac{\mu_0 N_f N_a k_{fn} k_{an} Rl}{p^2 g n^2} \cos(np\theta)$$

The winding factor can, for regular winding patterns, be expressed as the product of a *pitch* factor and a *breadth* factor, each of which can be estimated separately.

Pitch factor is found by considering the flux linked by a less-than-full pitched winding. Consider the situation in which radial magnetic flux density is

$$B_r = B_n \sin(np\phi - \omega t)$$

A winding with pitch α will link flux

$$\lambda = Nl \int_{\pi/2p - \alpha/2p}^{\pi/2p + \alpha/2p} B_n \sin(np\phi - \omega t) R d\phi$$

Pitch α refers to the angular displacement between sides of the coil, expressed in *electrical* radians. For a full-pitch coil $\alpha = \pi$.

The flux linked is

$$\lambda = \frac{2NlRB_n}{np} \sin\left(\frac{n\pi}{2}\right) \sin\left(\frac{n\alpha}{2}\right)$$

The *pitch* factor is seen to be

$$k_{pn} = \sin\frac{n\alpha}{2}$$

Now for *breadth* factor. This describes the fact that a winding may consist of a number of coils, each linking flux slightly out of phase with the others. A regular winding will have a number (say m) of coil elements, separated by *electrical* angle γ.

A full-pitch coil with one side at angle ξ will, in the presence of sinusoidal magnetic flux density, link flux

$$\lambda = Nl \int_{\xi/p}^{\pi/p - \xi/p} B_n \sin(np\phi - \omega t) R d\phi$$

This is readily evaluated to be

$$\lambda = \frac{2NlRB_n}{np} Re(e^{j(\omega t - n\xi)})$$

where complex number notation has been used for convenience in carrying out the rest of this derivation.

Now if the winding is distributed into m sets of slots and the slots are evenly spaced, the angular position of each slot will be

$$\xi_i = i\gamma - \frac{m-1}{2}\gamma$$

and the number of turns in each slot will be N/mp, so that actual flux linked will be

$$\lambda = \frac{2NlRB_n}{np} \frac{1}{m} \sum_{i=0}^{m-1} Re(e^{j(\omega t - n\xi_i)})$$

The *breadth* factor is then simply

$$k_b = \frac{1}{m} \sum_{i=0}^{m-1} e^{-jn(i\gamma - (m-1/2)\gamma)}$$

Note that this can be written as

$$k_b = \frac{e^{jn\gamma(m-1/2)}}{m} \sum_{i=0}^{m} e^{-jni\gamma}$$

Now, focus on that sum. Any coverging geometric sum has a simple sum

$$\sum_{i=0}^{\infty} x^i = \frac{1}{1-x}$$

and a truncated sum is

$$\sum_{i=0}^{m-1} = \sum_{i=0}^{\infty} - \sum_{i=m}^{\infty}$$

Then the useful sum can be written as

$$\sum_{i=0}^{m-1} e^{-jni\gamma} = (1 - e^{jnm\gamma}) \sum_{i=0}^{\infty} e^{-jni\gamma} = \frac{1 - e^{jnm\gamma}}{1 - e^{-jn\gamma}}$$

Now, the breadth factor is found

$$k_{bn} = \frac{\sin(nm\gamma/2)}{m\,\sin(n\gamma/2)}$$

5.17 Synchronous Machine-Simulation Models

5.17.1 Introduction

This section develops models useful for calculating the dynamic behavior of synchronous machines. It starts with a commonly accepted picture of the synchronous machine, assuming that the rotor can be fairly represented by three equivalent windings: one being the field and the other two, the d- and q-axis "damper" windings, representing the effects of rotor body, wedge chain, amortisseur and other current-carrying paths.

While a synchronous machine is assumed here, the results are fairly directly applicable to induction machines. Also, extension to situations in which the rotor representation must have more than one extra equivalent winding per axis should be straightforward.

5.17.2 Phase-variable model

To begin, assume that the synchronous machine can be properly represented by six equivalent windings. Four of these, the three armature-phase windings and the field winding, really are windings. The other two, representing the effects of distributed currents on the rotor, are referred to as "damper" windings. Fluxes are, in terms of currents

$$\begin{bmatrix} \underline{\lambda}_{ph} \\ \underline{\lambda}_{R} \end{bmatrix} = \begin{bmatrix} \underline{\underline{L}}_{ph} & \underline{\underline{M}} \\ \underline{\underline{M}}^{T} & \underline{\underline{L}}_{R} \end{bmatrix} \begin{bmatrix} \underline{I}_{ph} \\ \underline{I}_{R} \end{bmatrix} \tag{5.1}$$

where *phase* and *rotor* fluxes (and similarly, currents) are

$$\underline{\lambda}_{ph} = \begin{bmatrix} \lambda_{a} \\ \lambda_{b} \\ \lambda_{c} \end{bmatrix} \tag{5.2}$$

$$\underline{\lambda}_{R} = \begin{bmatrix} \lambda_{f} \\ \lambda_{kd} \\ \lambda_{kq} \end{bmatrix} \tag{5.3}$$

There are three inductance submatrices. The first of these describes armature winding inductances

$$\underline{\underline{L}}_{ph} = \begin{bmatrix} L_{a} & L_{ab} & L_{ac} \\ L_{ab} & L_{b} & L_{bc} \\ L_{ac} & L_{bc} & L_{c} \end{bmatrix} \tag{5.4}$$

where, for a machine that may have some saliency

$$L_a = L_{a0} + L_2 \cos 2\theta \qquad (5.5)$$

$$L_b = L_{a0} + L_2 \cos 2\left(\theta - \frac{2\pi}{3}\right) \qquad (5.6)$$

$$L_c = L_{a0} + L_2 \cos 2\left(\theta + \frac{2\pi}{3}\right) \qquad (5.7)$$

$$L_{ab} = L_{ab0} + L_2 \cos 2\left(\theta - \frac{\pi}{3}\right) \qquad (5.8)$$

$$L_{bc} = L_{ab0} + L_2 \cos 2\theta \qquad (5.9)$$

$$L_{ac} = L_{ab0} + L_2 \cos 2\left(\theta + \frac{\pi}{3}\right) \qquad (5.10)$$

Note that this last set of expressions assumes a particular form for the mutual inductances. This is seemingly restrictive, because it constrains the form of phase-to-phase mutual inductance variations with rotor position. The coefficient L_2 is actually the *same* in all six of these last expressions. As it turns out, this assumption does not really restrict the accuracy of the model very much.

The rotor inductances are relatively simply stated

$$\underline{\underline{L}}_R = \begin{bmatrix} L_f & L_{fkd} & 0 \\ L_{fkd} & L_{kd} & 0 \\ 0 & 0 & L_{kq} \end{bmatrix} \qquad (5.11)$$

And the stator-to-rotor mutual inductances are

$$\underline{\underline{M}} = \begin{bmatrix} M \cos \theta & L_{akd} \cos \theta & -L_{akq} \sin \theta \\ M \cos \left(\theta - \frac{2\pi}{3}\right) & L_{akd} \cos \left(\theta - \frac{2\pi}{3}\right) & -L_{akq} \sin \left(\theta - \frac{2\pi}{3}\right) \\ M \cos \left(\theta + \frac{2\pi}{3}\right) & L_{akd} \cos \left(\theta + \frac{2\pi}{3}\right) & -L_{akq} \sin \left(\theta + \frac{2\pi}{3}\right) \end{bmatrix}$$

$$(5.12)$$

5.17.3 Park's equations

The first step in the development of a suitable model is to *transform* the armature-winding variables to a coordinate system in which the rotor is stationary. We identify equivalent armature windings in the *direct* and *quadrature* axes. The *direct axis* armature winding is the equivalent of one of the phase windings, but aligned directly with the field. The *quadrature* winding is situated so that its axis *leads* the field winding by 90 *electrical* degrees. The transformation used to map the armature currents, fluxes and so forth onto the *direct* and *quadrature* axes is the celebrated *Park's Transformation,* named after Robert H. Park, an early investigator in transient behavior in synchronous machines. The mapping takes the form

$$\begin{bmatrix} u_d \\ u_q \\ u_0 \end{bmatrix} = \underline{u}_{dq} = \underline{\underline{T}}\underline{u}_{ph} = \underline{\underline{T}} \begin{bmatrix} u_a \\ u_b \\ u_c \end{bmatrix} \tag{5.13}$$

Where the transformation and its inverse are

$$\underline{\underline{T}} = \frac{2}{3} \begin{bmatrix} \cos\theta & \cos\left(\theta - \dfrac{2\pi}{3}\right) & \cos\left(\theta + \dfrac{2\pi}{3}\right) \\ -\sin\theta & -\sin\left(\theta - \dfrac{2\pi}{3}\right) & -\sin\left(\theta + \dfrac{2\pi}{3}\right) \\ \dfrac{1}{2} & \dfrac{1}{2} & \dfrac{1}{2} \end{bmatrix} \tag{15.14}$$

$$\underline{\underline{T}}^{-1} = \begin{bmatrix} \cos\theta & -\sin\theta & 1 \\ \cos\left(\theta - \dfrac{2\pi}{3}\right) & -\sin\left(\theta - \dfrac{2\pi}{3}\right) & 1 \\ \cos\left(\theta + \dfrac{2\pi}{3}\right) & -\sin\left(\theta + \dfrac{2\pi}{3}\right) & 1 \end{bmatrix} \tag{5.15}$$

This transformation maps *balanced* sets of phase currents into *constant* currents in the d-q frame. That is, if rotor angle is $\theta = \omega t + \theta_0$, and phase currents are

$$I_a = I \cos \omega t$$

$$I_b = I \cos\left(\omega t - \frac{2\pi}{3}\right)$$

$$I_c = I \cos\left(\omega t + \frac{2\pi}{3}\right)$$

then the transformed set of currents is

$$I_d = I \cos \theta_0$$

$$I_q = -I \sin \theta_0$$

Now, apply this transformation to Eq. (5.1) to express fluxes and currents in the armature in the d-q reference frame. To do this, extract the top line in (5.1)

$$\underline{\lambda}_{ph} = \underline{L}_{ph}\underline{I}_{ph} + \underline{M}\underline{I}_R \tag{5.16}$$

The transformed flux is obtained by premultiplying this whole expression by the transformation matrix. Phase current may be obtained from d-q current by multiplying by the inverse of the transformation matrix. Thus

$$\underline{\lambda}_{dq} = \underline{TL}_{ph}\underline{T}^{-1}\underline{I}_{dq} + \underline{TM}\underline{I}_R \tag{5.17}$$

The same process carried out for the lower line of (5.1) yields

$$\underline{\lambda}_R = \underline{M}^T\underline{T}^{-1}\underline{I}_{dq} + \underline{L}_R\underline{I}_R \tag{5.18}$$

Thus the fully transformed version of (5.1) is

$$\begin{bmatrix} \underline{\lambda}_{dq} \\ \underline{\lambda}_R \end{bmatrix} = \begin{bmatrix} \underline{L}_{dq} & \underline{L}_C \\ \frac{3}{2}\underline{L}_C^T & \underline{L}_R \end{bmatrix}\begin{bmatrix} \underline{I}_{dq} \\ \underline{I}_R \end{bmatrix} \tag{5.19}$$

If the conditions of (5.5) through (5.10) are satisfied, the inductance submatrices of (5.19) are of particularly simple form. (Please note that a substantial amount of algebra has been left out here!)

$$\underline{L}_{dq} = \begin{bmatrix} L_d & 0 & 0 \\ 0 & L_q & 0 \\ 0 & 0 & L_0 \end{bmatrix} \tag{5.20}$$

$$\underline{L}_C = \begin{bmatrix} M & L_{akd} & 0 \\ 0 & 0 & L_{akq} \\ 0 & 0 & 0 \end{bmatrix} \tag{5.21}$$

Note that (5.20) and (5.21) express three *separate* sets of apparently independent flux-current relationships. These may be re-cast into the following form

$$\begin{bmatrix} \lambda_d \\ \lambda_{kd} \\ \lambda_f \end{bmatrix} = \begin{bmatrix} L_d & L_{akd} & M \\ \frac{3}{2}L_{akd} & L_{kd} & L_{fkd} \\ \frac{3}{2}M & L_{fkd} & L_f \end{bmatrix}\begin{bmatrix} I_d \\ I_{kd} \\ I_f \end{bmatrix} \tag{5.22}$$

$$\begin{bmatrix} \lambda_q \\ \lambda_{kq} \end{bmatrix} = \begin{bmatrix} L_q & L_{akq} \\ \frac{3}{2}L_{akq} & L_{kq} \end{bmatrix} \begin{bmatrix} I_q \\ I_{kq} \end{bmatrix} \tag{5.23}$$

$$\lambda_0 = L_0 I_0 \tag{5.24}$$

Where the component inductances are

$$L_d = L_{a0} - L_{ab0} + \frac{3}{2} L_2 \tag{5.25}$$

$$L_q = L_{a0} - L_{ab0} - \frac{3}{2} L_2 \tag{5.26}$$

$$L_0 = L_{a0} + 2L_{ab0} \tag{5.27}$$

Note that the apparently restrictive assumptions embedded in (5.5) through (5.10) have resulted in the very simple form of (5.21) through (5.24) and which, in particular, result in three mutually independent sets of fluxes and currents. While one may be concerned about the restrictiveness of these expressions, note that the orthogonality between the d- and q-axes is not unreasonable. In fact, because these axes are orthogonal in *space,* it seems reasonable that they should not have mutual flux linkages. The principal consequence of these assumptions is the decoupling of the *zero-sequence* component of flux from the d- and q-axis components. It should be noted that departures from this form (that is, coupling between the "direct" and "zero" axes) must be through higher harmonic fields that will not couple well to the armature, so that any such coupling will be weak.

Next, armature voltage is, ignoring resistance, given by

$$\underline{V}_{ph} = \frac{d}{dt} \underline{\lambda}_{ph} = \frac{d}{dt} \underline{T}^{-1} \underline{\lambda}_{dq} \tag{5.28}$$

and that the *transformed* armature voltage must be

$$\underline{V}_{dq} = \underline{T}\underline{V}_{ph}$$

$$= \underline{T} \frac{d}{dt} (\underline{T}^{-1} \underline{\lambda}_{dq})$$

$$= \frac{d}{dt} \underline{\lambda}_{dq} + \left(\underline{T} \frac{d}{dt} \underline{T}^{-1} \right) \underline{\lambda}_{dq} \tag{5.29}$$

Much manipulation goes into reducing the second term of this, resulting in

$$T \frac{d}{dt} T^{-1} = \begin{bmatrix} 0 & -\dfrac{d\theta}{dt} & 0 \\ \dfrac{d\theta}{dt} & 0 & 0 \\ 0 & 0 & 0 \end{bmatrix} \tag{5.30}$$

This expresses the *speed voltage* that arises from a coordinate transformation. The two voltage-flux relationships that are affected are

$$V_d = \frac{d\lambda_d}{dt} - \omega\lambda_q \tag{5.31}$$

$$V_q = \frac{d\lambda_q}{dt} + \omega\lambda_d \tag{5.32}$$

where

$$\omega = \frac{d\theta}{dt} \tag{5.33}$$

5.17.4 Power and torque

Instantaneous *power* is given by

$$P = V_a I_a + V_b I_b + V_c I_c \tag{5.34}$$

Using the transformations given above, this can be shown to be

$$P = \frac{3}{2} V_d I_d + \frac{3}{2} V_q I_q + 3 V_0 I_0 \tag{5.35}$$

which, in turn, is

$$P = \omega \frac{3}{2} (\lambda_d I_q - \lambda_q I_d) + \frac{3}{2} \left(\frac{d\lambda_d}{dt} I_d - \frac{d\lambda_q}{dt} I_q \right) + 3 \frac{d\lambda_0}{dt} I_0 \tag{5.36}$$

Then, noting that *electrical* speed ω and shaft speed Ω are related by $\omega = p\Omega$ and that (5.36) describes electrical terminal power as the sum of shaft power and rate of change of stored energy, torque is given by

$$T = \frac{3}{2} p(\lambda_d I_q - \lambda_q I_d) \tag{5.37}$$

5.17.5 Per-unit normalization

The next thing to do is to investigate the way in which electric machine system are *normalized,* or put into what is called a *per-unit* system. The reason for this step is that, when the voltage, current, power and impedance are referred to normal operating parameters, the behavior characteristics of all types of machines become quite similar, giving us a better way of relating how a particular machine works to some reasonable standard. There are also numerical reasons for normalizing performance parameters to some standard.

The first step in normalization is to establish a set of *base* quantities. Normalize voltage, current, flux, power, impedance and torque by using base quantities for each of these. Note that the base quantities are *not* independent. In fact, for the armature, one need only specify three quantities: voltage (V_B), current (I_B) and frequency (ω_0). It is not customary to normalize time nor frequency. Once this is done for the armature circuits, the other base quantities are

- Base power

$$P_B = \frac{3}{2} V_B I_B$$

- Base impedance

$$Z_B = \frac{V_B}{I_B}$$

- Base flux

$$\lambda_B = \frac{V_B}{\omega_0}$$

- Base torque

$$T_B = \frac{p}{\omega_0} P_B$$

Note that base *voltage* and *current* are expressed as *peak* quantities. Base voltage is taken on a phase basis (line to neutral for a "wye"-connected machine), and base current is similarly taken on a phase basis, (line current for a "wye"-connected machine).

Normalized, or *per-unit* quantities are derived by dividing the *ordinary* variable (with units) by the corresponding *base*. For example, per-unit flux is

$$\psi = \frac{\lambda}{\lambda_B} = \frac{\omega_0 \lambda}{V_B} \tag{5.38}$$

In this derivation, per-unit quantities will usually be designated by lower case letters. Two notable exceptions are flux, where we use the letter ψ, and torque, where we will still use the upper case T and risk confusion.

Note that there will be *base* quantities for voltage, current and frequency for each of the different coils represented in our model. While it is reasonable to expect that the *frequency* base will be the same for all coils in a problem, the *voltage* and *current* bases may be different. Equation (5.22) becomes

$$
\begin{bmatrix} \psi_d \\ \psi_{kd} \\ \psi_f \end{bmatrix} =
\begin{bmatrix}
\dfrac{\omega_0 I_{dB}}{V_{db}} L_d & \dfrac{\omega_0 I_{kB}}{V_{db}} L_{akd} & \dfrac{\omega_0 I_{fB}}{V_{db}} M \\[3mm]
\dfrac{\omega_0 I_{dB}}{V_{kb}} \dfrac{3}{2} L_{akd} & \dfrac{\omega_0 I_{kB}}{V_{kb}} L_{kd} & \dfrac{\omega_0 I_{fB}}{V_{kdb}} L_{fkd} \\[3mm]
\dfrac{\omega_0 I_{dB}}{V_{fb}} \dfrac{3}{2} M & \dfrac{\omega_0 I_{kB}}{V_{fb}} L_{fkd} & \dfrac{\omega_0 I_{fB}}{V_{fb}} L_f
\end{bmatrix}
\begin{bmatrix} i_d \\ i_{kd} \\ i_f \end{bmatrix} \tag{5.39}
$$

where $i = I/I_B$ denotes *per-unit,* or normalized current.

Note that (5.39) may be written in simple form

$$
\begin{bmatrix} \psi_d \\ \psi_{kd} \\ \psi_f \end{bmatrix} =
\begin{bmatrix}
x_d & x_{akd} & x_{ad} \\
x_{akd} & x_{kd} & x_{fkd} \\
x_{ad} & x_{fkd} & x_f
\end{bmatrix}
\begin{bmatrix} i_d \\ i_{kd} \\ i_f \end{bmatrix} \tag{5.40}
$$

It is important to note that (5.40) *assumes* reciprocity in the normalized system. To wit, the following equations are implied

$$x_d = \omega_0 \frac{I_{dB}}{V_{dB}} L_d \tag{5.41}$$

$$x_{kd} = \omega_0 \frac{I_{kB}}{V_{kB}} L_{kd} \tag{5.42}$$

$$x_f = \omega_0 \frac{I_{fB}}{V_{fB}} L_f \tag{5.43}$$

$$x_{akd} = \omega_0 \frac{I_{kB}}{V_{dB}}$$

$$= \frac{3}{2} \omega_0 \frac{I_{dB}}{V_{kB}} \tag{5.44}$$

$$x_{ad} = \omega_0 \frac{I_{fB}}{V_{dB}} M$$

$$= \frac{3}{2} \omega_0 \frac{I_{dB}}{V_{fB}} M \tag{5.45}$$

$$x_{fkd} = \omega_0 \frac{I_{kB}}{V_{fB}} L_{fkd}$$

$$= \omega_0 \frac{I_{fB}}{V_{kB}} L_{fkd} \tag{5.46}$$

These in turn imply

$$\frac{3}{2} V_{dB} I_{dB} = V_{fB} I_{fB} \tag{5.47}$$

$$\frac{3}{2} V_{dB} I_{dB} = V_{kB} I_{kB} \tag{5.48}$$

$$V_{fB} I_{fB} = V_{kB} I_{kB} \tag{5.49}$$

These equations imply the same *power* base on all of the windings of the machine. This is so because the *armature* base quantities V_{dB} and I_{dB} are stated as *peak* values, while the *rotor* base quantities are stated as *dc* values. Thus power base for the *three-phase* armature is ³⁄₂ times the product of *peak* quantities, while the power base for the rotor is simply the product of those quantities.

The quadrature axis, which may have fewer equivalent elements than the direct axis and which may have different numerical values, still yields a similar structure. Without going through the details, the per-unit flux-current relationship for the q-axis is

$$\begin{bmatrix} \psi_q \\ \psi_{kq} \end{bmatrix} = \begin{bmatrix} x_q & x_{akq} \\ x_{akq} & x_{kq} \end{bmatrix} \begin{bmatrix} i_q \\ i_{kq} \end{bmatrix} \tag{5.50}$$

The voltage equations, including speed voltage terms, (5.31) and (5.32), may be augmented to reflect armature resistance

$$V_d = \frac{d\lambda_d}{dt} - \omega\lambda_q + R_a I_d \tag{5.51}$$

$$V_q = \omega\lambda_d + \frac{d\lambda_q}{dt} + R_a I_q \tag{5.52}$$

The *per-unit* equivalents of these are

$$v_d = \frac{1}{\omega_0}\frac{d\psi_d}{dt} - \frac{\omega}{\omega_0}\psi_q + r_a i_d \tag{5.53}$$

$$v_q = \frac{\omega}{\omega_0}\psi_d + \frac{1}{\omega_0}\frac{d\psi_q}{dt} + r_a i_q \tag{5.54}$$

where the per-unit armature resistance is just $r_a = R_a/Z_B$.

Note that none of the other circuits in this model have *speed voltage* terms, so their voltage expressions are exactly what one might expect

$$v_f = \frac{1}{\omega_0}\frac{d\psi_f}{dt} + r_f i_f \tag{5.55}$$

$$v_{kd} = \frac{1}{\omega_0}\frac{d\psi_{kd}}{dt} + r_{kd} i_{kd} \tag{5.56}$$

$$v_{kq} = \frac{1}{\omega_0}\frac{d\psi_{kq}}{dt} + r_{kq} i_{kq} \tag{5.57}$$

$$v_0 = \frac{1}{\omega_0}\frac{d\psi_0}{dt} + r_a i_0 \tag{5.58}$$

It should be noted that the *damper* winding circuits represent closed conducting paths on the rotor, so the two voltages v_{kd} and v_{kq} are always zero.

Per-unit torque is simply

$$T_e = \psi_d i_q - \psi_q i_d \tag{5.59}$$

Often, one needs to represent the dynamic behavior of the machine, including electromechanical dynamics involving rotor inertia. If J is the rotational inertia constant of the machine system, the rotor dynamics are described by the two ordinary differential equations

$$\frac{1}{p}J\frac{d\omega}{dt} = T^e + T^m \tag{5.60}$$

$$\frac{d\delta}{dt} = \omega - \omega_0 \tag{5.61}$$

where T^e and T^m represent *electrical* and *mechanical* torques in "ordinary" variables. The angle δ represents rotor phase angle with respect to some synchronous reference.

It is customary to define an "inertia constant" which is not dimensionless, but which nevertheless fits into the per-unit system of analysis. This is

$$H \equiv \frac{\text{Rotational kinetic energy at rated speed}}{\text{Base Power}} \qquad (5.62)$$

or

$$H = \frac{\frac{1}{2} J \left(\frac{\omega_0}{p}\right)^2}{P_B} = \frac{J \omega_0}{2 p T_B} \qquad (5.63)$$

Then the per-unit equivalent to (5.60) is

$$\frac{2H}{\omega_0} \frac{d\omega}{dt} = T_e + T_m \qquad (5.64)$$

where now T_e and T_m represent *per-unit* torques.

5.17.6 Equal mutual's base

In normalizing the differential equations that make up the model, a number of *base quantities* are used. For example, in deriving (5.40), the *per-unit* flux-current relationship for the *direct* axis, six base quantities appear V_B, I_B, V_{fB}, I_{fB}, V_{kB} and I_{kB}. Imposing reciprocity on (5.40) results in two constraints on these six variables, expressed in (5.47) through (5.49). Presumably the two armature base quantities will be fixed by machine rating. That leaves two more "degrees of freedom" in selection of base quantities. Note that the selection of base quantities will affect the reactance matrix in (5.40).

While there are different schools of thought on just how to handle these degrees of freedom, a commonly used convention is to employ what is called the *equal mutuals* base system. The two degrees of freedom are used to set the field and damper base impedances so that all three mutual inductances of (5.40) are equal

$$x_{akd} = x_{fkd} = x_{ad} \qquad (5.65)$$

The direct-axis flux-current relationship becomes

$$\begin{bmatrix} \psi_d \\ \psi_{kd} \\ \psi_f \end{bmatrix} = \begin{bmatrix} x_d & x_{ad} & x_{ad} \\ x_{ad} & x_{kd} & x_{ad} \\ x_{ad} & x_{ad} & x_f \end{bmatrix} \begin{bmatrix} i_d \\ i_{kd} \\ i_f \end{bmatrix} \qquad (5.66)$$

5.17.7 Equivalent circuit

The flux-current relationship of (5.66) is represented by the equivalent circuit of Fig. 5.25, if the "leakage" inductances are defined to be

$$x_{al} = x_d - x_{ad} \tag{5.67}$$

$$x_{kdl} = x_{kd} - x_{ad} \tag{5.68}$$

$$x_{fl} = x_f - x_{ad} \tag{5.69}$$

Many of the interesting features of the electrical dynamics of the synchronous machine may be discerned from this circuit. While a complete explication is beyond the scope of this chapter, it is possible to make a few observations.

The apparent inductance measured from the terminals of this equivalent circuit (ignoring resistance r_a) will be, in the frequency domain, of the form

$$x(s) = \frac{\psi_d(s)}{i_d(s)} = x_d \frac{P_n(s)}{P_d(s)} \tag{5.70}$$

Both the numerator and denominator polynomials in s will be second order. (Confirm this by writing an expression for terminal impedance). Since this is a "diffusion"-type circuit, having only resistances and inductances, all poles and zeros must be on the negative real axis of the "s-plane." The per-unit inductance is then

$$x(s) = x_d \frac{(1 + T'_d s)(1 + T''_d s)}{(1 + T'_{do} s)(1 + T''_{do} s)} \tag{5.71}$$

The two time constants T'_d and T''_d are the reciprocals of the *zeros* of the impedance, which are the *poles* of the admittance. These are called the *short-circuit* time constants.

The other two time constants T'_{do} and T''_{do} are the reciprocals of the

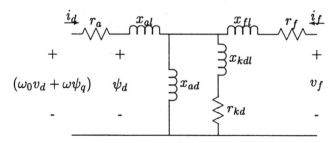

Figure 5.25 D-axis equivalent circuit for synchronous machine.

poles of the impedance, and so are called the *open-circuit* time constants.

This is cast as if there were two sets of well-defined time constants. These are the *transient* time constants T_d' and T_{do}', and the *subtransient* time constants T_d'' and T_{do}''. In many cases, these are indeed well separated, meaning that

$$T_d' \gg T_d'' \tag{5.72}$$

$$T_{do}' \gg T_{do}'' \tag{5.73}$$

If this is true, then the reactance is described by the pole-zero diagram shown in Fig. 5.26. Under this circumstance, the apparent terminal inductance has three distinct values, depending on frequency. These are the *synchronous* inductance, the *transient* inductance, and the *subtransient* inductance, given by

$$x_d' = x_d \frac{T_d'}{T_{do}'} \tag{5.74}$$

$$x_d'' = x_d' \frac{T_d''}{T_{do}''}$$

$$= x_d \frac{T_d'}{T_{do}'} \frac{T_d''}{T_{do}''} \tag{5.75}$$

A *Bode Plot* of the terminal reactance is shown in Fig. 5.27.

If the time constants are spread widely apart, they are given, approximately, by

$$T_{do}' = \frac{x_f}{\omega_0 r_f} \tag{5.76}$$

$$T_{do}'' = \frac{x_{kdl} + x_{fl}\|x_{ad}}{\omega_0 r_{kd}} \tag{5.77}$$

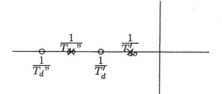

Figure 5.26 Pole-zero diagram for terminal inductance.

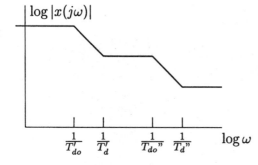

Figure 5.27 Frequency response of terminal inductance.

Finally, note that the three reactances are found simply from the model

$$x_d = x_{al} + x_{ad} \tag{5.78}$$

$$x_d' = x_{al} + x_{ad}\|x_{fl} \tag{5.79}$$

$$x_d'' = x_{al} + x_{ad}\|x_{fl}\|x_{kdl} \tag{5.80}$$

5.17.8 Statement of simulation model

Several simulation models may be derived, since the machine can be driven by either voltages or currents. Further, the expressions for permanent magnet machines are a bit different. So the first model is one in which the terminals are all constrained by *voltage*.

The state variables are the two stator fluxes ψ_d, ψ_q, two "damper" fluxes ψ_{kd}, ψ_{kq}, field flux ψ_f, rotor speed ω and torque angle δ. The most straightforward way of stating the model employs currents as auxiliary variables, and these are

$$\begin{bmatrix} i_d \\ i_{kd} \\ i_f \end{bmatrix} = \begin{bmatrix} x_d & x_{ad} & x_{ad} \\ x_{ad} & x_{kd} & x_{ad} \\ x_{ad} & x_{ad} & x_f \end{bmatrix}^{-1} \begin{bmatrix} \psi_d \\ \psi_{kd} \\ \psi_f \end{bmatrix} \tag{5.81}$$

$$\begin{bmatrix} i_q \\ i_{kq} \end{bmatrix} = \begin{bmatrix} x_q & x_{aq} \\ x_{aq} & x_{kq} \end{bmatrix}^{-1} \begin{bmatrix} \psi_q \\ \psi_{kq} \end{bmatrix} \tag{5.82}$$

Then the state equations are

$$\frac{d\psi_d}{dt} = \omega_0 v_d + \omega\psi_q - \omega_0 r_a i_d \tag{5.83}$$

$$\frac{d\psi_q}{dt} = \omega_0 v_q - \omega\psi_d - \omega_0 r_a i_q \tag{5.84}$$

$$\frac{d\psi_{kd}}{dt} = -\omega_0 r_{kd} i_{kd} \tag{5.85}$$

$$\frac{d\psi_{kq}}{dt} = -\omega_0 r_{kq} i_{kq} \tag{5.86}$$

$$\frac{d\psi_f}{dt} = -\omega_0 r_f i_f \tag{5.87}$$

$$\frac{d\omega}{dt} = \frac{\omega_0}{2H}(T_e + T_m) \tag{5.88}$$

$$\frac{d\delta}{dt} = \omega - \omega_0 \tag{5.89}$$

and, of course

$$T_e = \psi_d i_q - \psi_q i_d$$

5.17.8.1 Statement of parameters. Note that data for a machine may often be given in terms of the reactances $x_d, x_d', x_d'', T_{do}'$ and T_{do}'', rather than in terms of the elements of the equivalent circuit model. Note also that there are four inductances in the equivalent circuit, so we have to assume one. There is no loss in generality in doing so. Usually, one assumes a value for the stator leakage inductance, and if this is done, the translation is straightforward

$$x_{ad} = x_d - x_{al}$$

$$x_{fl} = \frac{x_{ad}(x_d' - x_{al})}{x_{ad} - x_d' + x_{al}}$$

$$x_{kdl} = \frac{1}{\dfrac{1}{x_d'' - x_{al}} - \dfrac{1}{x_{ad}} - \dfrac{1}{x_{fl}}}$$

$$r_f = \frac{x_{fl} + x_{ad}}{\omega_0 T_{do}'}$$

$$r_{kd} = \frac{x_{kdl} + x_{ad}\|x_{fl}}{\omega_0 T_{do}''}$$

5.17.8.2 Linearized model. Often, it becomes desirable to carry out a linearized analysis of machine operation, for example, to examine the damping of the swing mode at a particular operating point. What is done, then, is to assume a steady state operating point and examine the dynamics for "small" deviations from that operating point. The definition of "small" is really "small enough" that everything important appears in the first-order term of a Taylor series about the steady operating point.

Note that the expressions in the machine model are linear for the most part. There are, however, a few cases in which products of state variables cause us to do the expansion of the Taylor series. Assuming a steady state operating point $[\psi_{d0} \ \psi_{kd0} \ \psi_{f0} \ \psi_{q0} \ \psi_{kq0} \ \omega_0 \ \delta_0]$, the first-order (small-signal) variations are described by the following set of equations. First, since the flux-current relationship is linear

$$
\begin{bmatrix} i_{d1} \\ i_{kd1} \\ i_{f1} \end{bmatrix} = \begin{bmatrix} x_d & x_{ad} & x_{ad} \\ x_{ad} & x_{kd} & x_{ad} \\ x_{ad} & x_{ad} & x_f \end{bmatrix}^{-1} \begin{bmatrix} \psi_{d1} \\ \psi_{kd1} \\ \psi_{f1} \end{bmatrix} \tag{5.90}
$$

$$
\begin{bmatrix} i_{q1} \\ i_{kq1} \end{bmatrix} = \begin{bmatrix} x_q & x_{aq} \\ x_{aq} & x_{kq} \end{bmatrix}^{-1} \begin{bmatrix} \psi_{q1} \\ \psi_{kq1} \end{bmatrix} \tag{5.91}
$$

Terminal voltage will be, for operation against a voltage source

$$
V_d = V \sin \delta \quad V_q = V \cos \delta
$$

Then the differential equations governing the first-order variations are

$$
\frac{d\psi_{d1}}{dt} = \omega_0 V \cos \delta_0 \delta_1 + \omega_0 \psi_{q1} + \omega_1 \psi_{q0} - \omega_0 r_a i_{d1} \tag{5.92}
$$

$$
\frac{d\psi_{q1}}{dt} = -\omega_0 V \sin \delta_0 \delta_1 - \omega_0 \psi_{d1} - \omega_1 \psi_{d0} - \omega_0 r_a i_{q1} \tag{5.93}
$$

$$
\frac{d\psi_{kd1}}{dt} = -\omega_0 r_{kd} i_{kd1} \tag{5.94}
$$

$$
\frac{d\psi_{kq1}}{dt} = -\omega_0 r_{kq} i_{kq1} \tag{5.95}
$$

$$
\frac{d\psi_{f1}}{dt} = -\omega_0 r_f i_{f1} \tag{5.96}
$$

$$
\frac{d\omega_1}{dt} = \frac{\omega_0}{2H} (T_{e1} + T_{m1}) \tag{5.97}
$$

$$
\frac{d\delta_1}{dt} = \omega_1 \tag{5.98}
$$

$$T_e = \psi_{d0}i_{q1} + \psi_{d1}i_{q0} - \psi_{q0}i_{d1} - \psi_{q1}i_{d0}$$

5.17.8.3 Reduced order model for electromechanical transients. In many situations, the two armature variables contribute little to the dynamic response of the machine. Typically, the armature resistance is small enough that there is very little voltage drop across it and transients in the difference between armature flux and the flux that would exist in the "steady state" decay rapidly (or are not even excited). Further, the relatively short armature time constant makes for very short time steps. For this reason, it is often convenient, particularly when studying the relatively slow electromechanical transients, to omit the first two differential equations and set

$$\psi_d = v_q = V \cos \delta \qquad (5.99)$$

$$\psi_q = -v_d = -V \sin \delta \qquad (5.100)$$

The set of differential equations changes only a little when this approximation is made. Note, however, that it can be simulated with far fewer "cycles" if the armature time constant is short.

5.17.9 Current-driven model: connection to a system

The simulation expressions developed so far are useful in a variety of circumstances. They are, however, difficult to tie to network simulation programs because they use terminal voltage as an input. Generally, it is more convenient to use *current* as the input to the machine simulation and accept *voltage* as the output. Further, it is difficult to handle unbalanced situations with this set of equations.

An alternative to this set would be to employ the *phase* currents as state variables. Effectively, this replaces ψ_d, ψ_q and ψ_0 with i_a, i_b, and i_c. The resulting model will interface nicely with network simulations as we will show.

To start, write an expression for terminal flux on the d-axis:

$$\psi_d = x_d'' i_d + \psi_f \frac{x_{ad} \| x_{kdl}}{x_{ad} \| x_{kdl} + x_{fl}} + \psi_{kd} \frac{x_{ad} \| x_{fl}}{x_{ad} \| x_{fl} + x_{kdl}} \qquad (5.101)$$

and here, of course

$$x_d'' = x_{al} + x_{ad} \| x_{kdl} \| x_{fl}$$

This leads to a definition of "flux behind subtransient reactance"

$$\psi''_d = \frac{x_{ad}x_{kdl}\psi_f + x_{ad}x_{fl}\psi_{kd}}{x_{ad}x_{kdl} + x_{ad}x_{fl} + x_{kdl}x_{fl}} \qquad (5.102)$$

so that

$$\psi_d = \psi''_d + x''_d i_d$$

On the quadrature axis, the situation is essentially the same, but one step easier if there is only one quadrature-axis rotor winding

$$\psi_q = x''_q i_q + \psi_{kq}\frac{x_{aq}}{x_{aq} + x_{kql}} \qquad (5.103)$$

where

$$x''_q = x_{al} + x_{aq}\|x_{kql}$$

Very often these fluxes are referred to as "voltage behind subtransient reactance", with $\psi''_d = e''_q$ and $\psi''_q = -e''_d$. Then

$$\psi_d = x''_d i_d + e''_q \qquad (5.104)$$

$$\psi_q = x''_q i_q - e''_d \qquad (5.105)$$

Now, if i_d and i_q are determined, it is a bit easier to find the other currents required in the simulation. Write

$$\begin{bmatrix} \psi_{kd} \\ \psi_f \end{bmatrix} = \begin{bmatrix} x_{kd} & x_{ad} \\ x_{ad} & x_f \end{bmatrix}\begin{bmatrix} i_{kd} \\ i_f \end{bmatrix} + \begin{bmatrix} x_{ad} \\ x_{ad} \end{bmatrix} i_d \qquad (5.106)$$

and this inverts easily

$$\begin{bmatrix} i_{kd} \\ i_f \end{bmatrix} = \begin{bmatrix} x_{kd} x_{ad} \\ x_{ad} x_f \end{bmatrix}^{-1}\left(\begin{bmatrix} \psi_{kd} \\ \psi_f \end{bmatrix} - \begin{bmatrix} x_{ad} \\ x_{ad} \end{bmatrix} i_d\right) \qquad (5.107)$$

The quadrature-axis rotor current is simply

$$i_{kq} = \frac{1}{x_{kq}}\psi_{kq} - \frac{x_{aq}}{x_{kq}} i_q \qquad (5.108)$$

The torque equation is the same, but since it is usually convenient to assemble the fluxes behind subtransient reactance, it is possible to use

$$T_e = e''_q i_q + e''_d i_d + (x''_d - x''_q)i_d i_q \qquad (5.109)$$

Now it is necessary to consider terminal voltage. This is most conveniently cast in matrix notation. The vector of phase voltages is

$$\underline{v}_{ph} = \begin{bmatrix} v_a \\ v_b \\ v_c \end{bmatrix} \tag{5.110}$$

Then, with similar notation for phase flux, terminal voltage is, ignoring armature resistance

$$\underline{v}_{ph} = \frac{1}{\omega_0} \frac{d\underline{\psi}_{ph}}{dt}$$

$$= \frac{1}{\omega_0} \frac{d}{dt} \{\underline{\underline{T}}^{-1}\underline{\psi}_{dq}\} \tag{5.111}$$

Note that one may define the transformed vector of fluxes to be

$$\underline{\psi}_{dq} = \underline{\underline{x}}''\underline{i}_{dq} + \underline{e}'' \tag{5.112}$$

where the matrix of reactances shows orthogonality

$$\underline{\underline{x}}'' = \begin{bmatrix} x_d'' & 0 & 0 \\ 0 & x_q'' & 0 \\ 0 & 0 & x_0 \end{bmatrix} \tag{5.113}$$

and the vector of internal fluxes is

$$\underline{e}'' = \begin{bmatrix} e_q'' \\ -e_d'' \\ 0 \end{bmatrix} \tag{5.114}$$

Now, of course, $\underline{i}_{dq} = \underline{\underline{T}}\underline{i}_{ph}$, so that we may re-cast (5.111) as

$$\underline{v}_{ph} = \frac{1}{\omega_0} \frac{d}{dt} \{\underline{\underline{T}}^{-1}\underline{\underline{x}}''\underline{\underline{T}}\underline{i}_{ph} + \underline{\underline{T}}^{-1}\underline{e}''\} \tag{5.115}$$

Now it is necessary to make one assumption and one definition. The assumption, which is only moderately restrictive, is that subtransient saliency may be ignored. That is, we assume that $x_d'' = x_q''$. The definition separates the "zero-sequence" impedance into phase and neutral components

$$x_0 = x_d'' + 3x_g \tag{5.116}$$

Note that according to this definition, the reactance x_g accounts for any impedance in the neutral of the synchronous machine as well as mutual coupling between phases.

Then, the impedance matrix becomes

$$\underline{\underline{x}}'' = \begin{bmatrix} x_d' & 0 & 0 \\ 0 & x_d'' & 0 \\ 0 & 0 & x_d'' \end{bmatrix} + \begin{bmatrix} 0 & 0 & 0 \\ 0 & 0 & 0 \\ 0 & 0 & 3x_g \end{bmatrix} \tag{5.117}$$

In compact notation, this is

$$\underline{\underline{x}}'' = x_d'' \underline{\underline{I}} + \underline{\underline{x}}_g \tag{5.118}$$

where $\underline{\underline{I}}$ is the identity matrix.

Now the vector of phase voltages is

$$\underline{v}_{ph} = \frac{1}{\omega_0} \frac{d}{dt} \{x_d'' i_{ph} + \underline{\underline{T}}^{-1} \underline{\underline{x}}_g \underline{\underline{T}} i_{ph} + \underline{\underline{T}}^{-1} \underline{e}''\} \tag{5.119}$$

Note that in (5.119), multiplication by the identity matrix is already factored out. The next step is to carry out the matrix multiplication in the third term of (5.119). This operation turns out to produce a remarkably simple result

$$\underline{\underline{T}}^{-1} \underline{\underline{x}}_g \underline{\underline{T}} = x_g \begin{bmatrix} 1 & 1 & 1 \\ 1 & 1 & 1 \\ 1 & 1 & 1 \end{bmatrix} \tag{5.120}$$

The impact of this is that each of the three phase voltages has the same term, and that is related to the time derivative of the sum of the three currents, multiplied by x_g.

The third and final term in (5.119) describes voltages induced by rotor fluxes. It can be written as

$$\frac{1}{\omega_0} \frac{d}{dt} \{\underline{\underline{T}}^{-1} \underline{e}''\} = \frac{1}{\omega_0} \frac{d}{dt} \{\underline{\underline{T}}^{-1}\} \underline{e}'' + \frac{1}{\omega_0} \underline{\underline{T}}^{-1} \frac{d\underline{e}''}{dt} \tag{5.121}$$

Now, the time derivative of the inverse transform is

$$\frac{1}{\omega_0} \frac{d}{dt} \underline{\underline{T}}^{-1} = \frac{\omega}{\omega_0} \begin{bmatrix} -\sin(\theta) & -\cos(\theta) & 0 \\ -\sin\left(\theta - \frac{2\pi}{3}\right) & -\cos\left(\theta - \frac{2\pi}{3}\right) & 0 \\ -\sin\left(\theta + \frac{2\pi}{3}\right) & -\cos\left(\theta + \frac{2\pi}{3}\right) & 0 \end{bmatrix} \tag{5.122}$$

Now the three phase voltages can be extracted from all of this matrix algebra

$$v_a = \frac{x_d''}{\omega_0} \frac{di_a}{dt} + \frac{x_g}{\omega_0} \frac{d}{dt} (i_a + i_b + i_c) + e_a'' \qquad (5.123)$$

$$v_b = \frac{x_d''}{\omega_0} \frac{di_b}{dt} + \frac{x_g}{\omega_0} \frac{d}{dt} (i_a + i_b + i_c) + e_b'' \qquad (5.124)$$

$$v_c = \frac{x_d''}{\omega_0} \frac{di_c}{dt} + \frac{x_g}{v_0} \frac{d}{dt} (i_a + i_b + i_c) + e_c'' \qquad (5.125)$$

Where the internal voltages are

$$e_a'' = -\frac{\omega}{\omega_0} (e_q'' \sin(\theta) - e_d'' \cos(\theta))$$

$$+ \frac{1}{\omega_0} \cos(\theta) \frac{de_q''}{dt} + \frac{1}{\omega_0} \sin(\theta) \frac{de_d''}{dt} \qquad (5.126)$$

$$e_b'' = -\frac{\omega}{\omega_0} \left(e_q'' \sin \left(\theta - \frac{2\pi}{3} \right) - e_d'' \cos \left(\theta - \frac{2\pi}{3} \right) \right)$$

$$+ \frac{1}{\omega_0} \cos \left(\theta - \frac{2\pi}{3} \right) \frac{de_q''}{dt} + \frac{1}{\omega_0} \sin \left(\theta - \frac{2\pi}{3} \right) \frac{de_d''}{dt} \qquad (5.127)$$

$$e_c'' = -\frac{\omega}{\omega_0} \left(e_q'' \sin \left(\theta + \frac{2\pi}{3} \right) - e_d'' \cos \left(\theta + \frac{2\pi}{3} \right) \right)$$

$$+ \frac{1}{\omega_0} \cos \left(\theta + \frac{2\pi}{3} \right) \frac{de_q''}{dt} + \frac{1}{\omega_0} \sin \left(\theta + \frac{2\pi}{3} \right) \frac{de_d''}{dt} \qquad (5.128)$$

This set of expressions describes the equivalent circuit shown in Fig. 5.28.

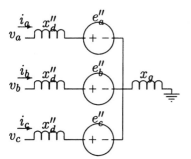

Figure 5.28 Equivalent network model.

5.17.10 Restatement of the model

The synchronous machine model which uses the three phase currents as state variables may now be stated in the form of a set of differential and algebraic equations

$$\frac{d\psi_{kd}}{dt} = -\omega_0 r_{kd} i_{kd} \tag{5.129}$$

$$\frac{d\psi_{kq}}{dt} = -\omega_0 r_{kq} i_{kq} \tag{5.130}$$

$$\frac{d\psi_f}{dt} = -\omega_0 r_f i_f \tag{5.131}$$

$$\frac{d\delta}{dt} = \omega - \omega_0 \tag{5.132}$$

$$\frac{d\omega}{dt} = \frac{\omega_0}{2H} (T_m + e_q'' i_q + e_d'' i_d) \tag{5.133}$$

where

$$\begin{bmatrix} i_{kd} \\ i_f \end{bmatrix} = \begin{bmatrix} x_{kd} & x_{ad} \\ x_{ad} & x_f \end{bmatrix}^{-1} \left(\begin{bmatrix} \psi_{kd} \\ \psi_f \end{bmatrix} - \begin{bmatrix} x_{ad} \\ x_{ad} \end{bmatrix} i_d \right)$$

and

$$i_{kq} = \frac{1}{x_{kq}} \psi_{kq} - \frac{x_{aq}}{x_{kq}} i_q$$

(It is assumed here that the difference between subtransient reactances is small enough to be neglected.)

The network interface equations are, from the network to the machine

$$i_d = i_a \cos(\theta) + i_b \cos\left(\theta - \frac{2\pi}{3}\right) + i_c \cos\left(\theta + \frac{2\pi}{3}\right) \tag{5.134}$$

$$i_q = -i_a \sin(\theta) - i_b \sin\left(\theta - \frac{2\pi}{3}\right) - i_c \sin\left(\theta + \frac{2\pi}{3}\right) \tag{5.135}$$

and, in the reverse direction, from the machine to the network

$$e''_a = -\frac{\omega}{\omega_0}\left(e''_q \sin(\theta) - e''_d \cos(\theta)\right)$$

$$+\frac{1}{\theta_0}\cos(\theta)\frac{de''_q}{dt} + \frac{1}{\omega_0}\sin(\theta)\frac{de''_d}{dt} \tag{5.136}$$

$$e''_b = -\frac{\omega}{\omega_0}\left(e''_q \sin\left(\theta - \frac{2\pi}{3}\right) - e''_d \cos\left(\theta - \frac{2\pi}{3}\right)\right)$$

$$+\frac{1}{\omega_0}\cos\left(\theta - \frac{2\pi}{3}\right)\frac{de''_q}{dt} + \frac{1}{\omega_0}\sin\left(\theta - \frac{2\pi}{3}\right)\frac{de''_d}{dt} \tag{5.137}$$

$$e''_c = -\frac{\omega}{\omega_0}\left(e''_q \sin\left(\theta + \frac{2\pi}{3}\right) - e''_d \cos\left(\theta + \frac{2\pi}{3}\right)\right)$$

$$+\frac{1}{\omega_0}\cos\left(\theta + \frac{2\pi}{3}\right)\frac{de''_q}{dt} + \frac{1}{\omega_0}\sin\left(\theta + \frac{2\pi}{3}\right)\frac{de''_d}{dt} \tag{5.138}$$

and, of course

$$\theta = \omega_0 t + \delta \tag{5.139}$$

$$e''_q = \psi''_d \tag{5.140}$$

$$e''_d = -\psi''_q \tag{5.141}$$

$$\psi''_d = \frac{x_{ad}x_{kdl}\psi_f + x_{ad}x_{fl}\psi_{kd}}{x_{ad}x_{kdl} + x_{ad}x_{fl} + x_{kdl}x_{fl}} \tag{5.142}$$

$$\psi''_q = \frac{x_{aq}}{x_{aq} + x_{kql}}\psi_{kq} \tag{5.143}$$

5.17.11 Network constraints

This model may be embedded in a number of networks. Different configurations will result in different constraints on currents. Consider, for example, the situation in which all of the terminal voltages are constrained, but perhaps by unbalanced (not entirely positive sequence) sources. In that case, the differential equations for the three phase currents would be

$$\frac{x''_d}{\omega_0}\frac{di_a}{dt} = (v_a - e''_a)\frac{x''_d + 2x_g}{x''_d + 3x_g}$$

$$- [(v_b - e''_b) + (v_c - e''_c)]\frac{x_g}{x''_d + 3x_g} \tag{5.144}$$

$$\frac{x_d''}{\omega_0}\frac{di_b}{dt} = (v_b - e_b'')\frac{x_d'' + 2x_g}{x_d'' + 3x_g}$$

$$- [(v_a - e_a'') + (v_c - e_c'')]\frac{x_g}{x_d'' + 3x_b} \qquad (5.145)$$

$$\frac{x_d''}{\omega_0}\frac{di_c}{dt} = (v_c - e_c'')\frac{x_d'' + 2x_g}{x_d'' + 3x_g}$$

$$- [(v_b - e_b'') + (v_a - e_a'')]\frac{x_g}{x_d'' + 3x_g} \qquad (5.146)$$

5.17.12   Example: line-line fault

This model is suitable for embedding into network analysis routines. It is also possible to handle many different situations directly. Consider, for example, the unbalanced fault represented by the network shown in Fig. 5.29. This shows a line-line fault situation, with one phase still connected to the network.

In this situation, one has only two currents to worry about, and their differential equations would be

$$\frac{di_b}{dt} = \frac{\omega_0}{2x_d''}(e_c'' - e_b'' - 2r_a i_b) \qquad (5.147)$$

$$\frac{di_a}{dt} = \frac{\omega_0}{x_d'' + x_g}(v_a - e_a'' - r_a i_a) \qquad (5.148)$$

and, of course, $i_c = -i_b$.

Note that included here are the effects of armature resistance, ignored in the previous section, but obviously important if the results are to be believed.

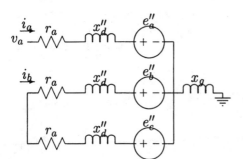

Figure 5.29 Line-to-line fault network model.

5.17.13   Permanent magnet machines

Permanent magnet machines are one state variable simpler than their wound-field counterparts. They may be accurately viewed as having *constant* field current. Assuming that we can define the internal (field) flux as

$$\psi_0 = x_{ad}i_{f0} \tag{5.149}$$

A reasonably simple expression for the rotor currents, in the case of a voltage driven machine becomes

$$\begin{bmatrix} i_d \\ i_{kd} \end{bmatrix} = \begin{bmatrix} x_d & x_{ad} \\ x_{ad} & x_{kd} \end{bmatrix}^{-1} \begin{bmatrix} \psi_d - \psi_0 \\ \psi_{kd} - \psi_0 \end{bmatrix} \tag{5.150}$$

$$\begin{bmatrix} i_q \\ i_{kq} \end{bmatrix} = \begin{bmatrix} x_q & x_{aq} \\ x_{aq} & x_{kq} \end{bmatrix}^{-1} \begin{bmatrix} \psi_q \\ \psi_{kq} \end{bmatrix} \tag{5.151}$$

The simulation model then has six states

$$\frac{d\psi_d}{dt} = \omega_0 v_d + \omega\psi_q - \omega_0 r_a i_d \tag{5.152}$$

$$\frac{d\psi_q}{dt} = \omega_0 v_q - \omega\psi_d - \omega_0 r_a i_q \tag{5.153}$$

$$\frac{d\psi_{kd}}{dt} = -\omega_0 r_{kd} i_{kd} \tag{5.154}$$

$$\frac{d\psi_{kq}}{dt} = -\omega_0 r_{kq} i_{kq} \tag{5.155}$$

$$\frac{d\omega}{dt} = \frac{\omega_0}{2H}(\psi_d i_q - \psi_q i_d + T_m) \tag{5.156}$$

$$\frac{d\delta}{dt} = \omega - \omega_0 \tag{5.157}$$

In the case of a current-driven machine, rotor currents required in the simulation are

$$i_{kd} = \frac{1}{x_{kd}}(\psi_{kd} - x_{ad}i_d - \psi_0) \tag{5.158}$$

$$i_{kq} = \frac{1}{x_{kq}}(\psi_{kq} - x_{aq}i_q) \tag{5.159}$$

Here, the "flux behind subtransient reactance" is, on the direct axis

$$\psi_d'' = \frac{x_{kdl}\psi_0 + x_{ad}\psi_{kd}}{x_{ad} + x_{kdl}} \tag{5.160}$$

and the subtransient reactance is

$$x_d'' = x_{al} + x_{ad}\|x_{kdl} \tag{5.161}$$

On the quadrature axis

$$\psi_q'' = \frac{x_{ad}\psi_{kq}}{x_{ad} + x_{kql}} \tag{5.162}$$

and

$$x_q'' = x_{al} + x_{aq}\|x_{kql} \tag{5.163}$$

In this case, there are only four state equations

$$\frac{d\psi_{kd}}{dt} = -\omega_0 r_{kd} i_{kd} \tag{5.164}$$

$$\frac{d\psi_{kq}}{dt} = -\omega_0 r_{kq} i_{kq} \tag{5.165}$$

$$\frac{d\omega}{dt} = \frac{\omega_0}{2H}(e_q'' i_q + e_d'' i_d + T_m) \tag{5.166}$$

$$\frac{d\delta}{dt} = \omega - \omega_0 \tag{5.167}$$

The interconnections to and from the network are the same as in the case of a wound-field machine: in the "forward" direction, from network to machine

$$i_d = i_a \cos(\theta) + i_b \cos\left(\theta - \frac{2\pi}{3}\right) + i_c \cos\left(\theta + \frac{2\pi}{3}\right) \tag{5.168}$$

$$i_q = -i_a \sin(\theta) - i_b \sin\left(\theta - \frac{2\pi}{3}\right) - i_c \sin\left(\theta + \frac{2\pi}{3}\right) \tag{5.169}$$

and, in the reverse direction, from the machine to the network

$$e_a'' = -\frac{\omega}{\omega_0} \left(e_q'' \sin(\theta) - e_d'' \cos(\theta) \right)$$

$$+\frac{1}{\omega_0} \cos(\theta) \frac{de_q''}{dt} + \frac{1}{\omega_0} \sin(\theta) \frac{de_d''}{dt} \qquad (5.170)$$

$$e_b'' = -\frac{\omega}{\omega_0} \left(e_q'' \sin\left(\theta - \frac{2\pi}{3}\right) - e_d'' \cos\left(\theta - \frac{2\pi}{3}\right) \right)$$

$$+\frac{1}{\omega_0} \cos\left(\theta - \frac{2\pi}{3}\right) \frac{de_q''}{dt} + \frac{1}{\omega_0} \sin\left(\theta - \frac{2\pi}{3}\right) \frac{de_d''}{dt} \qquad (5.171)$$

$$e_c'' = -\frac{\omega}{\omega_0} \left(e_q'' \sin\left(\theta + \frac{2\pi}{3}\right) - e_d'' \cos\left(\theta + \frac{2\pi}{3}\right) \right)$$

$$+\frac{1}{\omega_0} \cos\left(\theta + \frac{2\pi}{3}\right) \frac{de_q''}{dt} + \frac{1}{\omega_0} \sin\left(\theta + \frac{2\pi}{3}\right) \frac{de_d''}{dt} \qquad (5.172)$$

5.17.14   PM machines with no damper

PM machines without much rotor conductivity may often behave as if they have no damper winding at all. In this case, the model simplifies even further. Armature currents are

$$i_d = \frac{1}{x_d} (\psi_d - \psi_0) \qquad (5.173)$$

$$i_q = \frac{1}{x_q} \psi_q \qquad (5.174)$$

The state equations are

$$\frac{d\psi_d}{dt} = \omega_0 v_d + \omega \psi_q - \omega_0 r_a i_d \qquad (5.175)$$

$$\frac{d\psi_q}{dt} = \omega_0 v_q - \omega \psi_d - \omega_0 r_a i_q \qquad (5.176)$$

$$\frac{d\omega}{dt} = \frac{\omega_0}{2H} (\psi_d i_q - \psi_q i_d + T_m) \qquad (5.177)$$

$$\frac{d\delta}{dt} = \omega - \omega_0 \qquad (5.178)$$

In the case of no damper, the machine becomes quite simple. There is no "internal flux" on the quadrature axis. Further, there are no time derivatives of the internal flux on the d-axis. The only machine state equations are mechanical

$$\frac{d\omega}{dt} = \frac{\omega_0}{2H} (\psi_0 i_q + T_m) \tag{5.179}$$

$$\frac{d\delta}{dt} = \omega - \omega_0 \tag{5.180}$$

The "forward" network interface is as before

$$i_d = i_a \cos(\theta) + i_b \cos\left(\theta - \frac{2\pi}{3}\right) + i_c \cos\left(\theta + \frac{2\pi}{3}\right) \tag{5.181}$$

$$i_q = -i_a \sin(\theta) - i_b \sin\left(\theta - \frac{2\pi}{3}\right) - i_c \sin\left(\theta + \frac{2\pi}{3}\right) \tag{5.182}$$

and, in the reverse direction, from the machine to the network, things are simpler than before

$$e_q'' = -\frac{\omega}{\omega_0} \psi_0 \sin(\theta) \tag{5.183}$$

$$e_b'' = -\frac{\omega}{\omega_0} \psi_0 \sin\left(\theta - \frac{2\pi}{3}\right) \tag{5.184}$$

$$e_c'' = -\frac{\omega}{\omega_0} \psi_0 \sin\left(\theta + \frac{2\pi}{3}\right) \tag{5.185}$$

5.18 Standards

Synchronous motor standards fall in two broad categories, viz., standards for performance and standards for testing. In the U.S., National Electrical Manufacturers Association (NEMA) writes standards for performance, whereas The Institute of Electrical and Electronics Engineers (IEEE) writes standards for testing. The main standards that apply to synchronous motors are:

- NEMA MG1-1993, Rev 4, "Motors and Generators"
- IEEE Std 115-1995, "IEEE Guide: Test Procedures for Synchronous Machines"

- IEEE Std 522-1992, "IEEE Guide for Testing Turn-to-Turn Insulation on Form-Wound Stator Coils for Alternating Current Rotating Electric Machines"

NEMA MG1 is a standard for machines for general purpose applications. There are other U.S. standards that are industry-specific and deal with design and construction of machines pertinent to that particular industry. An example of this is American Petroleum Institute Standard API 546, which applies to synchronous motors for the petroleum and chemical industries, and addresses in some detail, design and manufacture in addition to performance. Such standards include issues other than performance because of the special needs of these industries. NEMA, on the other hand, writes standards that are, for the most part, applicable to all machines. Industry-specific standards generally use NEMA standards as the baseline and add to it to fit the needs of the specific industry.

Internationally, International Electrotechnical Commission (IEC) standards are available. These are developed by IEC member countries, and are expected to apply to machines everywhere, although country-specific standards such as the NEMA and IEEE standards in the U.S. also exist. IEC standards, like NEMA standards, are also general purpose standards.

5.19   NEMA MG1 Performance

NEMA MG1 applies to synchronous motors in sizes covered by Table 5.4.

5.19.1   Voltage, frequency, speed and power factor

NEMA synchronous motors are available for 50 and 60 Hz operation at power factors of 1.0 and 0.8. Standard service factor is 1.0, but

TABLE 5.4   Horsepower Ratings for Synchronous Motors

20	200	900	4000	12000	25000	60000
25	250	1000	4500	13000	27500	65000
30	300	1250	5000	14000	30000	70000
40	350	1500	5500	15000	32500	75000
50	400	1750	6000	16000	35000	80000
60	450	2000	7000	17000	37500	90000
75	500	2250	8000	18000	40000	100000
100	600	2500	9000	19000	45000	
125	700	3000	10000	20000	50000	
150	800	3500	11000	22500	55000	

motors at 1.15 service factor can also be specified. The available voltages are 460, 575, 2300, 4000, 4600, 6600 and 13200 volts. Since it is not possible to design all sizes of motors at all voltages and all speeds, NEMA identifies horsepower ranges available at various voltages, as well as speeds for all sizes. Table 5.5 gives horsepower assignments, and Table 5.6 available speeds.

TABLE 5.5   NEMA Horsepower Assignments

60 Hz voltage rating	Horsepower
460 or 575	100–600
2300	200–5000
4000 or 4600	200–10000
6600	1000–15000
13200	3500 and above

Copyright by NEMA. Used by permission.

TABLE 5.6   Available 60 Hz Speeds for Synchronous Motors

All ratings					
3600	600	327	225	138	95
1800	514	300	200	129	90
1200	450	277	180	120	86
900	400	257	164	109	80
720	360	240	150	100	

50 Hz speeds are 5/6 of the 60 Hz speeds

Copyright by NEMA. Used by permission.

5.19.2   Voltage and frequency variations

NEMA MG1 allows a variation of ±10% of rated voltage at rated frequency, ±5% of rated frequency at rated voltage, or a combination of rated voltage and frequency of 10%, provided that the frequency variation does not exceed ±5%. It is assumed the performance of the motor may differ from that at rated voltage and frequency conditions when operated at other than rated conditions.

5.19.3   Operating conditions

The usual or normal site operating conditions include the following:

- an ambient temperature in the range of 0°C to 40°C
- an altitude not exceeding 1000 m
- a location such that there is no serious interference with motor ventilation
- deviation factor of the supply voltage of less than 10%
- supply voltage balanced to within 1%

■ a grounded power system

5.19.4 Temperature rises

NEMA recognizes four insulation classes (A, B, F and H) for the windings. Table 5.7 gives allowable temperature rises for these four classes.

TABLE 5.7 NEMA Temperature Rise Limits in Degrees C

Notes	Class A	Class B	Class F	Class H
A	60	80	105	125
B	70	90	115	140
C	65	85	110	135
D	60	80	105	125

Note A: Rise by resistance. All ratings. Armature and field windings
Note B: Rise by RTD. 1500 HP and less for armature windings
Note C: Rise by RTD. Over 1500 HP at 7000 volts and less for armature windings
Note D: Rise by RTD. Over 1500 HP at over 7000 volts for armature windings.

Copyright by NEMA. Used by permission.

TABLE 5.8 Torque Values

Speed r/min	HP	Power factor	Locked rotor	Pull-in	Pull-out
500 to 800	200 and below	1.0	100	100	150
	150 and below	0.8	100	100	175
	250 to 1000	1.0	60	60	150
	200 to 1000	0.8	60	60	175
	1250 and larger	1.0	40	60	150
		0.8	40	60	175
450 and below	All ratings	1.0	40	30	150
		0.8	40	30	200

Copyright by NEMA. Used by permission.

5.19.5 Torques and starting

The minimum values of locked-rotor, pull-in and pull-out torques with rated voltage and frequency applied are given in Table 5.8. The required number of starts with normal load inertia are:

■ Two starts in succession, coasting to rest between starts, with the motor initially at ambient temperature

■ One start with the motor initially at a temperature not exceeding its rated load operating temperature

5.19.6   Overloads

NEMA motors have the capability to deliver higher than 150% torque for more than 15 seconds. They will withstand an excess current of 50% for 30 seconds. NEMA MG1 also requires an overspeed for two minutes of 20% for motors operating at higher than 1800 r/min, and 25% for all other motors.

5.19.7   Vibration

Vibration limits for completely assembled motors, running uncoupled are specified. The preferred vibration parameter is velocity in in/s or mm/s peak. Table 5.9 gives the vibration limits on the bearing housings of resiliently mounted motors. For rigidly mounted motors, the limits given in this Table are reduced to 80% of the table values.

Lower vibration limits are specified for special applications. Also included are shaft vibration limits by displacement using noncontacting probes. These both are available when specified.

TABLE 5.9   NEMA Unfiltered Vibration Limits

Speed (r/min)	Rotational frequency (Hz)	Peak velocity (in/s)
3600	60	0.15
1800	30	0.15
1200	20	0.15
900	15	0.12
720	12	0.09
600	10	0.08

Copyright by NEMA. Used by permission.

5.19.8   Surge withstand capabilities

Surge withstand capabilities for armature windings are specified by NEMA. These apply to armature windings only, and to motors using form-wound armature coils.

NEMA standard is a capability to withstand a steep-fronted surge of 2 per unit (pu) at a rise time of 0.1 to 0.2 μs and 4.5 pu at a rise time of 1.2 μs or slower. Option available is a surge of 3.5 per unit (pu) at a rise time of 0.1 to 0.2 μs and 5 pu at arise time of 1.2 μs or slower. One per unit is the peak of the rated line-to-ground voltage. For "green" coils, the test values are 65% and for resin rich coils, 80% during the manufacturing cycle.

The test methods and instrumentation are per IEEE Std 522. This latter standard establishes the test voltage levels and is a de facto standard for turn insulation.

5.20   International (IEC) Standards

IEC 34 series of standards apply to synchronous motors. A fair degree of harmonization exists between NEMA and IEC standards. Because of the needs of the international marketplace, the requirements for any specific design or performance parameter in the two standards are not always the same, although the intent is identical. NEMA standards however are somewhat more extensive than IEC standards in that a greater number of application related topics are addressed.

IEC does not have any requirements for power factor, service factor, speed and number of poles, and load inertia. A derating of the motor for altitudes higher than 1000 m is not required. It requires however, that the power supply be virtually balanced and sinusoidal with a harmonic voltage factor not exceeding 2%. Nine types of duty varying between continuous, standby and short time are identified (NEMA offers only continuous duty) and overspeed capability requirement is 20% for all speeds.

In some cases harmonization is not possible because of philosophical differences. For synchronous motors, a good example is the allowance of tolerances on performance by IEC. IEC performance for a given motor can sometimes be shown to be better than NEMA performance for the same motor. Another example is the surge withstand capability levels. NEMA requirements are based on some actual user and site testing experience, whereas the IEC requirements are a fixed percentage of the line voltage. For more information on standards, see the following references:

1. Ghai, Nirmal K., "Comparison of International and NEMA Standards for Salient Pole Synchronous Machines." Paper approved for publication in IEEE Transactions on Energy Conversion, 1998.
2. IEC 34-1, 1996, 10th Edition, "Rotating Electrical Machines, Part 1: Rating and Performance."
3. NEMA MG1-1993, Rev 4, "Motors and Generators."
4. IEEE Std 115-1995, "IEEE Guide: Test Procedures for Synchronous Machines."
5. IEEE Std 522-1992, "IEEE Guide for Testing Turn-to-Turn Insulation on Form-Wound Stator Coils for Alternating Current Rotating Electric Machines."

5.21   Testing of Synchronous Motors

Synchronous motor testing falls in two categories:

- Routine tests
- Prototype or complete tests

IEEE Std 115-1995 is the standard that applies to synchronous motors and contains instructions for conducting the tests to determine the performance characteristics of synchronous machines. This standard gives alternative methods for making many of the tests to enable the selection of a method appropriate for the size of motor under consideration. Where a method is superior to others, it is identified as the preferred method.

5.21.1   Routine tests

The main purpose of routine tests is to ensure that the motor is free from electrical or mechanical defects. Depending on the size of the motor, some or all of the following tests could constitute routine tests:

- Resistance of armature and field windings
- High potential test
- Polarity of field coils
- Insulation resistance
- Tests for short-circuited field coils
- Shaft currents
- Phase sequence
- Overspeed test
- Saturation curves
- Locked rotor current and torque
- Air-gap measurement
- Noise
- Vibration

NEMA MG1 includes the first two tests in the above list for all motors, and a field polarity check for unassembled motors or a no-load test for assembled motors.

5.21.2   Prototype tests

Prototype tests are performed to evaluate the complete performance of the motor. The following tests are included in addition to the routine tests:

- Measurement of segregated losses and efficiency
- Load excitation
- Temperature tests
- Speed-torque and pull-out torque tests
- Synchronous machine quantities

■ Sudden short-circuit tests

5.21.3   Resistance measurement

The stator, field winding and exciter winding resistances are usually measured using a digital bridge, or a calibrated ohmmeter if the resistance is greater than one ohm. The value is then corrected to 25°C for comparison with the expected value, or to any other temperature using the following relationship

$$R_s = R_t[(t_s + k)/(t_t + k)]$$

where R_s = winding resistance at temperature t_s
    R_t = winding temperature at temperature t_t
    k = characteristic constant for winding material
       = 234.5 for copper.

t_s and t_t are in °C.

5.21.4   Polarity of field coils

This test can be made by means of a small permanent magnet mounted in such a manner that it can rotate and reverse its direction freely. For the test, the field winding is energized with a low value of current. The correct polarity is indicated by the magnet reversing direction as it is passed from pole to pole.

5.21.5   High potential test

This test is usually made after all other tests have been completed. The test voltage is applied to each phase of the stator winding, the field winding and the exciter windings in sequence, with all windings not under test and other metal parts grounded. The leads of each winding are connected together for the test. The test voltage can be an ac voltage at power frequency or, for the stator windings, a dc voltage equal to 1.7 times the rms value of the power frequency voltage. The duration of the test for each winding is one minute. The test voltages specified by NEMA are:

stator windings	twice the line voltage plus 1000 volts with a minimum of 1500 volts.
field windings	10 times the excitation voltage with a minimum of 1500 volts for rated excitation voltage of ≤500 volts and 4000 volts plus twice the excitation voltage for rated excitation voltages of greater than 500 volts.
exciter armature winding	same as for field windings

exciter field winding   10 times the excitation voltage with a minimum of 1500 volts for rated excitation voltage of ≤500 volts and 4000 volts plus twice the excitation voltage for rated excitation voltages greater than 500 volts.

5.21.6   Insulation resistance

Insulation resistance is the resistance of the winding's insulation. Its measurement is useful as a long term maintenance tool. Measured frequently during the life of the motor, it provides an indication of winding deterioration and potential need for preventive maintenance.

For performing this test, all accessories with leads located at the machine terminals are disconnected from the motor and their leads connected to each other and to the motor frame. The test is made on each phase with other phases grounded. The usual voltages for making this test are 500 or 1000 volts dc for machines operating at 7000 volts or less, and 2500 volts dc for higher voltages. The most commonly used testing device is an insulation resistance tester with a self-contained constant dc voltage source and a direct reading cross-coil type ohmmeter. The test voltage is applied to both ends of the winding under test.

The minimum acceptable value of insulation resistance in megohms is equal to the rated rms line voltage in kilovolts plus one at 40°C.

5.21.7   Tests for short-circuited field coils

This test is made to determine if any of the field winding turns are short circuited. A good test method is to pass constant amplitude alternating current through the entire field winding and to measure the voltage across each field coil. Compared to a good coil, a field coil with a short-circuited turn or turns will have a substantially lower voltage drop across it while a good coil adjacent to the coil with the short-circuited turn will have a voltage drop somewhat less than that across a good coil because of the reduced flux in the short-circuited coil. A comparison of the voltage drops across all coils can thus be used to determine the coil with the short-circuited turns.

5.21.8   Shaft currents

Unbalances in the magnetic circuits can create flux linkages with the rotating systems which can produce a potential difference between the shaft ends. This potential difference, if large enough, can produce a circulating current through the bearing systems resulting in premature bearing failures unless the circuit is interrupted by insulation. With the motor running at rated voltage and frequency, the voltage between the shaft ends is measured using an electronic voltmeter. A voltage of less than 100 millivolts for anti-friction bearings and 200 millivolts for sleeve bearings should create no problems for the motor. For higher voltages, one bearing should be insulated to interrupt the current flow and eliminate bearing currents.

5.21.9 Phase sequence

This test is made to insure that the motor terminals have been marked correctly for the required direction of rotation. The test is performed by starting the motor from its normal power source and observing the direction of rotation.

5.21.10 Overspeed test

This test is normally made with the motor unexcited. The motor rotor is driven at the specified overspeed and its vibration performance observed for any sign of distress. The motor may be dismantled, if necessary, to look for any signs of damage.

NEMA standard requires an overspeed capability for two minutes of 20% of rated speed for motors running at speeds higher than 1800 r/min and 20% for all others.

5.21.11 Vibration

The normal test entails reading vibration at the bearing housing with the motor running uncoupled and on no-load at rated voltage and frequency. The limits are established in NEMA MG1. (See above.)

5.21.12 Bearing temperature rise

This test is made by operating the motor unloaded for at least two hours while monitoring the bearing temperature. The test is continued until the bearing temperature stabilizes. A good indication of temperature stability is when there is less than 1°C rise between two consecutive readings taken 30 minutes apart.

5.21.13 Measurement of saturation curves, segregated losses and efficiency

Efficiency is the ratio of the motor output power and the motor input power.

$$\text{Efficiency} = (\text{output})/(\text{input})$$

or $\qquad\qquad\quad = (\text{output})/(\text{output} + \text{losses})$

or, approximately $\qquad = (\text{input} - \text{losses})/(\text{input})$

It can thus be calculated by a knowledge of power input and power output, or of power output and losses, or power input and losses.

The losses in the synchronous motor consist of the following:

- Stator I^2R loss
- Field I^2R loss
- Core loss
- Friction and windage loss
- Stray load loss

The stator and field I^2R losses can be calculated using the stator and field currents and the resistance of the stator and field windings at the operating temperature.

Four different methods can be used for the determination of other losses and saturation curves. One of these, the electrical input method, will be described here. Two tests are required for this. These are the open-circuit test and the short-circuit test.

5.21.13.1 Open-circuit test. The open-circuit test yields the friction and windage loss, and the core loss for the motor. Open-circuit or no-load saturation curve is also obtained from this test.

The machine is run as a synchronous motor at approximately unity power factor and readings of power input, field current and armature current taken at the following voltage points:

- Four points below 60% voltage
- Two points between 60% and 90% voltage
- Four between 90% and 110% voltage
- Two points above 110% voltage, with one of these at approximately 120% voltage.

From this test, the open-circuit saturation curve which is the relationship between armature voltage and field current can be plotted, as well as the core loss and friction and windage loss curve (Fig. 5.30).

Since the motor is unloaded for this test, the power input is the motor loss for the test. For each voltage point, the difference between

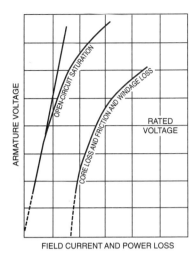

ARMATURE VOLTAGE

OPEN-CIRCUIT SATURATION

CORE LOSS AND WINDAGE LOSS

CORE LOSS AND FRICTION AND WINDAGE LOSS

RATED
VOLTAGE

FIELD CURRENT AND POWER LOSS

Figure   5.30 Open-circuit   and
core-loss curves.

the total loss and the stator and field I^2R losses for the measured value of stator and field currents and winding resistances at the test temperature is the core loss plus friction and windage loss. A plot of this loss against the stator voltage when extrapolated down to the point where it intercepts the x-axis gives the friction and windage loss. Subtracting the friction and windage loss as well as the I^2R losses from power inputs at all voltage points will give the core-loss curve.

5.21.13.2   Short-circuit test. The short-circuit test gives the stray load losses for the motor. For this test, the machine is operated as a motor at a fixed voltage, preferably about one-third the rated voltage or at a lower value for which stable operation can be obtained. The armature current is varied by changing the field current in about six steps between 25% and 125% of rated armature current.

The total power input under the short-circuit test consists of the core loss, the friction and windage loss, the field and stator I^2R loss, and the stray load loss. A curve can now be plotted showing the relationship between the total loss and the square of the armature current or voltage (Fig. 5.31). The intercept of this curve with the power loss axis gives the friction, windage and core losses. Subtracting these losses from the total power input at any stator current gives the short-circuit loss for that current. Subtracting the stator I^2R loss from the short-circuit loss at the test temperature gives the stray load loss for that armature current.

The curve showing the relationship of stator current to field current is over-excited part of the zero power factor V-curve (see Fig. 5.32). A

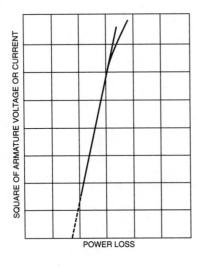

Figure 5.31 Determination of friction windage and core loss.

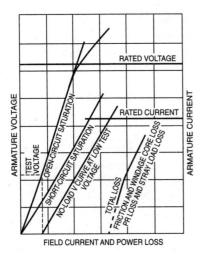

Figure 5.32 Saturation and loss curves.

curve parallel to this and passing through the origin is the short-circuit saturation curve for the motor.

5.21.13.3 Load excitation. The load excitation or the field current under load conditions can be determined for any load, voltage, frequency and power factor by loading the motor and measuring the field current. When the motor is too large to be so loaded, the field current can be obtained by the potier reactance method.

In the potier reactance method, the voltage back of the potier reactance is determined as in Fig. 5.33, or from the equation

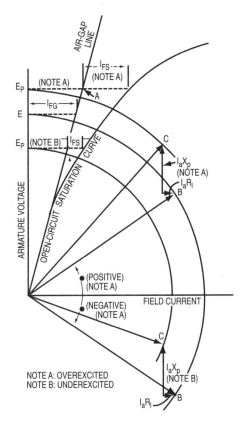

Figure 5.33 Determination of voltage back of potier reactance.

NOTE A: OVEREXCITED
NOTE B: UNDEREXCITED

$$E_p = \sqrt{[(E \cos \phi - I_a R_a)^2 + (E \sin \phi + I_a X_p)^2]}$$

where X_p = Potier reactance
E = Stator terminal voltage
I_a = Stator current
R_a = Stator resistance,
all in per unit
ϕ = Power factor angle, positive for over-excited, negative for under-excited operation

$I_a X_p$ is obtained from Fig. 5.34.

The potier reactance can be calculated graphically as shown in Fig. 5.34. In this figure, line ab is parallel to the air-gap line and line bc is the product of the stator current and the potier reactance. The potier reactance is obtained by dividing this product by the stator current. Or the stator leakage reactance can be used in place of potier reactance.

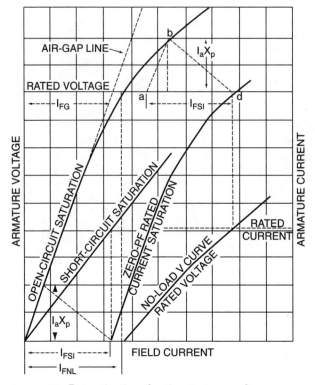

Figure 5.34 Determination of potier reactance voltage.

The field excitation I_{FS} is then calculated from Fig. 5.35 or Fig. 5.36 for which I_{FG} and I_{FS} are obtained from Fig. 5.33 and I_{FSI} from Fig. 5.34. Or the following equation may be used

$$I_{FL} = I_{FS} + \sqrt{[(I_{FG} + I_{FSI} \sin \phi)^2 + (I_{FSI} \cos \phi)^2]}$$

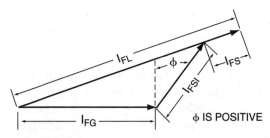

Figure 5.35 Determination of load excitation for over-excited generator.

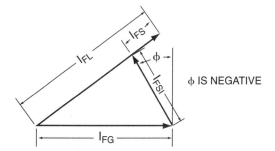

φ IS NEGATIVE

Figure 5.36 Determination of load excitation for under-excited generator.

where ϕ = power factor angle, positive for over-excited and negative for under-excited operation

I_{FG} = field current from the air-gap line at rated voltage

I_{FSI} = field current at rated current from short-circuit curve

I_{FS} = the difference between the field currents on open-circuit curve and the air-gap line for voltage E_p

5.21.14 Temperature tests

Temperature tests are made to determine the temperature rise of the stator and field windings when running under load. These tests can be made using more than one method, the actual method selected depending on the capabilities of the test facility and the size of the motor.

The preferred test method is to load the motor at specified load, voltage, stator current, power factor and frequency until stable temperatures are achieved. For large motors, this method is usually not feasible and another method, the zero power factor test, is more commonly used. In this method, the motor is operated at no load, zero power factor as a synchronous condenser, maintaining appropriate conditions of stator current, voltage and frequency. Since the voltage back of the potier reactance under this test condition is greater than at rated power factor, the terminal voltage must be reduced so that this voltage is the same as its value under rated load and power factor conditions. For the zero power factor test, the field winding losses and hence the field winding temperature rise will differ considerably from those at normal operating conditions. The rise must therefore be corrected for the specified field current.

5.21.15 Other tests

For more information on the tests described above, tests for synchronous machine reactances and time constants, as well as other tests, see IEEE Std 115-1995.

5.22 Bibliography

References for synchronous motors

1. Bose, B. K.; *Adjustable Speed AC Drives;* IEEE Press, 1980.
2. Concordia, C.; *Synchronous Machines;* New York, Wiley, 1951.
3. Fitzgerald, A. E, Kingsley, C., Jr., and Kusko, A.; *Electric Machinery,* 3d ed.; New York, McGraw-Hill Book Company, 1971.
4. IEEE Std. 115 − 1965 *Test Procedures for Synchronous Machines.*
5. IEEE Std. 421 − 1972 *Criteria and Definition for Excitation Systems for Synchronous Machines.*
6. Kostenko, M., and Piotrovsky, L.; *Electric Machines,* vol. 2; Moscow, MIR Publications, 1974.
7. Miller, T. J.; *Brushless Permanent-Magnet and Reluctance Motor Drives,* Oxford University Press, 1989.
8. NEMA Std. MS1−*Motors and Generators.*
9. Sarma, Mulukutla, S.; *Synchronous Machines;* New York, Gordon & Breach, 1979.
10. Say, M. G.; *Alternating Current Machines;* New York, Wiley, 1976.
11. Synchronous Machines (General); ANSI C50.10, Synchronous Motors ANSI C50.11.

Permanent Magnet-Synchronous (Brushless) Motors

J. Kirtley

6.1 Introduction

This section introduces the design evaluation of permanent magnet motors, with an eye toward servo and drive applications. It is organized in the following manner. First, three different geometrical arrangements for permanent magnet motors are described:

1. Surface-mounted magnets, conventional stator.

2. Surface-mounted magnets, air-gap stator winding.

3. Internal magnets (flux-concentrating).

The section then includes a qualitative discussion of these geometries. Also examined is the elementary rating parameters of the machine and how to arrive at a rating, how to estimate the torque and power vs. speed capability of the motor, how the machine geometry can be used to estimate both the elementary rating parameters; and finally, the parameters used to make more detailed estimates of the machine performance.

6.2 Motor Morphologies

There are, of course, many ways of building permanent magnet motors. However, only a few will be considered in this section. Actually, once these are understood, rating evaluations of most other geometrical arrangements should be fairly straightforward. It should be understood that the "rotor inside" vs. "rotor outside" distinction is in fact trivial, with very few exceptions.

6.2.1 Surface-magnet machines

Figure 6.1 shows the basic *magnetic* morphology of the motor with magnets mounted on the surface of the rotor and an otherwise conventional stator winding. This sketch does not show some of the important mechanical aspects of the machine, such as the means for fastening the permanent magnets to the rotor, and so one should look at it with a bit of caution. In addition, this sketch and the other sketches to follow are not necessarily to a scale that would result in workable machines.

This figure shows an axial section of a four-pole ($p = 2$) machine. The four magnets are mounted on a cylindrical rotor "core", or shaft, made of ferromagnetic material. Typically, this would simply be a steel shaft. In some applications, the magnets may be simply bonded to the steel. For applications in which a glue joint is not satisfactory (e.g. for high-speed machines), some sort of rotor-banding or retaining-ring structure is required.

The stator winding of this machine is "conventional", very much like that of an induction motor, consisting of wires located in slots in the

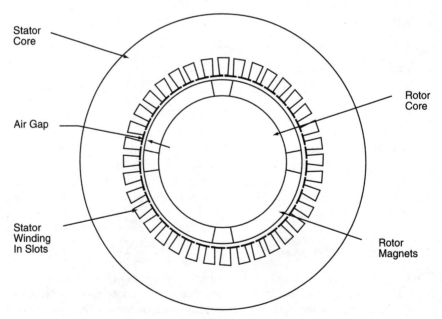

Figure 6.1   Axial view of a surface mount motor.

surface of the stator core. The stator core itself is made of laminated ferromagnetic material (probably silicon iron sheets). The character and thickness of these sheets are determined by operating frequency and efficiency requirements. They are required to carry alternating magnetic fields, so they must be laminated to reduce eddy current losses.

This sort of machine is simple in construction. Note that the operating magnetic flux density in the air-gap is nearly the same as in the magnets, so that this sort of machine cannot have air-gap flux densities higher than that of the remanent flux density of the magnets. If low-cost ferrite magnets are used, this means relatively low induction and consequently relatively low efficiency and power density. (Note the qualifier "relatively" here!). Beware, however, that with modern, high-performance permanent magnet materials in which remanent flux densities can be on the order of 1.2 T, air-gap working flux densities can be on the order of 1 T. With the requirement for slots to carry the armature current, this may be, anyway, a practical limit for air-gap flux density.

It is also important to note that the magnets in this design are really in the "air gap" of the machine, and therefore are exposed to all of the time- and space-harmonics of the stator winding MMF. Because some permanent magnets have electrical conductivity (particularly the higher performance magnets), any asynchronous fields will tend to produce eddy currents and consequent losses in the magnets.

6.2.2   Interior magnet or flux-concentrating machines

Interior magnet designs have been developed to counter several apparent or real shortcomings of surface mount motors:

- Flux-concentrating designs allow the flux density in the air-gap to be higher than the flux density in the magnets themselves.

- In interior magnet designs, there is some degree of shielding of the magnets from high order space harmonic fields by the pole pieces.

- There are control advantages to some types of interior magnet motors, to be discussed shortly.

- Some types of internal magnet designs have (or are claimed to have) structural advantages over surface-mount magnet designs.

The geometry of one type of internal magnet motor is shown (crudely) in Fig. 6.2. The permanent magnets are oriented so that their magnetization is azimuthal. They are located between wedges of

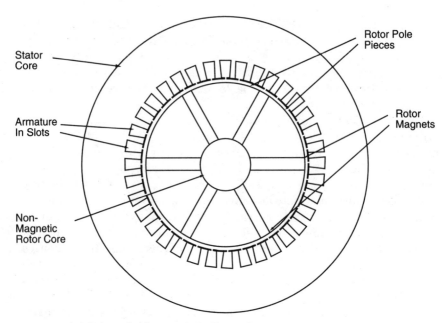

Figure 6.2   Axial view of a flux concentrating motor.

magnetic material (the pole pieces) in the rotor. Flux passes through these wedges, going radially at the air-gap, then azimuthally through the magnets. The central core of the rotor must be non-magnetic, to prevent "shorting out" the magnets. No structure is shown at all in this drawing, but quite obviously this sort of rotor is a structural challenge. A six-pole machine is shown. Typically, one does not expect flux-concentrating machines to have small pole numbers, because it is difficult to get more area inside the rotor than around the periphery. On the other hand, a machine built in this way, but without substantial flux concentration, will still have saliency and magnet shielding properties.

At first sight, these machines appear to be quite complicated to analyze, and that judgement seems to hold up.

6.2.3   Air-gap armature windings

Shown in Fig. 6.3 is a surface-mounted magnet machine with an air-gap, or surface armature winding. Such machines take advantage of the fact that modern, permanent magnet materials have very low permeabilities and that, therefore, the magnetic field produced is relatively insensitive to the size of the air-gap of the machine. It is possible to eliminate the stator teeth and use all of the periphery of the air-gap for windings.

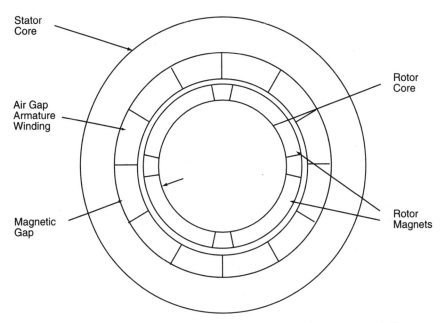

Figure 6.3 Axial view of a permanent magnet (PM) motor with an air-gap winding.

Not shown in this figure is the structure of thee armature winding. This is not an issue in "conventional" stators, since the armature is contained in slots in the iron stator core. The use of an air-gap winding gives opportunities for economy of construction, new armature winding forms such as helical windings, elimination of "cogging" torques and (possibly) higher power densities.

6.3   Zeroth Order-Rating

In determining the rating of a machine, consider two separate sets of parameters. The first set, the elementary rating parameters, consist of the machine inductances, internal flux linkage and stator resistance. From these and a few assumptions about base and maximum speed, it is possible to get a first estimate of the rating and performance of the motor. More detailed performance estimates, including efficiency in sustained operation, require estimation of other parameters.

6.3.1   Voltage and current: round rotor

To get started, consider the equivalent circuit shown in Fig. 6.4. This is actually the equivalent circuit which describes all *round rotor* synchronous machines. It is directly equivalent only to some of the ma-

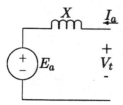

Figure 6.4 Synchronous machine equivalent circuit.

chines dealt with here, but serves to illustrate one or two important points.

Shown in Fig. 6.4 is the equivalent circuit of a single phase of the machine. Most motors are three-phase, but it is not difficult to carry out most of the analysis for an arbitrary number of phases. The circuit shows an internal voltage E_a and a reactance X which, together with the terminal current I, determine the terminal voltage V. In this picture, armature resistance is ignored. If the machine is running in the sinusoidal steady state, the major quantities are of the form

$$E_a = \omega\lambda_a \cos(\omega t + \delta)$$

$$V_t = V \cos \omega t$$

$$I_a = I \cos(\omega t - \psi)$$

The machine is in synchronous operation if the internal and external voltages are at the same frequency and have a constant (or slowly changing) phase relationship (δ). The relationship between the major variables may be visualized by the phasor diagram shown in Fig. 6.5. The internal voltage is just the time-derivative of the internal flux from the permanent magnets, and the voltage drop in the machine

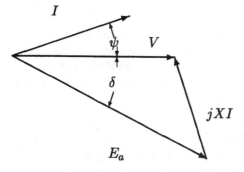

Figure 6.5 Phasor diagram for a synchronous machine.

reactance is also the time-derivative of flux produced by armature current in the air-gap and in the "leakage" inductances of the machine. By convention, the angle ψ is positive when current I lags voltage V and the angle δ is positive, then internal voltage E_a leads terminal voltage V. So both of these angles have negative signs in the situation shown in Fig. 6.5.

If there are q phases, the *time-average* power produced by this machine is simply

$$P = \frac{q}{2} \, VI \, \cos \psi$$

For most polyphase machines operating in what is called "balanced" operation (all phases doing the same thing with uniform phase differences between phases), torque (and consequently power) are approximately constant. Since we have ignored power dissipated in the machine armature, it must be true that power absorbed by the internal voltage source is the same as terminal power, or

$$P = \frac{q}{2} \, E_a I \, \cos (\psi + \delta)$$

Since in the steady state

$$P = \frac{\omega}{p} \, T$$

where T is torque and ω/p is mechanical rotational speed, torque can be derived from the terminal quantities by simply

$$T = p \, \frac{q}{2} \, \lambda_a I \, \cos (\psi + \delta)$$

In principal, then, to determine the torque and hence power rating of a machine, it is only necessary to determine the internal flux, the terminal current capability, and the speed capability of the rotor. In fact, it is *almost* that simple. Unfortunately, the model shown in Fig. 6.4 is not quite complete for some of the motors under discussion, so one more level into machine theory must be considered.

6.3.2 Two-reaction theory

The material in this subsection is framed in terms of three-phase ($q = 3$) machine theory, but it is actually generalizable to an arbitrary number of phases. Suppose a machine whose three-phase armature can be characterized by *internal* fluxes and inductance which may, in general, not be constant but is a function of rotor position. Note that the simple model we presented in the previous subsection does not conform to this picture, because it assumes a constant terminal inductance. In that case, the relationship between fluxes and currents is

$$\underline{\lambda}_{ph} = \underline{\underline{L}}_{ph}\underline{I}_{ph} + \underline{\lambda}_R \tag{6.1}$$

where $\underline{\lambda}_R$ is the set of internally produced fluxes (from the permanent magnets) and the stator winding may have both self and mutual inductances.

Now, it is useful to do a transformation on these stator fluxes in the following way: each armature quantity, including flux, current and voltage, is projected into a coordinate system that is fixed to the rotor. This is often called the *Park's Transformation*. For a three-phase machine, it

$$\begin{bmatrix} u_d \\ u_q \\ u_0 \end{bmatrix} = \underline{u}_{dq} = \underline{\underline{T}}\,\underline{u}_{ph} = \underline{\underline{T}} \begin{bmatrix} u_a \\ u_b \\ u_c \end{bmatrix} \tag{6.2}$$

Where the transformation and its inverse are

$$\underline{\underline{T}} = \frac{2}{3} \begin{bmatrix} \cos\theta & \cos\left(\theta - \dfrac{2\pi}{3}\right) & \cos\left(\theta + \dfrac{2\pi}{3}\right) \\ -\sin\theta & -\sin\left(\theta - \dfrac{2\pi}{3}\right) & -\sin\left(\theta + \dfrac{2\pi}{3}\right) \\ \dfrac{1}{2} & \dfrac{1}{2} & \dfrac{1}{2} \end{bmatrix} \tag{6.3}$$

$$\underline{\underline{T}}^{-1} = \begin{bmatrix} \cos\theta & -\sin\theta & 1 \\ \cos\left(\theta - \dfrac{2\pi}{3}\right) & -\sin\left(\theta - \dfrac{2\pi}{3}\right) & 1 \\ \cos\left(\theta + \dfrac{2\pi}{3}\right) & -\sin\left(\theta + \dfrac{2\pi}{3}\right) & 1 \end{bmatrix} \tag{6.4}$$

It is easy to show that balanced polyphase quantities in the stationary, or phase-variable frame, translate into *constant* quantities in the so-called "d–q" frame. For example

$$I_a = I \cos \omega t$$

$$I_b = I \cos \left(\omega t - \frac{2\pi}{3} \right)$$

$$I_c = I \cos \left(\omega t + \frac{2\pi}{3} \right)$$

$$\theta = \omega t + \theta_0$$

maps to

$$I_d = I \cos \theta_0$$

$$I_q = -I \sin \theta_0$$

Now, if $\theta = \omega t + \theta_0$, the transformation coordinate system is chosen correctly and the "d-" axis will correspond with the axis on which the rotor magnets are making positive flux. That happens if, when $\theta = 0$, phase A links maximum positive flux from the permanent magnets. If this is the case, the *internal* fluxes are

$$\lambda_{aa} = \lambda_f \cos \theta$$

$$\lambda_{ab} = \lambda_f \cos \left(\theta - \frac{2\pi}{3} \right)$$

$$\lambda_{ac} = \lambda_f \cos \left(\theta + \frac{2\pi}{3} \right)$$

The fluxes in the d-q frame are

$$\underline{\lambda}_{dq} = \underline{\underline{L}}_{dq} \underline{I}_{dq} + \underline{\lambda}_R = \underline{\underline{TL}}_{ph} \underline{\underline{T}}^{-1} \underline{I}_{dq} + \underline{\lambda}_R \qquad (6.5)$$

Two things should be noted here. The first is that, if the coordinate system has been chosen as described above, the flux induced by the rotor is, in the d-q frame, simply

$$\underline{\lambda}_R = \begin{bmatrix} \lambda_f \\ 0 \\ 0 \end{bmatrix}$$

$$(6.6)$$

That is, the magnets produce flux *only* on the d-axis.

The second thing to note is that, under certain assumptions, the inductances in the d-q frame are *independent of rotor position* and have no mutual terms. That is

$$\underline{\underline{L}}_{dq} = \underline{\underline{TL}}_{ph}\underline{\underline{T}}^{-1} = \begin{bmatrix} L_d & 0 & 0 \\ 0 & L_q & 0 \\ 0 & 0 & L_0 \end{bmatrix} \qquad (6.7)$$

The assertion that inductances in the d-q frame are constant is actually questionable, but it is close enough to being true and analyses that use it have proven to be close enough to being correct that it (the assertion) has held up to the test of time. In fact, the deviations from independence on rotor position are small. Independence of axes (that is, absence of mutual inductances in the d-q frame) is correct because the two axes are physically orthogonal. The third, or "zero" axis in this analysis is generally ignored. It does not couple to anything else and has neither flux nor current. Note that the direct- and quadrature-axis inductances are, in principal, straightforward to compute. They are:

- Direct axis. The inductance of one of the armature phases (corrected for the fact of multiple phases) with the rotor aligned with the axis of the phase.

- Quadrature axis. The inductance of one of the phases with the rotor aligned 90 electrical degrees away from the axis of that phase.

Next, armature voltage is, ignoring resistance, given by

$$\underline{V}_{ph} = \frac{d}{dt}\,\underline{\lambda}_{ph} = \frac{d}{dt}\,\underline{\underline{T}}^{-1}\underline{\lambda}_{dq} \qquad (6.8)$$

and that the *transformed* armature voltage must be

$$\underline{V}_{dq} = \underline{\underline{T}}\underline{V}_{ph}$$

$$= \underline{\underline{T}}\frac{d}{dt}\,(\underline{\underline{T}}^{-1}\underline{\lambda}_{dq}) \qquad (6.9)$$

$$= \frac{d}{dt}\,\underline{\lambda}_{dq} + \left(\underline{\underline{T}}\frac{d}{dt}\,\underline{\underline{T}}^{-1}\right)\underline{\lambda}_{dq}$$

The second term in this expresses "speed voltage". A good deal of straightforward but tedious manipulation yields

$$\underline{\underline{T}}\frac{d}{dt}\,\underline{\underline{T}}^{-1} = \begin{bmatrix} 0 & -\dfrac{d\theta}{dt} & 0 \\ \dfrac{d\theta}{dt} & 0 & 0 \\ 0 & 0 & 0 \end{bmatrix} \qquad (6.10)$$

The direct- and quadrature-axis voltage expressions are then

$$V_d = \frac{d\lambda_d}{dt} - \omega\lambda_q \qquad (6.11)$$

$$V_q = \frac{d\lambda_q}{dt} + \omega\lambda_d \qquad (6.12)$$

where

$$\omega = \frac{d\theta}{dt}$$

Instantaneous *power* is given by

$$P = V_a I_a + V_b I_b + V_c I_c \qquad (6.13)$$

Using the transformations given above, this can be shown to be

$$P = \frac{3}{2} V_d I_d + \frac{3}{2} V_q I_q + 3 V_0 I_0 \qquad (6.14)$$

which, in turn, is

$$P = \omega \frac{3}{2}(\lambda_d I_q - \lambda_q I_d) + \frac{3}{2}\left(\frac{d\lambda_d}{dt} I_d + \frac{d\lambda_q}{dt} I_q\right) + 3 \frac{d\lambda_0}{dt} I_0 \quad (6.15)$$

Noting that $\omega = p\Omega$ and that (6.15) describes electrical terminal power as the sum of shaft power and rate of change of stored energy, one may deduce that torque is given by

$$T = \frac{q}{2} p(\lambda_d I_q - \lambda_q I_d) \qquad (6.16)$$

Expression (6.15) states a generalization to a q-phase machine, even though the derivation given here was carried out for the $q = 3$ case. Of course three-phase machines are by far the most common case. Machines with higher numbers of phases behave in the same way (and this generalization is valid for most purposes), but there are more rotor variables analogous to "zero axis".

Now, noting that, in general, L_d and L_q are not necessarily equal

$$\lambda_d = L_d I_d + \lambda_f \qquad (6.17)$$

$$\lambda_q = L_q I_q \qquad (6.18)$$

then torque is given by

$$T = p\,\frac{q}{2}\,(\lambda_f + (L_d - L_q)I_d)I_q \tag{6.19}$$

6.3.3 Finding torque capability

For high performance drives, assume that the power supply, generally an inverter, can supply currents in the correct spatial relationship to the rotor to produce torque in some reasonably effective fashion. Shown in this section is how to determine the required values of I_d and I_q in order to produce a required torque (or if the torque is limited by either voltage or current). This is the essence of what is known as "field-oriented control", or putting stator currents in the correct location *in space* to produce the required torque.

The objective in this section is, given the elementary parameters of the motor, to find the capability of the motor to produce torque. There are three things to consider here:

1. Armature current is limited, generally by heating.
2. A second limit is the voltage capability of the supply, particularly at high speed.
3. If the machine is operating within these two limits, one should consider the optimal placement of currents (that is, how to get the most torque per unit of current to minimize losses).

Often the discussion of current placement is carried out using the I_d, I_q plane as a tool to visualize what is going on. Operation in the steady state implies a single point on this plane. A simple illustration is shown in Fig. 6.6. The thermally-limited armature current capability is represented as a circle around the origin, since the magnitude of armature current is just the length of a vector from the origin in this space. Since in general, for permanent magnet machines with buried magnets, $L_d < L_q$, so the optimal operation of the machine will be with negative I_d. How to determine this optimum operation will be shown shortly.

Finally, an ellipse describes the *voltage* limit. To start, consider what would happen if the terminals of the machine were to be short-circuited so that $V = 0$. If the machine is operating at sufficiently high speed so that armature resistance is negligible, armature current would be simply

$$I_d = -\frac{\lambda_f}{L_d}$$

$$I_q = 0$$

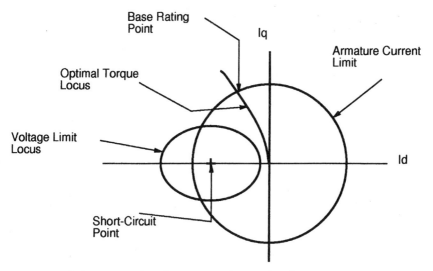

Figure 6.6 Limits to operation of a permanent magnet motor.

Now, loci of constant flux turn out to be ellipses around this point on the plane. Since terminal flux is proportional to voltage and inversely proportional to frequency, if the machine is operating with a given terminal voltage, the ability of that voltage to command current in the I_d, I_q plane is an ellipse whose size "shrinks" as speed increases.

To simplify the mathematics involved in this estimation, normalize reactances, fluxes, currents and torques. First, define the *base* flux to be $\lambda_b = \lambda_f$ and the *base* current I_b to be the armature capability. Then define two *per-unit* reactances

$$x_d = \frac{L_d I_b}{\lambda_b} \tag{6.20}$$

$$x_q = \frac{L_q I_b}{\lambda_b} \tag{6.21}$$

Next, define the *base torque* to be

$$T_b = p\frac{q}{2}\lambda_b I_b$$

and then, given *per-unit* currents i_d and i_q, the *per-unit* torque is simply

$$t_e = (1 - (x_q - x_d)i_d)i_q \tag{6.22}$$

It is fairly straightforward (but a bit tedious) to show that the locus of current-optimal operation (that is, the largest torque for a given current magnitude or the smallest current magnitude for a given torque) is along the curve

$$i_d = -\sqrt{\frac{i_a^2}{2} + 2\left(\frac{1}{4(x_q - x_d)}\right)^2 - \frac{1}{2(x_q - x_d)}\sqrt{\left(\frac{1}{4(x_q - x_d)}\right)^2 + \frac{i_a^2}{2}}} \quad (6.23)$$

$$i_q = -\sqrt{\frac{i_a^2}{2} - 2\left(\frac{1}{4(x_q - x_d)}\right)^2 + \frac{1}{2(x_q - x_d)}\sqrt{\left(\frac{1}{4(x_q - x_d)}\right)^2 + \frac{i_a^2}{2}}} \quad (6.24)$$

The "rating point" will be the point along this curve when $i_a = 1$, or where this curve crosses the armature capability circle in the i_d, i_q plane. It should be noted that this set of expressions only works for salient machines. For non-salient machines, of course, torque-optimal current is on the q-axis. In general, for machines with saliency, the "per-unit" torque will *not* be unity at the rating (so that the rated, or "base speed" torque is not the "base" torque), but

$$T_r = T_b \times t_e \quad (6.25)$$

where t_e is calculated at the rating point (that is, $i_a = 1$ and i_d and i_q as per (6.23) and (6.24)).

For sufficiently low speeds, the power-electronic drive can command the optimal current to produce torque up to rated. However, for speeds higher than the "base speed", this is no longer true. Define a per-unit terminal flux

$$\psi = \frac{V}{\omega \lambda_b}$$

Operation at a given flux magnitude implies

$$\psi^2 = (1 + x_d i_d)^2 + (x_q i_q)^2$$

which is an ellipse in the i_d, i_q plane. The base speed is that speed at which this ellipse crosses the point where the optimal current curve crosses the armature capability. Operation at the highest attainable torque (for a given speed) generally implies d-axis currents that are higher than those on the optimal current locus. What is happening here is the (negative) d-axis current serves to reduce effective machine flux and hence voltage which is limiting q-axis current. Thus, operation above the base speed is often referred to as "flux weakening".

The strategy for picking the correct trajectory for current in the i_d, i_q plane depends on the value of the per-unit reactance x_d. For values of $x_d > 1$, it is possible to produce *some* torque at *any* speed. For values of $x_d < 1$, there is a speed for which no point in the armature-current capability is within the voltage-limiting ellipse, so that useful torque has gone to zero. Generally, the maximum torque operating point is the intersection of the armature-current limit and the voltage-limiting ellipse

$$
i_d = \frac{x_d}{x_q^2 - x_d^2} - \sqrt{\left(\frac{x_d}{x_q^2 - x_d^2}\right)^2 + \frac{x_q^2 - \psi^2 + 1}{x_q^2 - x_d^2}} \tag{6.26}
$$

$$
i_q = \sqrt{1 - i_d^2} \tag{6.27}
$$

It may be that there is no intersection between the armature capability and te voltage-limiting ellipse. If this is the case and if $x_d < 1$, torque capability at the given speed is zero. If, on the other hand, $x_d > 1$, it may be that the intersection between the voltage-limiting ellipse and the armature-current limit is *not* the maximum torque point. To find out, we calculate the maximum torque point on the voltage-limiting ellipse. This is done in the usual way by differentiating torque with respect to i_d while holding the relationship between i_d and i_q to be on the ellipse. The algebra is a bit messy, and results in

$$
i_d = -\frac{3x_d(x_q - x_d) - x_d^2}{4x_d^2(x_q - x_d)} \tag{6.28}
$$

$$
- \sqrt{\left(\frac{3x_d(x_q - x_d) - x_d^2}{4x_d^2(x_q - x_d)}\right)^2 + \frac{(x_q - x_d)(\psi^2 - 1) + x_d}{2(x_q - x_d)x_d^2}}
$$

$$
i_q = \frac{1}{x_q}\sqrt{\psi^2 - (1 + x_d i_d)^2} \tag{6.29}
$$

Ordinarily, it is probably easiest to compute (6.28) and (6.29) first, then test to see if the currents are outside the armature capability. If they are, use (6.26) and (6.27).

These expressions give us the capability to estimate the torque-speed curve for a machine. As an example, the machine described by the parameters cited in Table 6.1 is a (nominal) 3HP, 4-pole, 3000 RPM machine.

The rated operating point turns out to have the attributes shown in Table 6.2. The loci of operation in the I_d, I_q plane is shown in Fig. 6.7. The armature-current limit is shown only in the second and third quadrants, so it shows up as a semicircle. The two ellipses correspond with the rated point (the larger ellipse) and with a speed that is three times rated (9000 RPM). The torque-optimal current locus can be seen

TABLE 6.1   Data for the Example Machine

D- Axis Inductance	2.53 mHy
Q- Axis Inductance	6.38 mHy
Internal Flux	58.1 mWb
Armature Current	30 A

TABLE 6.2   Operating Characteristics of Example Machine

Per-Unit D-Axis Current At Rating Point	i_d	-.5924
Per-Unit Q-Axis Current At Rating Point	i_q	.8056
Per-Unit D-Axis Reactance	x_d	1.306
Per-Unit Q-Axis Reactance	x_q	3.294
Rated Torque (Nm)	T_r	9.17
Terminal Voltage at Base Point (V)		97

running from the origin to the rating point, and the higher speed operating locus follows the armature-current limit. Figure 6.8 shows the torque/speed and power/speed curves. Note that this sort of machine only approximates "constant power" operation at speeds above the "base" or rating-point speed.

6.4   Parameter Estimation

Because there are a number of different motor geometries to consider and because they share parameters in a not-too-orderly fashion, this section will have a number of sub-parts. First, calculate flux linkage, then reactance.

6.4.1   Flux linkage

Given a machine which may be considered to be uniform in the axial direction, flux linked by a single, full-pitched coil which spans an angle from zero to π/p is

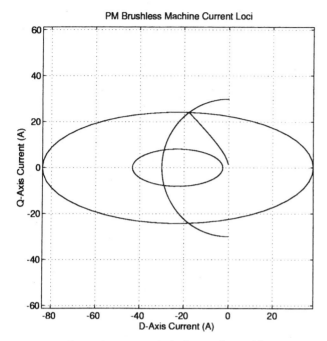

Figure 6.7   Operating current loci of example machine.

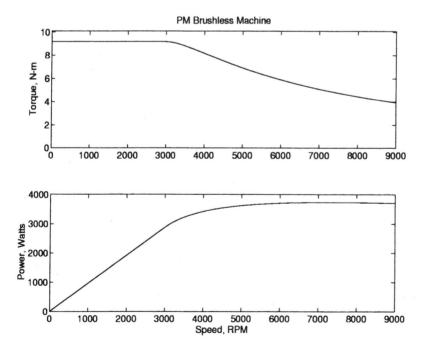

Figure 6.8   Torque- and power-speed capability.

$$\phi = \int_0^{\pi/p} B_r Rld\phi$$

where B_r is the radial flux through the coil. And, if B_r is sinusoidally distributed, this will have a peak value of

$$\phi_p = \frac{2RlB_r}{p}$$

Now, if the actual winding has N_a turns, and using the pitch and breadth factors derived in Appendix 1, the total flux linked is simply

$$\lambda_f = \frac{2RlB_1 N_a k_w}{p} \tag{6.30}$$

where

$$k_w = k_p k_b$$

$$k_p = \sin \frac{\alpha}{2}$$

$$k_b = \frac{\sin m \dfrac{\gamma}{2}}{m \sin \dfrac{\gamma}{2}}$$

The angle α is the *pitch* angle

$$\alpha = 2\pi p \frac{N_p}{N_s}$$

where N_p is the coil span (in slots) and N_s is the total number of slots in the stator. The angle γ is the slot electrical angle

$$\gamma = \frac{2\pi p}{N_s}$$

Now, what remains to be found is the space-fundamental magnetic flux density B_1. In the Appendix 3, it is shown that, for magnets in a surface-mount geometry, the magnetic field at the surface of the magnetic gap is

$$B_1 = \mu_0 M_1 k_g \tag{6.31}$$

where the space-fundamental magnetization is

$$M_1 = \frac{B_r}{\mu_0} \frac{4}{\pi} \sin \frac{p\theta_m}{2}$$

where B_r is remanent flux density of the permanent magnets and θ_m is the magnet angle. The factor that describes the geometry of the magnetic gap depends on the case. For magnets inside and $p \neq 1$

$$k_g = \frac{R_s^{p-1}}{R_s^{2p} - R_i^{2p}} \left(\frac{p}{p+1} (R_2^{p+1} - R_1^{p+1}) + \frac{p}{p-1} R_i^{2p} (R_1^{1-p} - R_2^{1-p}) \right)$$

For magnets inside and $p = 1$

$$k_g = \frac{1}{R_s^2 - R_i^2} \left(\frac{1}{2} (R_2^2 - R_1^2) + R_i^2 \log \frac{R_2}{R_1} \right)$$

For the case of magnets outside and $p \neq 1$

$$k_g = \frac{R_i^{p-1}}{R_s^{2p} - R_i^{2p}} \left(\frac{p}{p+1} (R_2^{p+1} - R_1^{p+1}) + \frac{p}{p-1} R_s^{2p} (R_1^{1-p} - R_2^{1-p}) \right)$$

and for magnets outside and $p = 1$

$$k_g = \frac{1}{R_s^2 - R_i^2} \left(\frac{1}{2} (R_2^2 - R_1^2) + R_s^2 \log \frac{R_2}{R_1} \right)$$

where R_s and R_i are the outer and inner magnetic boundaries, respectively, and R_2 and R_1 are the outer and inner boundaries of the magnets.

Note that for the case of a small gap, in which both the *physical* gap g and the magnet thickness h_m are both much less than rotor radius, it is straightforward to show that all of the above expressions approach what one would calculate using a simple, one-dimensional model for the permanent magnet

$$k_g \longrightarrow \frac{h_m}{g + h_m}$$

This is the whole story for the winding-in-slot, narrow air-gap, surface magnet machine. For air-gap armature windings, it is necessary to take into account the radial dependence of the magnetic field.

6.4.2 Air-gap armature windings

With no windings in slots, the conventional definition of winding factor becomes difficult to apply. If, however, each of the phase belts of the winding occupies an angular extent θ_w, then the equivalent to (6.31) is

$$k_w = \frac{\sin p \, \dfrac{\theta_w}{2}}{p \, \dfrac{\theta_w}{2}}$$

Next, assume that the "density" of conductors within each of the phase belts of the armature winding is uniform, so that the density of turns as a function of radius is

$$N(r) = \frac{2N_a r}{R_{wo}^2 - R_{wi}^2}$$

This just expresses the fact that there is more azimuthal room at larger radii, so with uniform density, the number of turns as a function of radius is linearly dependent on radius. Here, R_{wo} and R_{wi} are the outer and inner radii, respectively, of the winding.

Now it is possible to compute the flux linked due to a magnetic field distribution

$$\lambda_f = \int_{R_{wi}}^{R_{wo}} \frac{2lN_a k_w r}{p} \frac{2r}{R_{wo}^2 - R_{wi}^2} \mu_0 H_r(r) dr \qquad (6.32)$$

Note the form of the magnetic field as a function of radius expressed in (6.80) and (6.81) of Appendix 2. For the "winding outside" case, it is

$$H_r = A(r^{p-1} + R_s^{2p} r^{-p-1})$$

Then, a winding with all its turns concentrated at the outer radius $r = R_{wo}$ would link flux

$$\lambda_c = \frac{2lR_{wo} k_w}{p} \mu_0 H_r(R_{wo}) = \frac{2lR_{wo} k_w}{p} \mu_0 A \, (R_{wo}^{p-1} + R_s^2 R_{wo}^{-p-1})$$

Carrying out (6.32), it is possible to express the flux linked by a thick winding in terms of the flux that would have been linked by a radially concentrated winding at its outer surface by

$$k_t = \frac{\lambda_f}{\lambda_c}$$

where, for the winding outside, $p \neq 2$ case

$$k_t = \frac{2}{(1 - x^2)(1 + \xi^{2p})} \left(\frac{(1 - x^{2+p})\xi^{2p}}{2 + p} + \frac{1 - x^{2-p}}{2 - p} \right) \qquad (6.33)$$

where $\xi = R_{wo}/R_s$ and $x = R_{wi}/R_{wo}$. In the case of winding outside and $p = 2$

$$k_t = \frac{2}{(1 - x^2)(1 + \xi^{2p})} \left(\frac{(1 - x^4)\xi^4}{4} - \log x \right) \qquad (6.34)$$

In a very similar way, define a winding factor for a thick winding in which the reference radius is at the inner surface (Note: this is done because the inner surface of the inside winding is likely to be coincident with the inner ferromagnetic surface, as the outer surface of the outer winding is likely to be coincident with the outer ferromagnetic surface.) For $p \neq 2$

$$k_t = \frac{2x^{-p}}{(1 - x^2)(1 + \eta^{2p})} \left(\frac{1 - x^{2+p}}{2 + p} + (\eta x)^{2p} \frac{1 - x^{2-p}}{2 - p} \right) \qquad (6.35)$$

and for $p = 2$

$$k_t = \frac{2x^{-2}}{(1 - x^2)(1 + \eta^{2p})} \left(\frac{1 - x^4}{4} - (\eta x)^4 \log x \right) \qquad (6.36)$$

where $\eta = R_i/R_{wi}$

So, in summary, the flux linked by an air-gap armature is given by

$$\lambda_f = \frac{2RlB_1 N_a k_w k_t}{p} \qquad (6.37)$$

where B_1 is the flux density at the outer radius of the physical winding (for outside-winding machines) or at the inner radius of the physical winding (for inside-winding machines). Note that the additional factor k_t is a bit more than one (it approaches unity for thin windings), so that, for small pole numbers and windings that are not too thick, it is almost correct and in any case "conservative" to take it to be one.

6.4.3 Interior magnet motors

For the flux-concentrating machine, it is possible to estimate air-gap flux density using a simple reluctance model.

The air-gap permeance of one pole piece is

$$\mathscr{P}_{ag} = \mu_0 l \frac{R\theta_p}{g}$$

where θ_p is the angular width of the pole piece.

And the incremental permeance of a magnet is

$$\wp_m = \mu_0 \frac{h_m l}{w_m}$$

The magnet sees a *unit permeance* consisting of its own permeance in series with one-half of each of two-pole pieces (in parallel)

$$\wp_u = \frac{\wp_{ag}}{\wp_m} = \frac{R\theta_p}{4g} \frac{w_m}{h_m}$$

Magnetic flux density in the *magnet* is

$$B_m = B_0 \frac{\wp_u}{1 + \wp_u}$$

And then flux density in the *air-gap* is

$$B_g = \frac{2h_m}{R\theta_p} B_m = B_0 \frac{2h_m w_m}{4gh_m + R\theta_p w_m}$$

The space fundamental of that can be written as

$$B_1 = \frac{4}{\pi} \sin \frac{p\theta_p}{2} B_0 \frac{w_m}{2g} \gamma_m$$

where

$$\gamma_m = \frac{1}{1 + \dfrac{w_m}{g} \dfrac{\theta_p}{4} \dfrac{R}{h_m}}$$

The flux linkage is then computed as before

$$\lambda_f = \frac{2RlB_1 N_a k_w}{p} \tag{6.38}$$

6.4.4 Winding inductances

The next important set of parameters to compute are the d- and q-axis inductances of the machine. Considered are three separate cases, the winding-in-slot, surface magnet case which is magnetically "round", or non-salient, the air-gap winding case and the flux-concentrating case which is salient or has different direct- and quadrature-axis inductances.

6.4.4.1 Surface magnets, windings-in-slots. In this configuration, there is no saliency, so that $L_d = L_q$. There are two principal parts to inductance, the air-gap inductance and slot-leakage inductance. Other components, including end-turn leakage, may be important in some configurations, and they would be computed in the same way as for an induction machine. As is shown in Appendix 1, the fundamental part of air-gap inductance is

$$L_{d1} = \frac{q}{2} \frac{4}{\pi} \frac{\mu_0 N_a^2 k_w^2 l R_s}{p^2 (g + h_w)} \tag{6.39}$$

Here, g is the magnetic gap, including the physical rotational gap and any magnet retaining means that might be used. h_m is the magnet thickness.

Since the magnet thickness is included in the air-gap, the air-gap permeance may not be very large, so that slot-leakage inductance may be important. To estimate this, assume that the slot shape is rectangular, characterized by the following dimensions:

h_s height of the main portion of the slot

w_s width of the top of the main portion of the slot

h_d height of the slot depression

w_d slot depression opening

Of course, not all slots are rectangular. In fact, in most machines, the slots are trapezoidal in shape to maintain the cross-sections that are radially uniform. However, only a very small error (a few percent) is incurred in calculating slot permeance if the slot is assumed to be rectangular and the *top* width is used (that is the width closest to the air-gap). Then the slot permeance is, per unit length

$$\mathscr{P} = \mu_0 \left(\frac{1}{3} \frac{h_s}{w_s} + \frac{h_d}{w_d} \right)$$

Assume for the rest of this discussion a standard winding, with m slots in each phase-belt (this assumes, then, that the total number of slots is $N_s = 2pqm$), and each slot holds two half-coils. (A half-coil is one side of a coil which, of course, is wound in two slots). If each coil has N_c turns (meaning $N_a = 2pmN_c$), then the contribution to phase self-inductance of *one* slot is, if both half-coils are from the same phase, $4l\mathscr{P}N_c^2$. If the half-coils are from different phases, then the contribution to self inductance is $l\mathscr{P}N_c^2$ and the magnitude of the contribution to mutual inductance is $l\mathscr{P}N_c^2$. (Some caution is required here. For three-phase windings, the mutual inductance is negative, as are the senses of the currents in the two other phases. Thus, the impact of "mutual leakage" is to increase the reactance. This will be true for other numbers of phases as well, even if the algebraic sign of the mutual leakage inductance is positive, in which case so will be the sense of the other phase current.)

Two other assumptions are made here. The standard one is that the winding "coil throw", or span between sides of a coil, is $N_s/2p - N_{sp}$. N_{sp} is the coil "short pitch". The other is that each phase-belt will overlap with, at most, two other phases: the ones on either side in sequence. This last assumption is immediately true for three-phase windings (because there *are* only two other phases). It is also likely to be true for any reasonable number of phases.

Noting that each phase occupies $2p(m - N_{sp})$ slots with both coil halves in the same slot and $4pN_{sp}$ slots in which one coil half shares a slot with a different phase, we can write down the two components of slot-leakage inductance, self and mutual

$$L_{as} = 2pl[(m - N_{sp})(2N_c)^2 + 2N_{sp}N_c^2]$$

$$L_{am} = 2plN_{sp}N_c^2$$

For a three-phase machine, then, the total slot-leakage inductance is

$$L_a = L_{as} + L_{am} = 2pl9N_c^2(4m - N_{sp})$$

For a uniform, symmetric winding with an odd number of phases, it is possible to show that the effective slot-leakage inductance is

$$L_a = L_{as} - 2L_{am} \cos \frac{2\pi}{q}$$

Total synchronous inductance is the sum of air-gap and leakage components. So far, this is

$$L_d = L_{d1} + L_a$$

6.4.4.2 Air-gap armature windings. It is shown in Appendix 2 that the inductance of a single-phase of an air-gap winding is:

$$L_a = \sum_n LL_{np}$$

where the harmonic components are

$$L_k = \frac{8}{\pi} \frac{\mu_0 l k_{wn}^2 N_a^2}{k(1-x^2)^2} \left[\frac{(1-x^{2-k}\gamma^{2k})(1-x^{2+k})}{(4-k^2)(1-\gamma^{2k})} \right.$$

$$+ \frac{\xi^{2k}(1-x^{k+2})^2}{(2+k)^2(1-\gamma^{2k})} + \frac{\xi^{-2k}(1-x^{2-k})^2}{(2-k)^2(\gamma^{-2k}-1)}$$

$$+ \left. \frac{(1-\gamma^{-2k}x^{2+k})(1-x^{2-k})}{(4-k^2)(\gamma^{-2k}-1)} - \frac{k}{4-k^2} \frac{1-x^2}{2} \right]$$

and the following shorthand coefficients are used

$$x = \frac{R_{wi}}{R_{wo}}$$

$$\gamma = \frac{R_i}{R_s}$$

$$\xi = \frac{R_{wo}}{R_s}$$

This fits into the conventional inductance framework

$$L_n = \frac{4}{\pi} \frac{\mu_0 N_a^2 R_s L k_{wn}^2}{N^2 p^2 g} k_a$$

The "thick armature" coefficient is

$$k_a = \frac{2gk}{R_{wo}} \frac{1}{(1-x^2)^2} \left[\frac{(1-x^{2-k}\gamma^{2k})(1-x^{2+k})}{(4-k^2)(1-\gamma^{2k})} \right.$$

$$+ \frac{\xi^{2k}(1-x^{k+2})^2}{(2+k)^2(1-\gamma^{2k})} + \frac{\xi^{-2k}(1-x^{2-k})^2}{(2-k)^2(\gamma^{-2k}-1)}$$

$$+ \left. \frac{(1-\gamma^{-2k}x^{2+k})(1-x^{2-k})}{(4-k^2)(\gamma^{-2k}-1)} - \frac{k}{4-k^2} \frac{1-x^2}{2} \right]$$

where $k = np$ and $g = R_s - R_i$ is the conventionally defined "air-gap". In the case of $p = 2$, the fundamental component of k_a is

$$k_a = \frac{2gk}{R_{wo}} \frac{1}{(1-x^2)^2} \left[\frac{1-x^4}{8} - \frac{2\gamma^4 + x^4(1-\gamma^4)}{4(1-\gamma^4)} \log x \right.$$

$$+ \left. \frac{\gamma^4}{\xi^4(1-\gamma^4)} (\log x)^2 + \frac{\xi^4(1-x^4)^2}{16(1-\gamma^4)} \right]$$

If the aspect ratio R_i/R_s is not too far from unity, neither is k_a.

For a q-phase winding a good approximation to the inductance is given by just the first space harmonic term, or

$$L_d = \frac{q}{2} \frac{4}{\pi} \frac{\mu_0 N_a^2 R_s L k_{wn}^2}{n^2 p^2 g} k_a$$

6.4.4.3 Internal magnet motor. The permanent magnets will have an effect on reactance because the magnets are in the main flux path of the armature. Further, they affect direct and quadrature reactances differently, so that the machine will be salient. Actually, the effect on the direct axis will likely be greater, so that this type of machine will exhibit "negative" saliency: the quadrature-axis reactance will be larger than the direct-axis reactance.

A full-pitch coil aligned with the direct axis of the machine would produce flux density

$$B_r = \frac{\mu_0 N_a I}{2g \left(1 + \dfrac{R \theta_p}{4g} \dfrac{w_m}{h_m}\right)}$$

Note that only the pole area is carrying useful flux, so that the space fundamental of radial-flux density is

$$B_1 = \frac{\mu_0 N_a I}{2g} \frac{4}{\pi} \frac{\sin \dfrac{p \theta_m}{2}}{1 + \dfrac{w_m}{h_m} \dfrac{R \theta_p}{4g}}$$

Then, since the flux linked by the winding is

$$\lambda_a = \frac{2 R I N_a k_w B_1}{p}$$

The d-axis inductance, including mutual phase coupling, is (for a q-phase machine)

$$L_d = \frac{q}{2} \frac{4}{\pi} \frac{\mu_0 N_a^2 R l k_w^2}{p^2 g} \gamma_m \sin \frac{p \theta_p}{2}$$

The quadrature axis is quite different. On that axis, the armature does *not* tend to push flux through the magnets, so they have only a minor effect. What effect they *do* have is due to the fact that the magnets produce a space in the active air-gap. Thus, while a full-pitch coil aligned with the quadrature axis will produce an air-gap flux density

$$B_r = \frac{\mu_0 NI}{g}$$

the space fundamental of that will be

$$B_1 = \frac{\mu_0 NI}{g} \frac{4}{\pi} \left(1 - \sin \frac{p\theta_t}{2} \right)$$

where θ_t is the angular width taken out of the pole by the magnets. So that the expression for quadrature-axis inductance is

$$L_q = \frac{q}{2} \frac{4}{\pi} \frac{\mu_0 N_a^2 R l k_w^2}{p^2 g} \left(1 - \sin \frac{p\theta_t}{2} \right)$$

6.5 Current Rating and Resistance

The last part of machine rating is its current capability. This is heavily influenced by cooling methods, for the principal limit on current is the heating produced by resistive dissipation. Generally, it is possible to do first-order design estimates by assuming a current density that can be handled by a particular cooling scheme. Then, in an air-gap winding

$$N_a I_a = (R_{wo}^2 - R_{wi}^2) \frac{\theta_{we}}{2} J_a$$

and note that, usually, the armature fills the azimuthal space in the machine

$$2q\,\theta_{we} = 2\pi$$

For a winding in slots, nearly the same thing is true: if the rectangular slot model holds true

$$2qN_a I_a = N_s h_s w_s J_s$$

where J_s denotes *slot* current density. Characterize the total slot area by a "space factor" λ_s which is the ratio between total slot area and the annulus occupied by the slots. For the rectangular slot model

$$\lambda_s = \frac{N_s h_s w_s}{\pi(R_{wo}^2 - R_{wi}^2)}$$

where $R_{wi} = R + h_d$ and $R_{wo} = R_{wi} + h_s$ in a normal, stator-outside winding. In this case, $J_a = J_s \lambda_s$ and the two types of machines can be evaluated in the same way.

It would seem apparent that one would want to make λ_s as large as possible, to permit high currents. The limit on this is that the magnetic teeth between the conductors must be able to carry the air-gap flux, and making them too narrow would cause them to saturate. The peak of the time fundamental magnetic field in the teeth is, for example

$$B_t = B_1 \frac{2\pi R}{N_s w_t}$$

where w_t is the width of a stator tooth

$$w_t = \frac{2\pi(R + h_d)}{N_s} - w_s$$

so that

$$B_t \approx \frac{B_1}{1 - \lambda_s}$$

6.5.1 Resistance

Winding resistance may be estimated as the length of the stator conductor divided by its area and its conductivity. The length of the stator conductor is

$$l_c = 2lN_a f_e$$

where the "end winding factor" f_e is used to take into account the extra length of the end turns (which is usually *not* negligible). The *area* of each turn of wire is, for an air-gap winding

$$A_w = \frac{\theta_{we}}{2} \frac{R_{wo}^2 - R_{wi}^2}{N_a} \lambda_w$$

where λ_w, the "packing factor" relates the area of conductor to the total area of the winding. The resistance is then just

$$R_a = \frac{4lN_a^2}{\theta_{we} (R_{wo}^2 - R_{wi}^2)\lambda_w \sigma}$$

and, of course, σ is the conductivity of the conductor.

For windings in slots, the expression is almost the same, simply substituting the total slot area

$$R_a = \frac{2qlN_a^2}{N_s h_s w_s \lambda_w \sigma}$$

The end turn allowance depends strongly on how the machine is made. One way of estimating what it might be is to assume that the end turns follow a roughly circular path from one side of the machine to the other. The radius of this circle would be, very roughly, R_w/p, where R_w is the average radius of the winding

$$R_w \approx (R_{wo} + R_{wi})/2$$

Then the end-turn allowance would be

$$f_e = 1 + \frac{\pi R_w}{pl}$$

6.6 Appendix 1: Air-Gap Winding Inductance

A simple two-dimensional model is used to estimate the magnetic fields and then inductances of an air-gap winding. The principal limiting assumption here is that the winding is uniform in the $\bar{z}$ direction, which means it is long in comparison with its radii. This is generally not true, nevertheless, the answers obtained are not too far from being correct. The *style* of analysis used here can be carried into a three-dimensional, or quasi three-dimensional domain to get much more precise answers, at the expense of a very substantial increase in complexity.

The coordinate system to be used is shown in Fig. 6.9. To maintain generality, four radii are defined: R_i and R_s are ferromagnetic bound-

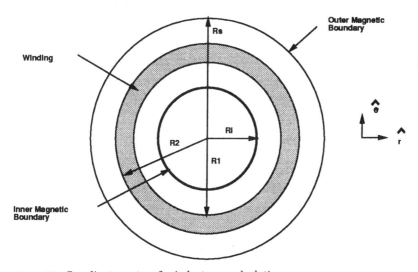

Figure 6.9 Coordinate system for inductance calculation.

aries, and would, of course, correspond with the machine shaft and the stator core. The winding itself is carried between radii R_1 and R_2, which correspond with radii R_{wi} and R_{wo} in the body of the text. It is assumed that the armature is carrying a current in the z direction, and that this current is uniform in the radial dimension of the armature. If a single phase of the armature is carrying current, that current will be

$$J_{z0} = \frac{N_a I_a}{\frac{\theta_{we}}{2}(R_2^2 - R_1^2)}$$

over the annular wedge occupied by the phase. The resulting distribution can be fourier analyzed, and the *n-th* harmonic component of this will be (assuming the coordinate system has been chosen appropriately)

$$J_{zn} = \frac{4}{n\pi} J_{z0} \sin n \frac{\theta_{we}}{2} = \frac{4}{\pi} \frac{N_a I_a}{R_2^2 - R_1^2} k_{wn}$$

where the *n-th* harmonic winding factor is

$$k_{wn} = \frac{\sin n \frac{\theta_{we}}{2}}{n \frac{\theta_{we}}{2}}$$

and note that θ_{we} is the *electrical* winding angle

$$\theta_{we} = p\theta_w$$

It is easier to approach this problem using vector potential. Since the divergence of flux density is zero, it is possible to let the magnetic flux density be represented by the curl of a vector potential

$$\bar{B} = \nabla \times \bar{A}$$

Taking the curl of *that*

$$\nabla \times (\nabla \times \bar{A}) = \mu_0 \bar{J} = \nabla\nabla \cdot \bar{A} - \nabla^2 \bar{A}$$

and using the coulomb gage

$$\nabla \cdot \overline{A} = 0$$

A reasonably tractable partial differential equation in the vector potential is

$$\nabla^2 \overline{A} = -\mu_0 \overline{J}$$

Assuming there is only a z-directed component of $\overline{J}$, that one component, in circular cylindrical coordinates is

$$\frac{1}{r} \frac{\partial}{\partial r} r \frac{\partial A_z}{\partial r} + \frac{1}{r^2} \frac{\partial^2}{\partial \theta^2} A_z = -\mu_0 J_z \qquad (6.40)$$

For this problem, all variables will be varying sinusoidally with angle, so angular dependence is $e^{jk\theta}$. Thus

$$\frac{1}{r} \frac{\partial}{\partial r} r \frac{\partial A_z}{\partial r} - \frac{k^2}{r^2} A_z = -\mu_0 J_z$$

This is a three-region problem. Note the regions as

$$\text{i} \quad R_i < r < R_1$$

$$\text{w} \quad R_1 < r < R_2$$

$$\text{o} \quad R_2 < r < R_s$$

For i and o, the current density is zero and an appropriate solution to (6.40) is

$$A_z = A_+ r^k + A_- r^{-k}$$

In the region of the winding, w, a particular solution must be used in addition to the homogenous solution, and

$$A_z = A_+ r^k + A_- r^{-k} + A_p$$

where, for $k \neq 2$

$$A_p = -\frac{\mu_0 J_z r^2}{4 - k^2}$$

or, if $k = 2$

$$A_p = -\frac{\mu_0 J_z r^2}{4}\left(\log r - \frac{1}{4}\right)$$

And, of course, the two pertinent components of the magnetic flux density are

$$B_r = \frac{1}{r}\frac{\partial A_z}{\partial \theta}$$

$$B_\theta = -\frac{\partial A_z}{\partial r}$$

Next, it is necessary to match boundary conditions. There are six free variables and correspondingly there must be six of these boundary conditions. They are the following:

- At the inner and outer magnetic boundaries, $r = R_i$ and $r = R_s$, the azimuthal magnetic field must vanish.
- At the inner and outer radii of the winding itself, $r = R_1$ and $r = R_2$, *both* radial and azimuthal magnetic field must be continuous.

These conditions may be summarized by

$$kA_+^i R_i^{k-1} - kA_-^i R_i^{-k-1} = 0$$

$$kA_+^o R_s^{k-1} - kA_-^o R_s^{-k-1} = 0$$

$$A_+^w R_2^{k-1} + A_-^w R_2^{-k-1} - \frac{\mu_0 J_z R_2}{4 - k^2} = A_+^o R_2^{k-1} + A_-^o R_2^{-k-1}$$

$$-kA_+^w R_2^{k-1} + kA_-^w R_2^{-k-1} + \frac{2\mu_0 J_z R_2}{4 - k^2} = -kA_+^o R_2^{k-1} + kA_-^o R_2^{-k-1}$$

$$A_+^w R_1^{k-1} + A_-^w R_1^{-k-1} - \frac{\mu_0 J_z R_1}{4 - k^2} = A_+^i R_1^{k-1} + A_-^i R_1^{-k-1}$$

$$-kA_+^w R_1^{k-1} + kA_-^w R_1^{-k-1} + \frac{2\mu_0 J_z R_1}{4 - k^2} = -kA_+^i R_1^{k-1} + kA_-^i R_1^{-k-1}$$

This derivation is carried out here only for the case of $k \neq 2$. The $k = 2$ case may be obtained by substituting its particular solution at the beginning or by using L'Hopital's rule on the final solution. This set may be solved to yield, for the winding region

$$A_z = \frac{\mu_0 J_z}{2k} \left[\left(\frac{R_s^{2k} R_2^{2-k} - R_i^{2k} R_1^{2-k}}{(2-k)(R_s^{2k} - R_i^{2k})} + \frac{R_2^{2+k} - R_1^{2+k}}{(2+k)(R_s^{2k} - R_i^{2k})} \right) r^k \right.$$

$$+ \left(\frac{R_2^{2-k} - R_1^{2-k}}{(2-k)(R_i^{-2k} - R_s^{-2k})} + \frac{R_s^{-2k} R_2^{2+k} - R_i^{-2k} R_1^{2+k}}{(2+k)(R_i^{-2k} - R_s^{-2k})} \right) r^{-k}$$

$$\left. - \frac{2k}{4-k^2} r^2 \right]$$

Now, the inductance linked by any single, full-pitched loop of wire located with one side at azimuthal position θ and radius r is

$$\lambda_i = 2l A_z(r,\theta)$$

To extend this to the whole winding, integrate over the area of the winding the incremental flux linked by each element times the turns density. This is, for the n-th harmonic of flux linked

$$\lambda_n = \frac{4l k_{wn} N_a}{R_2^2 - R_1^2} \int_{R_1}^{R_2} A_z(r) r \, dr$$

Making the appropriate substitutions for current into the expression for vector potential, this becomes

$$\lambda_n = \frac{8}{\pi} \frac{\mu_0 l k_{wn}^2 N_a^2 I_a}{k(R_2^2 - R_1^2)^2} \left[\left(\frac{R_s^{2k} R_2^{2-k} - R_i^{2k} R_1^{2-k}}{(2-k)(R_s^{2k} - R_i^{2k})} + \frac{R_2^{2+k} - R_1^{2+k}}{(2+k)(R_2^{2k} - R_i^{2k})} \right) \right.$$

$$\frac{R_2^{k+2} - R_1^{k+2}}{k+2}$$

$$+ \left(\frac{R_2^{2-k} - R_1^{2-k}}{(2-k)(R_i^{-2k} - R_s^{-2k})} + \frac{R_s^{-2k} R_2^{2+k} - R_i^{-2k} R_1^{2+k}}{(2+k)(R_i^{-2k} - R_s^{-2k})} \right)$$

$$\left. \frac{R_2^{2-k} - R_1^{2-k}}{2-k} - \frac{2k}{4-k^2} \frac{R_2^4 - R_1^4}{4} \right]$$

6.7 Appendix 2:   Permanent Magnet Field Analysis

This section is a field analysis of the kind of radially magnetized, permanent magnet structures commonly used in electric machinery. It is a fairly general analysis, which will be suitable for use with either surface or in-slot windings, and for the magnet inside or the magnet outside case.

This is a two-dimensional layout suitable for situations in which field variation along the length of the structure is negligible.

6.7.1 Layout

The assumed geometry is shown in Fig. 6.10. Assumed iron (highly permeable) boundaries are at radii R_i and R_s. The permanent magnets, assumed to be polarized radially and alternately (i.e. North-South), are located between radii R_1 and R_2. We assume there are p pole pairs ($2p$ magnets) and that each magnet subsumes an electrical angle of θ_{me}. The electrical angle is just p times the physical angle, so that if the magnet angle were $\theta_{me} = \pi$, the magnets would be touching.

If the magnets are arranged so that the radially-polarized magnets are located around the azimuthal origin ($\theta = 0$), the space fundamental of magnetization is

$$\overline{M} = \bar{i}_r M_0 \cos p\theta$$

where the fundamental magnitude is

$$M_0 = \frac{4}{\pi} \sin \frac{\theta_{me}}{2} \frac{B_{\text{rem}}}{\mu_0}$$

and B_{rem} is the remanent magnetization of the permanent magnet.

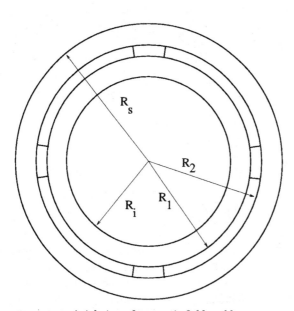

Figure 6.10 Axial view of magnetic field problem.

Since there is no current anywhere in this problem, it is convenient to treat magnetic field as the gradient of a scalar potential

$$\overline{H} = -\nabla \psi$$

The divergence of this is

$$\nabla^2 \psi = -\nabla \cdot \overline{H}$$

Since magnetic *flux* density is divergence-free

$$\nabla \cdot \overline{B} = 0$$

we have

$$\nabla \cdot \overline{H} = -\nabla \cdot \overline{M}$$

or

$$\nabla^2 \psi = \nabla \cdot \overline{M} = \frac{1}{r} M_0 \cos p\theta$$

Now, if we let the magnetic scalar potential be the sum of *particular* and *homogeneous* parts

$$\psi = \psi_p + \psi_h$$

where $\nabla^2 \psi_h = 0$, then

$$\nabla^2 \psi_p = \frac{1}{r} M_0 \cos p\theta$$

We can find a suitable solution to the *particular* part of this in the region of magnetization by trying

$$\psi_p = C r^\gamma \cos p\theta$$

Carrying out the Laplacian equation on this

$$\nabla^2 \psi_p = C r^{\gamma-2} (\gamma^2 - p^2) \cos p\theta \frac{1}{r} M_0 \cos p\theta$$

which works if $\gamma = 1$, in which case

$$\psi_p = \frac{M_0 r}{1 - p^2} \cos p\theta$$

Of course, this solution holds only for the region of the magnets: $R_1 < r < R_2$, and is zero for the regions outside of the magnets.

A suitable *homogeneous* solution satisfies Laplace's equation, $\nabla^2 \psi_h = 0$, and is, in general, of the form

$$\psi_h = Ar^p \cos p\theta + Br^{-p} \cos p\theta$$

Then we may write a trial *total* solution for the flux density as

$$R_i < r < R_1 \quad \psi = (A_1 r^p + B_1 r^{-p}) \cos p\theta$$

$$R_1 < r < R_2 \quad \psi = \left(A_2 r^p + B_2 r^{-p} + \frac{M_0 r}{1 - p^2}\right) \cos p\theta$$

$$R_2 < r < R_s \quad \psi = (A_3 r^p + B_3 r^{-p}) \cos p\theta$$

The boundary conditions at the inner and outer (assumed infinitely permeable) boundaries at $r = R_i$ and $r = R_s$ require that the azimuthal field vanish, or $\partial \psi / \partial \theta = 0$, leading to

$$B_1 = -R_i^{2p} A_1$$

$$B_3 = -R_s^{2p} A_3$$

At the magnet inner and outer radii, H_θ and B_r must be continuous. These are

$$H_\theta = -\frac{1}{r} \frac{\partial \psi}{\partial \theta}$$

$$B_r = \mu_0 \left(-\frac{\partial \psi}{\partial r} + M_r\right)$$

These become, at $r = R_1$

$$-pA_1 (R_1^{p-1} - R_i^{2p} R_1^{-p-1}) = -p (A_2 R_1^{p-1} + B_2 R_1^{-p-1}) - p \frac{M_0}{1 - p^2}$$

$$-pA_1 (R_1^{p-1} + R_i^{2p} R_1^{-p-1}) = -p (A_2 R_1^{p-1} - B_2 R_1^{-p-1})$$

$$-\frac{M_0}{1 - p^2} + M_0$$

and at $r = R_2$

$$-pA_3(R_2^{p-1} - R_s^{2p}R_2^{-p-1}) = -p\,(A_2R_2^{p-1} + B_2R_2^{-p-1})$$

$$- p\,\frac{M_0}{1 - p^2}$$

$$-pA_3(R_2^{p-1} + R_s^{2p}R_2^{-p-1}) = -p(A_2R_2^{p-1} - B_2R_2^{-p-1})$$

$$- \frac{M_0}{1 - p^2} + M_0$$

Some small-time manipulation of these yields

$$A_1(R_1^p - R_i^{2p}R_1^{-p}) = A_2R_1^p + B_2R_1^{-p} + R_1\frac{M_0}{1 - p^2}$$

$$A_1(R_1^p + R_i^{2p}R_1^{-p}) = A_2R_1^p - B_2R_1^{-p} + pR_1\frac{M_0}{1 - p^2}$$

$$A_3(R_2^p - R_s^{2p}R_2^{-p}) = A_2R_2^p + B_2R_2^{-p} + R_2\frac{M_0}{1 - p^2}$$

$$A_3(R_2^p + R_s^{2p}R_2^{-p}) = A_2R_2^p - B_2R_2^{-p} + pR_2\frac{M_0}{1 - p^2}$$

Taking sums and differences of the first and second and then third and fourth of these, we obtain

$$2A_1R_1^p = 2A_2R_1^p + R_1M_0\frac{1 + p}{1 - p^2}$$

$$2A_1R_i^{2p}R_1^{-p} = -2B_2R_1^{-p} + R_1M_0\frac{p - 1}{1 - p^2}$$

$$2A_3R_2^p = 2A_2R_2^p + R_2M_0\frac{1 + p}{1 - p^2}$$

$$2A_3R_s^{2p}R_2^{-p} = -2B_2R_2^{-p} + R_2M_0\frac{p - 1}{1 - p^2}$$

and then multiplying through by appropriate factors (R_2^p and R_1^p) and then taking sums and differences of *these*

$$(A_1 - A_3) R_1^p R_2^p = (R_1 R_2^p - R_2 R_1^p) \frac{M_0}{2} \frac{p+1}{1-p^2}$$

$$(A_1 R_i^{2p} - A_3 R_s^{2p}) R_1^{-p} R_2^{-p} = (R_1 R_2^{-p} - R_2 R_1^{-p}) \frac{M_0}{2} \frac{p-1}{1-p^2}$$

Dividing through by the appropriate groups

$$A_1 - A_3 = \frac{R_1 R_2^p - R_2 R_1^p}{R_1^p R_2^p} \frac{M_0}{2} \frac{1+p}{1-p^2}$$

$$A_1 R_i^{2p} - A_3 R_s^{2p} = \frac{R_1 R_2^{-p} - R_2 R_1^{-p}}{R_1^{-p} R_2^{-p}} \frac{M_0}{2} \frac{p-1}{1-p^2}$$

and then, by multiplying the top equation by R_s^{2p} and subtracting

$$A_1(R_s^{2p} - R_i^{2p}) = \left(\frac{R_1 R_2^p - R_2 R_1^p}{R_1^p R_2^p} \frac{M_0}{2} \frac{1+p}{1-p^2} \right) R_s^{2p}$$

$$-\frac{R_1 R_2^{-p} - R_2 R_1^{-p}}{R_1^{-p} R_2^{-p}} \frac{M_0}{2} \frac{p-1}{1-p^2}$$

This is readily solved for the field coefficients A_1 and A_3

$$A_1 = -\frac{M_0}{2(R_s^{2p} - R_i^{2p})}$$

$$\times \left(\frac{p+1}{p^2-1} (R_1^{1-p} - R_2^{1-p}) R_s^{2p} + \frac{p-1}{p^2-1} (R_2^{1+p} - R_1^{1+p}) \right)$$

$$A_3 = -\frac{M_0}{2(R_s^{2p} - R_i^{2p})}$$

$$\times \left(\frac{1}{1-p} (R_1^{1-p} - R_2^{1-p}) R_i^{2p} - \frac{1}{1+p} (R_2^{1+p} - R_1^{1+p}) \right)$$

Now, noting that the scalar potential is, in Region 1 (radii less than the magnet)

$$\psi = A_1(r^p - R_i^{2p} r^{-p}) \cos p\theta \quad r < R_1$$

$$\psi = A_3(r^p - R_s^{2p} r^{-p}) \cos p\theta \quad r > R_2$$

and noting that $p(p + 1)/(p^2 - 1) = p/(p - 1)$ and $p(p - 1)/(p^2 - 1) = p/(p + 1)$, magnetic field is

$$r < R_1$$

$$H_r = \frac{M_0}{2(R_s^{2p} - R_i^{2p})} \left(\frac{p}{p-1} (R_1^{1-p} - R_2^{1-p}) R_s^{2p} \right.$$

$$\left. + \frac{p}{p+1} (R_2^{1+p} - R_1^{1+p}) \right) (r^{p-1} + R_i^{2p} r^{-p-1}) \cos p\theta$$

$$r > R_2$$

$$H_r = \frac{M_0}{2(R_s^{2p} - R_i^{2p})} \left(\frac{p}{p-1} (R_1^{1-p} - R_2^{1-p}) R_i^{2p} \right.$$

$$\left. + \frac{p}{p+1} R_2^{1+p} - R_1^{1+p}) \right) (r^{p-1} + R_s^{2p} r^{-p-1}) \cos p\theta$$

The case of $p = 1$ appears to be a bit troublesome here, but is easily handled by noting that

$$\lim_{p \to 1} \frac{p}{p-1} (R_1^{1-p} - R_2^{1-p}) = \log \frac{R_2}{R_1}$$

Now there are a number of special cases to consider.
For the iron-free case, $R_i \to 0$ and $R_2 \to \infty$, this becomes, simply, for $r < R_1$

$$H_r = \frac{M_0}{2} \frac{p}{p-1} (R_1^{1-p} - R_2^{1-p}) r^{p-1} \cos p\theta$$

Note that for the case of $p = 1$, the limit of this is

$$H_r = \frac{M0}{2} \log \frac{R_2}{R_1} \cos \theta$$

and for $r > R_2$

$$H_r = \frac{M_0}{2} \frac{p}{p+1} (R_2^{p+1} - R_1^{p+1}) r^{-(p+1)} \cos p\theta$$

For the case of a machine with iron boundaries and windings in slots, we are interested in the fields at the boundaries. In such a case, usually, either $R_i = R_1$ or $R_s = R_2$. The fields are, at the outer boundary: $r = R_s$

$$H_r = M_0 \frac{R_s^{p-1}}{R_s^{2p} - R_i^{2p}} \left(\frac{p}{p+1} (R_2^{p+1} - R_1^{p+1}) \right.$$

$$\left. + \frac{p}{p-1} R_i^{2p} (R_1^{1-p} - R_2^{1-p}) \right) \cos p\theta$$

or at the inner boundary: $r = R_i$

$$H_r = M_0 \frac{R_i^{p-1}}{R_s^{2p} - R_i^{2p}}$$

$$\times \left(\frac{p}{p+1} (R_2^{p+1} - R_1^{p+1}) + \frac{p}{p-1} R_s^{2p} (R_1^{1-p} - R_2^{1-p}) \right) \cos p\theta$$

Direct Current Motors

J. Kirtley and N. Ghai

7.1 Introduction

Virtually all electric machines and all practical electric machines employ some form of rotating or alternating field/current system to produce torque. While it is possible to produce a "true DC" machine (e.g. the "Faraday Disk"), for practical reasons such machines have not reached application and are not likely to. The AC machine is operated from an alternating voltage source. Indeed, this is one of the principal reasons for employing AC in power systems.

The first electric machines employed a mechanical switch, in the form of a carbon brush/commutator system, to produce this rotating field. While the widespread use of power electronics is making "brushless" motors (which are really just synchronous machines) more popular and common, commutator machines are still economically very important. They are relatively cheap, particularly in small sizes, and they tend to be rugged and simple.

Commutator machines are found in a very wide range of applications. The starting motor on all automobiles is a commutator machine. Many of the other electric motors in automobiles, from the little motors that drive the outside rear-view mirrors to the motors that drive the windshield wipers are permanent magnet commutator machines. The large traction motors that drive subway trains and diesel/electric locomotives are DC commutator machines (although induction machines are making some inroads here). Many common appliances use "universal" motors: series connected commutator motors adapted to AC.

A schematic picture ("cartoon") of a commutator type machine is shown in Fig. 7.1. The armature of this machine is on the rotor (this is the part that handles the electric power), and the current is fed to the armature through the brush/commutator system. The interaction magnetic field is provided by a field winding. A permanent magnet field is applicable here.

Assume that the interaction magnetic flux density averages B_r, and that there are C_a conductors underneath the poles at any one time, and if there are m parallel paths, then one may estimate torque produced by the machine by

$$T_e = \frac{C_a}{m} R\ell B_r I_a$$

where R and ℓ are rotor radius and length, respectively, and I_a is terminal current. Note that C_a is not necessarily the total number of conductors, but rather the total number of *active* conductors (that is, conductors underneath the pole and therefore subject to the interaction field). If N_f is the number of field turns per pole, the interaction field is

$$B_r = \mu_0 \frac{N_f I_f}{g}$$

leading to a simple expression for torque in terms of the two currents

$$T_e = G I_a I_f$$

where G is now the motor coefficient (units of N-m/ampere squared)

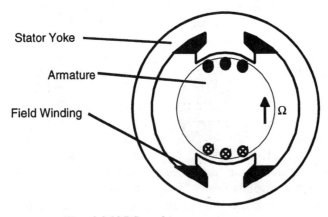

Figure 7.1 Wound-field DC machine geometry.

$$G = \mu_0 \frac{C_a}{m} \frac{N_f}{g} R\ell$$

Reviewing this using Faraday's Law

$$\nabla \times \vec{E} = -\frac{\partial \vec{B}}{\partial t}$$

Integrating both sides and noting that the area integral of a curl is the edge integral of the quantity

$$\oint \vec{E} \cdot d\vec{\ell} = -\iint \frac{\partial \vec{B}}{\partial t}$$

In the case in which the edge of the contour is moving (see Fig. 7.2), this is a bit more convenient to use if

$$\frac{d}{dt} \iint \vec{B} \cdot \vec{n} \, da = \iint \frac{\partial \vec{B}}{\partial t} \cdot \vec{n} \, da + \oint \vec{v} \times \vec{B} \cdot d\vec{\ell}$$

where $\vec{v}$ is the velocity of the contour. This is a convenient way of noting the apparent electric field within a moving object (as in the conductors in a DC machine)

$$\vec{E}' = \vec{E} + \vec{v} \times \vec{B}$$

Note that the armature conductors are moving through the magnetic field produced by the stator (field) poles, and one can ascribe to them an axially directed electric field

$$E_z = -R\Omega B_r$$

If the armature conductors are arranged as described above, with C_a conductors in m parallel paths underneath the poles and with a mean active radial magnetic field of B_r, voltage induced in the stator conductors is

$$E_b = \frac{C_a}{m} R\Omega B_r$$

Note that this is only the voltage induced by motion of the armature conductors through the field and does not include brush or conductor resistance. If the expression for effective magnetic field is included, the back voltage is

$$E_b = G\Omega I_f$$

which leads to the conclusion that newton-meters per ampere squared equals volt seconds per ampere. This stands to reason if one examines electric power into the interaction and mechanical power out

$$P_{em} = E_b I_a = T_e \Omega$$

A more complete model of this machine would include the effects of armature, brush and lead resistance, so that in steady state operation

$$V_a = R_a I_a + G \Omega I_f$$

This corresponds with the equivalent circuit shown in Fig. 7.3.

Now, consider this machine with its armature connected to a voltage source and its field operating at steady current, so that

$$I_a = \frac{V_a - G \Omega I_f}{R_a}$$

Then torque, electric power in and mechanical power out are

$$T_e = G I_f \frac{V_a - G \Omega I_f}{R_a}$$

$$P_e = V_a \frac{V_a - G \Omega I_f}{R_a}$$

$$P_m = G \Omega I_f \frac{V_a - G \Omega I_f}{R_a}$$

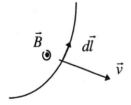

Figure 7.2 Motion of a contour through a magnetic field produces flux change and electric field in the moving contour.

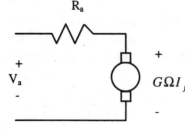

Figure 7.3 DC machine equivalent circuit.

Now, note that these expressions define three regimes defined by rotational speed, as is illustrated in Fig. 7.4. The two "break points" are at zero speed and at the "zero torque" speed

$$\Omega_0 = \frac{V_a}{GI_f}$$

For $0 < \Omega < \Omega_0$, the machine is a motor: electric power in and mechanical power out are both positive. For higher speeds: $\Omega_0 < \Omega$, the machine is a generator, with electrical power in and mechanical power out being both negative. For speeds less than zero, electrical power in is positive and mechanical power out is negative. There are few needs to operate machines in this regime, short of some types of "plugging" or emergency braking in traction systems.

7.2 Connections

The previous section described a mode of operation of a commutator machine usually called "separately excited," in which field and armature circuits are controlled separately. This mode of operation is used in some types of traction applications in which the flexibility it affords is useful. For example, some traction applications apply voltage control in the form of "choppers" to separately excited machines, as is shown in Fig. 7.5.

Note that the "zero torque speed" is dependent on armature voltage and on field current. For high torque at low speed, one would operate the machine with high field current and enough armature voltage to produce the requisite current. As speed increases so does back voltage, and field current may need to be reduced. At any steady operating speed, there will be some optimum mix of field and armature currents to produce the required torque. As is often done for braking, one could re-connect the armature of the machine to a braking resistor and turn

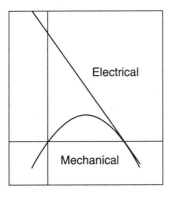

Figure 7.4 DC machine operating regimes.

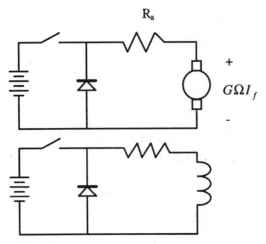

Figure 7.5 Two-chopper, separately excited machine hookup.

the machine into a generator. Braking torque is controlled by field current.

A subset of the separately excited machine is the shunt connection in which armature and field are supplied by the same source, in parallel. This connection is not widely used anymore: it does not yield any meaningful ability to control speed, and the simple applications to which it used to be used are mostly being handled by induction machines.

Another connection which is still widely used in the series connection, shown in Fig. 7.6, in which the field winding is sized so that its normal operating current level is the same as normal armature current and the two windings are connected in series. Then

$$I_a = I_f = \frac{V}{R_a + R_f + G\Omega}$$

And then torque is

$$T_e = \frac{GV^2}{(R_a + R_f + G\Omega)^2}$$

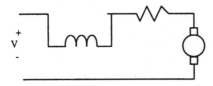

Figure 7.6 Series connection.

It is important to note that this machine has no "zero-torque" speed, leading to the possibility that an unloaded machine might accelerate to dangerous speeds. This is particularly true because the commutator, made of pieces of relatively heavy material tied together with non-conductors, is not very strong.

Speed control of series-connected machines can be achieved with voltage control and many appliances using this type of machine use choppers or phase-controlled rectifiers. An older form of control used in traction applications was the series dropping resistor: obviously not a very efficient way of controlling the machine and not widely used (except in old equipment, of course).

A variation on this class of machine is the very widely used "universal motor," in which the stator and rotor (field and armature) of the machine are both constructed to operate with alternating current. This means that both the field and armature are made of laminated steel. Note that such a machine will operate just as it would have with direct current, with the only addition being the reactive impedance of the two windings. Working with RMS quantities

$$\underline{I} = \frac{\underline{V}}{R_a + R_f + G\Omega + j\omega\,(L_a + L_f)}$$

$$T_e = \frac{|V|^2}{(R_a + R_f + G\Omega)^2 + (\omega L_a + \omega L_f)^2}$$

where ω is the electrical supply frequency. Note that, unlike other AC machines, the universal motor is not limited in speed to the supply frequency. Appliance motors typically turn substantially faster than the 3,600 RPM limit of AC motors, and this is one reason why they are so widely used: with the high rotational speeds, it is possible to produce more power per unit mass (and more power per dollar).

7.3 Commutator

The commutator is what makes the DC machine work. There are still aspects of how the brush and commutator system work that are poorly understood. However, this section makes some attempt to show a bit of what the brush/commutator system does.

To start, look at the picture shown in Fig. 7.7. Represented are a pair of poles (shaded) and a pair of brushes. Conductors make a group of closed paths. The current from one of the brushes takes two parallel paths. It is possible to follow one of those paths around a closed loop, under each of the two poles (remember that the poles are of opposite polarity) to the opposite brush. Open commutator segments (most of them) do not carry the current into or out of the machine.

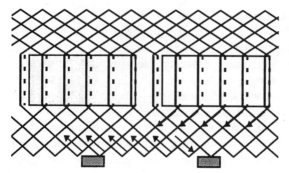

Figure 7.7 Commutator and current paths.

A commutation interval occurs when the current in one coil must be reversed, as is shown in Fig. 7.8. In the simplest form, this involves a brush bridging between two commutator segments, shorting out that coil. The resistance of the brush causes the current to decay. When the brush leaves the leading segment, the current in the leading coil must reverse.

This section does not attempt to fully explore the commutation process in this type of machine, but note a few things. *Resistive* commutation is the process relied upon in small machines. When the current in one coil must be reversed (because it has left one pole and is approaching the other), that coil is shorted by the brushes. The brush resistance causes the current in the coil to decay. Then the leading commutator segment leaves the brush and the current MUST reverse (the trailing coil has current in it), often resulting in sparking.

In larger machines, the commutation process would involve too much sparking, which causes brush wear and noxious gases (ozone) that promote corrosion, etc. In these cases, it is common to use sepa-

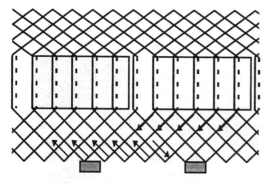

Figure 7.8 Commutator at commutation.

rate commutation interpoles. These are separate, usually narrow or seemingly vestigal pole pieces which carry armature current. They are arranged in such a way that the flux from the interpole drives current in the commutated coil in the proper direction. Remember that the coil being commutated is located physically between the active poles and the interpole is therefore in the right spot to influence commutation. The interpole is wound with armature current (it is in series with the main brushes). It is easy to see that the interpole must have a flux density proportional to the current to be commutated. Since the speed with which the coil must be commutated is proportional to rotational velocity and so is the voltage induced by the interpole, if the right number of turns are put around the interpole, commutation can be made to be quite accurate.

7.4 Compensation

The analysis of commutator machines often ignores armature reaction flux. Obviously these machines **do** produce armature reaction flux, in quadrature with the main field. Normally, commutator machines are highly salient and the quadrature inductance is lower than direct-axis inductance, but there is still flux produced. This adds to the flux density on one side of the main poles (possibly leading to saturation). To make the flux distribution more uniform, and therefore to avoid this saturation effect of quadrature axis flux, it is common in very highly-rated machines to wind compensation coils: essentially mirror-images of the armature coils, but this time wound in slots in the surface of the field poles. Such coils will have the same number of ampere-turns as the armature. Normally, they have the same number of turns and are connected directly in series with the armature brushes. What they do is to almost exactly cancel the flux produced by the armature coils, leaving only the main flux produced by the field winding. One might think of these coils as providing a reaction torque, produced in exactly the same way as main torque is produced by the armature.

7.5 Standards

DC motor standards fall into two categories, viz., standards for performance and standards for test. For test, IEEE Std 113 applies. For performance, the applicable U.S. standard is NEMA MG1-1993, Revision 4.

In NEMA MG1, DC motors are divided in two categories by size: those with outputs of up to and including 1.25 hp per r/min and those with higher outputs. Because of their smaller size, motors in the former category (the medium DC motors) have their size and mounting dimensions standardized, and their performance is defined more precisely. Motors in the latter category (the large DC motors) do not have standardized dimensions and the performance is more generalized. This is so because larger motors are not mass produced and are usually designed to fit the needs of specific applications.

7.6 NEMA Performance for Medium DC Motors

The following performances are included in NEMA MG1:

- Operating conditions
- Temperature rise of windings
- Rated voltage variations
- Speed variations
- Momentary overload

- Commutation
- Overspeed
- Operation on rectified AC current
- Operation on variable voltage
- Vibration

7.6.1 Operating conditions

The usual or normal site operating conditions include the following:

- An ambient temperature in the range of 0°C to 40°C.
- An altitude not exceeding 1000 m.
- A location such that there is no serious interference with motor ventilation.
- Installation on a rigid mounting surface.

7.6.2 Temperature rise

For continuous operation and for motors in totally enclosed fan-cooled, and all other enclosures, permissible temperature rises by resistance in degrees centigrade for classes A, B, F and H insulation are given in Table 7.1. For short time-rated motors, higher temperature rises are permitted. These temperature rises are dependent on the time rating of the motor and are lower for longer time ratings.

TABLE 7.1 Temperature Rise

Insulation Class	A	B	F	H
Temperature Rise	70	100	130	155

7.6.3 Rated voltage variations

NEMA requires that the motors operate successfully with a voltage variation of +10% from the rated field and armature DC voltages, and of ±10% of rated AC line voltage for motors operating on rectified power supply, although the performance is not expected to be necessarily the same as that established for operation at rated voltage.

7.6.4 Speed variations due to load

Speed variation from rated load to no load for straight shunt-wound, stabilized shunt-wound or permanent magnet DC motors when operated at rated armature voltage, with ambient temperature per the standard, and the windings at constant temperature is allowed in accordance with Table 7.2.

7.6.5 Base speed variations due to heating

For straight shunt-wound, stabilized shunt-wound and permanent magnet motors, NEMA limits the speed variation from base speed between ambient temperature to that attained at rated load armature and field voltage following a run of specified duration to the values given in Table 7.3.

7.6.6 Variation from rated base speed

When operated at rated load and voltage and at full field, with windings at constant temperature, the speed variation allowed by NEMA is ±7.5%.

TABLE 7.2   Speed Variation

Horsepower	Percent speed regulation at base speed
<3	25
3–50	20
51–100	15
>100	10

TABLE 7.3   Percent Variation of Rated Base Speed

Enclosure type	A	B	F	H
Open	10	15	20	25
Totally Enclosed	15	20	25	30

7.6.7 Momentary overloads

NEMA DC motors are capable of 50% armature over-current for one minute at rated voltage. For adjustable speed motors this capability is required for all speeds within the speed range.

7.6.8 Successful commutation

Successful commutation is defined as operation in normal service at rated load with no serious damage to the commutator or brushes due to sparking that might require abnormal maintenance. It is recognized that some visible sparking is not evidence of unsuccessful commutation.

7.6.9 Overspeed

For one minute duration, all DC motors are required to be capable of overspeeds as follows:

1. *Shunt-wound:* 25% above the highest rated speed, or 15% above the no-load speed, whichever is greater.
2. *Compound-wound motors with ≤35% speed regulation:* 25% above the highest rated speed, or 15% above the corresponding no-load speed, whichever is greater, but not exceeding 50% above the highest rated speed.
3. *Series-wound and compound wound with >35% speed regulation:* 10% above the maximum safe operating speed.

7.6.10 Operation on rectified alternating currents

When small DC motors intended for use on adjustable voltage electronic power supplies are operated from rectified power sources, they are required to be designed or selected for this application. NEMA suggests that successful operation in this application is possible if the combination results in a form factor at rated load equal to or less than the motor-rated form factor. For some typical power supplies, the recommended rated motor form factors are given in Table 7.4. The form factor of the current is the ratio of the rms value of the current to its average value.

7.6.11 Operation on variable voltage power supply

NEMA cautions that the temperature rise of the motors operated at full-load torque and reduced speed and reduced armature voltage will

TABLE 7.4 Recommended Form Factors

Power source	Armature current form factor range	Recommended motor form factor
Single Phase Thyristor		
Half Wave	1.86–2.0	2
Half Wave	1.71–1.85	1.85
Half or Full Wave	1.51–1.7	1.7
Full Wave	1.41–1.5	1.5
Full Wave	1.31–1.4	1.4
Full Wave	1.21–1.3	1.3
Three Phase Thyristor		
Half Wave	1.11–1.2	1.2
Full Wave	1.00–1.1	1.1

increase requiring a reduction in load torque. With some rectifier circuits, the current ripple at rated current also increases as the armature voltage is reduced requiring a further reduction in load torque. Under these conditions, motors are capable of operating successfully at only 67% of rated torque at 50% of rated speed.

7.6.12 Vibration

See Section 7.7.10 on large DC motors.

7.6.13 Other performance

For additional performance, rating charts, etc., see NEMA MG1-1993, Rev 4.

7.7 NEMA Performance for Large DC Motors

7.7.1 Motor types

Three broad categories of DC motor are recognized in this standard. These are general industrial motors, metal-rolling mill motors and reversing hot-mill motors.

7.7.1.1 General industrial. These motors are designed for all industrial service other than metal rolling mills, and may, when specified, operate at speeds higher than the base speed by field weakening.

7.7.1.2 Metal rolling-mill. These motors apply to metal rolling-mill service other than reversing hot-mill applications. They may be specified for single or either direction of rotation. These motors differ from the general industrial motors in that they have special requirements for continuous overload capability, heavy mechanical construction, high momentary overloads and close speed regulation.

Metal rolling-mill motors are further divided into two categories: the class N and the class S motor. Class N motors are designed for a given base speed, but when specified, may be designed for operation at higher speeds by field weakening. Class S motors are designed for speeds still higher than those for class N (by field weakening), and therefore require special mechanical construction.

7.7.1.3 Reversing hot-mill. These motors are designed specifically for reversing hot-mill applications. They are characterized by no continuous overloads, mechanical construction suitable for rapid speed reversals and sudden application of heavy loads, and higher momentary overload capability.

7.7.2 Continuous overloads

General industrial and reversing hot-mill motors have no continuous overload capability. Metal rolling-mill motors are capable of carrying 115% of rated load throughout the rated speed range, and 125% of rated load for two hours at rated voltage throughout the rated speed range following continuous operation at rated load. At 115% load, the temperature rises specified for the applicable insulation class will be higher than at rated load and other performance characteristics may differ from those at rated load. At 125% load for two hours, the temperature rise permitted for the applicable insulation class is not exceeded, but other performance characteristics may differ.

7.7.3 Momentary overloads

All motors are capable of carrying momentary overloads for one minute of magnitudes given in Tables 7.5, 7.6, and 7.7.

7.7.4 Operating conditions

The usual or normal site operating conditions include the following:

- An ambient temperature in the range of 0°C to 40°C.

TABLE 7.5 Overloads for General Industrial Motors

Percent of base speed	Percent of rated horsepower load	
	Occasionally applied	Frequently applied
100	150	140
200	150	130
≥300	140	125

TABLE 7.6 Overload for Metal Rolling-Mill Motors

Percent of base speed	Percent of rated horsepower load	
	Occasionally applied	Frequently applied
100	200	175
200	200	160
≥300	175	125

TABLE 7.7 Overload for Reversing Hot-Mill Motors

Percent base of speed	Occasionally applied		Frequently applied	
	% of rated base speed torque	% of rated horsepower	% of rated base speed torque	% of rated horsepower
93	275	256	. . .	. . .
95	. . .	. . .	225	214
125	199	248.5	166	207.5
150	162	242.5	135	202
175	135	236.5	112	196.5
200	115	230	95.5	191
225	99.5	224	82.5	185.5
250	87.5	218	72	180
275	77	212	63.5	174.5
300	68.5	206	56.3	169

- An altitude not exceeding 1000 m.
- A location such that there is no serious interference with motor ventilation.

7.7.5 Overspeed

All NEMA motors are required to have an overspeed capability of 25% above rated full-load speed for a period of two minutes.

7.7.6 Variation from rated voltage

NEMA DC motors are designed to operate successfully at rated load at up to 110% of rated DC armature or field voltage, or both, provided that maximum speed is not exceeded, but the performance within this variation may not necessarily be in accordance with the standard. The motors are also required to be capable of withstanding repetitive transient voltages of 160% of rated voltage and random transients of 200% of rated voltage.

7.7.7 Reversal time of reversing hot-mill motors

Maximum time typically required to reverse direction of rotation at no load and with suitable controls are given in Table 7.8.

7.7.8 Successful commutation

Successful commutation is defined as operation in normal service at rated load with no serious damage to the commutator or brushes due to sparking that might require abnormal maintenance. It is recognized that some visible sparking is not evidence of unsuccessful commutation.

7.7.9 Operation on rectified alternating currents

Large DC motors to NEMA standards are designed to operate from a direct-current source and their performance will vary in every material respect when operated from a rectified alternating-current supply. The difference in performance will be more marked when the rectifier pulse number is less than six or when the rectifier current is phase-controlled to produce an output voltage of 85% or less of the maximum possible rectified output voltage.

7.7.10 Vibration

Vibration limits for completely assembled motors, running uncoupled are specified. The preferred vibration parameter is velocity in in/s or mm/s peak. Table 7.9 gives the vibration limits on the bearing hous-

TABLE 7.8 Reversal Times for Reversing Hot-Mill Motors

Motor reverse speed (forward and reverse) (percent of base speed)	Reversal Time (seconds)
Horsepower × Base Speed (r/min) ≤ 250,000 and Speed Ratio ≤ 2	
100	1.5
150	2.5
200	4.0
Horsepower × Base Speed (r/min) > 250,000 or Speed Ratio > 2	
100	2
150	3
200	5
240	7
300	12

TABLE 7.9 NEMA Unfiltered Vibration Limits

Speed (r/min)	Rotational frequency (Hz)	Peak velocity (in/s)
3600	60	0.15
1800	30	0.15
1200	20	0.15
900	15	0.12
720	12	0.09
600	10	0.08

ings of resiliently mounted motors. The standard is 0.15 in/s. For rigidly mounted motors, the limits given in this Table are reduced to 80% of the table values.

7.7.11 Temperature rise

Permissible temperature rises in NEMA MG1 for various insulation classes are a function not only of the class of insulation, but also of the type of enclosure and machine ventilation. They also vary with the type of service. Tables 7.10 and 7.11 give the temperature-rise limits for large DC motors.

7.7.12 Other performance

For more information, see NEMA MG1-1993, Rev 4.

TABLE 7.10 Temperature-Rise Limits in Degrees Celsius for General Industrial Service

			General industrial service							
			Semi-enclosed continuous rated 100% Load				Totally-enclosed continuous rated 100% load			
		Method of temperature determination	Insulation class				Insulation class			
Item	Machine part		A	B	F	H	A	B	F	H
1	Armature windings and all other windings other than those given in items 2 and 3	Thermometer	50	70	90	110	55	75	95	115
		Resistance	70	100	130	155	70	100	130	155
2	Multilayer field windings	Resistance	70	100	130	155	70	100	130	155
3	Single-layer field windings with exposed uninsulated surfaces and bare copper windings	Thermometer	60	80	105	130	65	85	110	135
		Resistance	70	100	130	155	70	100	130	155
4	Commutator and collector rings	Thermometer	65	85	105	125	65	85	105	125

TABLE 7.11 Temperature-Rise Limits in Degrees Celsius for Mill Service

			Metal rolling mill service								
			Metal rolling mills						Reversing hot mills		
			Forced-ventilated or totally-enclosed water–air-cooled						Forced-ventilated or totally-enclosed water–air-cooled		
			Continuous rated 100% load			2 Hours 125% Load			Continuous rated 100% load		
			Insulation class			Insulation class			Insulation class		
Item	Machine part	Method of temperature determination	B	F	H	B	F	H	B	F	H
1	Armature windings and all other windings other than those given in items 2 and 3	Thermometer	40	60	75	55	75	95	50	70	90
		Resistance	60	90	110	80	110	135	70	100	130
2	Multilayer field windings	Resistance	70	100	120	80	110	135	70	100	130
3	Single-layer windings with exposed uninsulated surfaces and bare copper windings	Thermometer	50	70	90	65	85	110	60	80	105
		Resistance	60	90	110	80	110	135	70	100	130
4	Commutator and collector rings	Thermometer	55	75	90	65	85	105	65	85	105

7.8 References for DC Motors

Anderson, E. P.; *Electric Motors;* New York, Macmillan, 1991.

Dewan, S., Slemon, G. R., and Straughen, A.; *Power Semi-Conductor Drives;* New York, Wiley, 1984.

Fitzgerald, A. E., Kingsley, C., and Kusko, A.; Electric Machinery; New York, McGraw-Hill Book Company, 1971.

Lightband, D. A., and Bicknell, D. A.; *The Direct Current Traction Motor;* London, Business Books Ltd., 1970.

Kostenko, M., and Piotrovsky, L.; *Electrical Machines,* vol. 1; Moscow MIR Publishers, 1974.

Kusko, A.; *Solid State – DC Motor Drives;* Cambridge, MA. MIT Press, 1969.

Nasar, S. A., and Unnewehr, L. E.; *Electromechanics and Electric Machines;* New York, Wiley, 1979.

NEMA Standard MS1—*Motors and Generators.*

Say, M. G., and Taylor, E. O.; *Direct Current Machines;* New York, Wiley, 1980.

Smeaton, R. W.; *Motor Applications and Maintenance Handbook;* New York, McGraw-Hill Book Company, 1969.

Sokira, T. J., and Jaffe, W.; *Brushless DC Motors,* Blue Ridge Summit, Pa., TAB Books, 1989.

8

Other Types of Electric Motors and Related Apparatus

J. Kirtley

8.1 Induction Generators

Any induction motor, if driven above its synchronous speed when connected to an ac power source, will deliver power to the external circuit. The generator operation is easily visualized from the equivalent circuit of Fig. 8.1, corresponding to negative slip. The induction generator must always take reactive power from the load or the line for excitation and for the I^2X losses. For this reason, the induction generator can only operate in parallel with an electric power system or independently with a load supplemented by capacitors. For independent operation, the speed must be increased with load to maintain constant frequency; the voltage is controlled with the capacitors.

An induction generator delivers an instantaneous three-phase short-circuit current equal to the terminal voltage divided by its

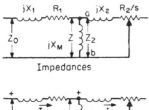

Impedances

Voltages and currents

Figure 8.1 Equivalent circuit of polyphase induction motor.

locked-rotor impedance. Its rate of decay is much faster than that of a synchronous generator of the same rating, corresponding to the subtransient time constant T'_{do}; sustained short-circuit current is zero.

The virtue of the induction generator is its ability to self-synchronize when the stator circuit is closed to a power system. At one time induction generators were used for small, unattended hydro stations. Today, induction generators are being used in a similar manner for wind turbines and cogeneration units. They have also been used for high-speed, high-frequency generators, because of their squirrel-cage rotor construction.

8.2   Synchronous Induction Motors

There are three types of motors that can start and run as induction motors yet can lock into the supply frequency and run as synchronous motors as well. They are (1) the wound-rotor motor with dc exciter; (2) the permanent-magnet (PM) synchronous motor; (3) the reluctance-synchronous motor. The latter two types are used today primarily with adjustable-frequency inverter power supplies. In Europe, wound-rotor induction motors have often been provided with low-voltage dc exciters that supply direct current to the rotor, making them operate as synchronous machines. With secondary rheostats for starting, such a motor gives the low starting current and high torque of the wound-rotor induction motor and an improved power factor under load. Several different forms of these synchronous induction motors have been proposed, but they have not shown any net advantage over usual salient-pole synchronous or induction machines and are very seldom used in the United States. The PM synchronous motor is shown in Fig. 8.2a. The construction is the same as that of an ordinary squirrel-cage motor (either single or polyphase), except that the depth of rotor core below the squirrel-cage bars is very shallow, just enough to carry the rotor flux under locked-rotor conditions. Inside this shallow rotor core is placed a permanent magnet, fully magnetized. The rotor core serves as a keeper, so that the rotor is not demagnetized by removing it from the stator. In starting, the rotor flux is confined to the laminated core. As the speed rises, the rotor frequency decreases and the rotor flux builds up, creating a pulsating torque with the field of the magnet, as when a synchronous motor is being synchronized after the dc field has been applied. As the motor approaches full speed, therefore, the ac impressed field locks into step with the field of the magnet and the machine runs as a synchronous motor. The absence of rotor I^2R loss, the synchronous speed operation, and the high efficiency and power factor make the motor very attractive for special applications, such as high-frequency spinning motors. When many such motors are supplied from a high-frequency source, the kVA requirements are reduced to perhaps 50% of those needed for usual induction motor types, with consequent large savings.

If the rotor surface of a P-pole squirrel-cage motor is cut away at symmetrically spaced points, forming P salient poles, the motor will accelerate to full speed as an induction motor and then lock into step and operate as a synchronous motor. The synchronizing torque is due to the change in reluctance and, therefore, in stored magnetic energy, when the air-gap flux moves from the low- into the high-reluctance region. Such motors are often used in small-horsepower sizes, when synchronous operation is required, but they have inherently low pull-out torque and low power factor, and also poor efficiency, and therefore require larger frames than the same horsepower induction motor. The PM synchronous motor has superior performance in every way, except possibly cost. A cross section of the reluctance-synchronous motor is shown in Fig. 8.2b. These motors are available up to about 5 hp.

If the number of rotor salients is nP, instead of P, and if the P-pole motor winding is arranged to also produce a field of $(n + 1)P$ or $(n - 1)P$ poles, the motor may lock into step at a subsynchronous speed and run as a subsynchronous motor. For the P-pole fundamental mmf, acting on the varying rotor permeance, will create $(n + 1)P$ and $(n - 1)P$ pole fields from this base, and these will lock into step with the independently produced $(n - 1)P$- or $(n + 1)P$-pole field, when the rotor speed is such as to make the two harmonic fields turn at the same speed in the same direction.

It is difficult to provide much torque in such subsynchronous motors, and their use is therefore limited to very small sizes, such as may be used in small timer or instrument motors.

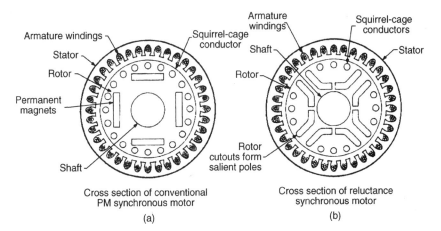

Cross section of conventional
PM synchronous motor
(a)

Cross section of reluctance
synchronous motor
(b)

Figure 8.2 Cross section of (a) a permanent magnet (PM) synchronous motor and (b) a reluctance-synchronous motor.

8.3   Linear Motors

Linear induction motors (LIMs) have been built in fractional-horse-power ratings for such applications as moving drapes, and up to several thousand horsepower for driving tracked air-cushion transit vehicles on a guideway. Other applications include moving freight cars in yards, driving people-mover vehicles, and providing reciprocating motion for machine tools. LIMs are built like rotary induction motors with distributed multipole polyphase windings placed in the slots of a plane laminated status as shown in Fig. 8.3. When the windings are excited by a polyphase voltage of frequency f, an air-gap space flux wave is propagated along the length of the stator at a velocity of $v = 2fp$, where p is the pole pitch. The rotor consists of an aluminum or copper sheet, which is propelled by the field with a slip velocity to provide the required thrust. LIMs are either double-sided, with two facing stators operating on a single rotor, or single-sided, with the rotor sheet backed by a moving or stationary magnetic return path. The magnetic force density normal to the stator surface is considerable compared to the tangential force density that moves the rotor, which requires that the stator be well-braced mechanically to maintain constant air-gap distances over the surface of the stator. The typical tangential force density is about 3 lb/in^2 for air-cooled windings, where the normal force density is about 30 lb/in^2.

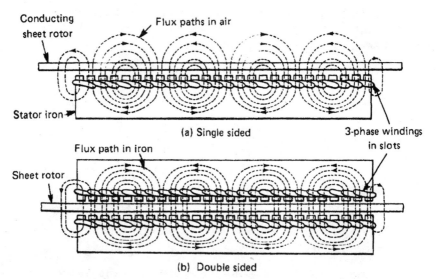

Figure 8.3   Single-sided and double-sided linear induction motors (LIMs) with sheet rotor.

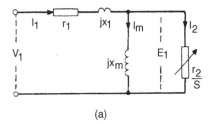

(a)

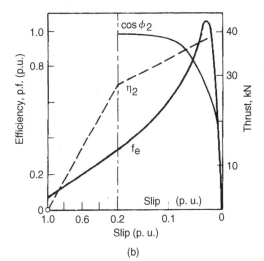

Slip (p. u.)

(b)

Figure 8.4   (a) Equivalent circuit of a double-sided LIM and (b) the characteristic curves of a typical LIM.

The magnetic air-gap of a double-sided LIM is the thickness of the sheet rotor plus the clearance between the rotor and the stators on either side. Whereas most rotary induction motors are built with an air-gap of 0.025 to 0.1 mils, the air-gap in the LIM is 0.25 to 1.5 in. For this reason, the magnetizing reactance of the LIM is lower than that of an equivalent rotary induction motor. Also the stator leakage reactance is higher. The equivalent circuit of the LIM is shown in Fig. 8.4a. Figure 8.4b shows the thrust-slip power factor and efficiency curves of a double-sided LIM.[1] This LIM has an air gap of 1.47 in, a rotor sheet thickness of 0.25 in, and a stator length of 9.8 in. The two, three-phase windings of the stator are excited at 173 Hz from inverters to produce a linear synchronous velocity of 395 ft/s. Speed control

---

[1] M. G. Say; *Alternating Current Machines;* New York, John Wiley & Sons, 1983.

and breaking of LIMs is done in the same way as in the rotary induction motors.

8.4 High-Frequency Motors

For high-speed tools and for spinning of rayon and other threads, a variety of interesting motor constructions have been developed. Normally these are two-pole, three-phase motors, with special high-frequency power supply of 90, 120, or 180 Hz, giving operating speeds between 500 and 10,500 r/min and up to 25 hp. In textile applications, the motors usually drive individual spinning buckets, which are subject to considerable unbalance due to uneven building up of thread, etc. The continual starting and stopping for loading and unloading the buckets requires the motors to carry unbalance reliability through the entire speed range, necessitating careful design of mounting flexibility and shaft stiffness. Most usual applications, however, are in woodworking and similar industries, where separate motor stators and rotors are supplied to the tool manufacturers for building into their particular devices. These motors were powered from high-frequency alternators, but are now powered by adjustable-frequency solid-state inverters.

Three-phase 400-Hz power systems, used on large airplanes, have led to the development of 400-Hz motors with speeds of 12,000 and 24,000 r/min, having weights averaging 2 lb/hp for motors of 1 to 15 hp with 5-min ratings. These motors are open, with an external fan to force air over the windings.

8.5 Stepper Motors

The primary characteristic of a stepper motor is its ability to rotate a prescribed small angle (step) in response to each control pulse applied to its windings. Below about 200 pulses per second, the motor rotates in discrete steps in synchrony with the pulses; at higher frequencies up to 16,000 pulses per second, the motor slews without stopping between pulses. Although motors are available for step angles of 90 to 0.180°, the common step is 1.8°. Stepper motors are categorized as permanent-magnet rotor (PM), variable reluctance (VR), or hybrid (PM-VR). The rotor of the PM aligns itself with the energized stator poles as shown in Fig. 8.5b. The rotor turns until the poles are aligned at each step. The PM-VR hybrid shown in Fig. 8.5c has a high skew rate yet retains holding torque when the power is turned off. Motors can be made to rotate in half-steps to increase accuracy. Performance of stepper motors is described by two types of curves: the pull-out torque vs. speed curve, as shown in Fig. 8.6; and the holding torque

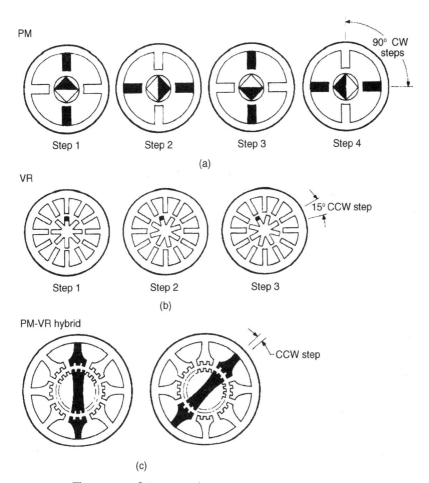

Figure 8.5 Three types of stepper motors.

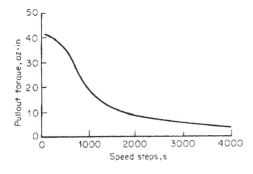

Figure 8.6 Pull-out torque vs. speed for a four-phase 5° step VR step motor running at half steps (2½°).

angle curves, as shown in Fig. 8.7. Stepper motors are available with holding torques up to 4000 oz · in.

8.6 Hysteresis Motors

By constructing the secondary core of an induction motor of hardened magnet steel, in place of the usual annealed low-loss silicon-steel laminations, the secondary hysteresis can be greatly magnified, producing effective synchronous motor action. Such hysteresis motors, having smooth rotor surfaces without secondary teeth or windings, give extremely uniform torque, are practically noiseless, and give substantially the same torque from standstill all the way up to synchronous speed. A hysteresis motor is a true synchronous motor, with its load torque produced by an angular shift between the axis of rotating primary mmf and the axis of secondary magnetization. When the load torque exceeds the maximum hysteresis torque, the secondary magnetization axis slips on the rotor, giving the same effect as a friction brake set for a fixed torque.

Despite the interesting characteristics of this type of motor, it is limited to small sizes, because of the inherently small torque derivable from hysteresis losses. Only moderate flux densities are practicable, owing to the excessive excitation losses required to produce high densities in hard magnet steel, and, therefore, about 20 W/lb of rotor magnet steel represents the maximum useful synchronous power on 60 Hz. Hysteresis motors have found an important use for phonograph-motor drives, their synchronous speed enabling a governor to be dispensed with and freedom from tone waver to be secured.

The Telechron motor, which is so widely used for operating electric clocks, also operates on the hysteresis-motor principle. In the Telechron motor, a two-pole rotating field is produced in a cylindrical air space, and into this space is introduced a sealed thin-metal cylinder containing a shaft carrying one or more hardened magnet-steel disks, driving a gear train. The 60-Hz magnetic field causes the steel disks to revolve at 3600 r/min, driving through the gears a low-speed shaft,

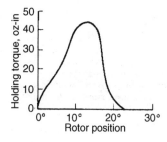

Figure 8.7 Holding torque vs. rotor position for a 3-phase, VR step motor, 24 steps per revolution, bi-directional, 1600 steps/s.

usually 1 r/min, which merges from the sealed cylinder through a closely fitting bushing designed to minimize oil leakage. Although the magnetic field has to cross a very considerable air-gap length and pass through the thin walls of the metal cylinder, the power required to drive a well-designed clock is so small that sample output is obtained with only about 2 W input for ordinary household-clock sizes.

The hysteresis motor has been displaced for phonograph and tape-reel drives by the transistor-driven brushless dc motor. It has been displaced for electric clocks by solid-state circuits with digital readout.

8.7 Alternating-Current Commutator Motors

8.7.1 Classification[2]

As compared with the induction motor, the ac commutator motor possesses two of the advantages of the dc motor: a wide speed range without sacrifice of efficiency, and superior starting ability. In the induction motor, the starting torque is limited by the small space-phase displacement between the air-gap flux and the induced secondary current and by magnetic saturation of the flux paths. In the ac commutator motor, on the other hand, the air-gap flux and current are held at the optimum space-phase displacement by proper location of the brush axis, and the secondary current is not limited by magnetic saturation, giving high torque per ampere at starting. Furthermore, the series commutator motor may be operated far above the induction-motor synchronous speed, giving high power output per unit of weight.

Alternating-current commutator motors may be grouped into two classes:

1. Those motors in which the resultant mmf providing the flux increases with the load. When operated from a source of constant voltage, the speed of such motors decreases with increasing load. They are termed *series motors* from the similarity of their characteristics to those of series-wound dc motors. The speed at any given load may be varied by changing the applied votage or, in some cases, by shifting the brushes.

2. Those motors in which the resultant mmf providing the flux is substantially constant irrespective of the load. For operation from a source of constant voltage, the speed of such motors is approximately constant. The speed may, however, be increased or decreased (inde-

---

[2] C. W. Olliver; "The A-C Commutator Motor"; Princeton, N.J., D. Van Nostrand Company, Inc., 1927.

pendently of the load) by increasing or decreasing the voltage at the terminals of the motor, by brush shifting, or by the provision of suitably disposed and connected auxiliary coils. Such motors are termed *shunt motors.*

Alternating-current commutator motors are either single-phase or polyphase. A unique characteristic of all single-phase motors is a double line-frequency pulsation of the torque produced, corresponding to the sinusoidal variation twice each cycle of the single-phase power supplied. This torque pulsation is partly transmitted to the load, causing small speed pulsations and necessitating special coupling and mounting designs to minimize vibration and fatigue stresses.

Polyphase commutator motors have the advantage of better inherent commutating ability, due in part to the need for shifting the rotor current only 60° in time phase at each brush stud for a six-phase motor of 30° for 12 phases, as compared with 180° shift for a single-phase or dc machine. Single-phase motors are generally limited to sizes below about 10 hp, except for railway applications.

With the advent of solid-state devices, the ac commutator motor is being displaced by the thyristor-rectifier-powered dc motors and inverter-fed induction motors, at less cost and superior performance. The dc motor does not have the difficulties of commutation, the requirement for extra windings, and shifting brush arrangements, and can be built on an unlaminated frame. The induction motor has no commutator and can run at the high speeds of the commutator motor.

8.7.2 Single-phase straight series motor

An ordinary dc series motor, if constructed with a well-laminated field circuit, will operate (although unsatisfactorily) if connected to a suitable source of single-phase alternating current. Since the armature is in series with the field, the periodic reversals of current in the armature will correspond with simultaneous reversals in the direction of the flux, and consequently the torque will always be in the same direction. But the inductance of the motor will be so great that the current will lag far behind the voltage, and the motor will have a very low power factor. The entire amount of armature flux produced along the brush axis generates a reactive voltage in the armature, which must be overcome by the applied voltage, without performing any useful function whatever.

When the motor is first thrown in the circuit, and before the armature has moved from rest, the field constitutes the primary of a transformer and sends flux through the armature core. Those armature turns which at that instant are short-circuited under the brushes act as short-circuited secondary coils and are traversed by heavy currents which serve no useful purpose whatever and occasion serious heating. When the armature starts to revolve, these short-circuited turns are opened as they pass out from under the brushes and are replaced by other turns which are momentarily short-circuited and then opened. These interruptions of heavy currents are accompanied by serious sparking, since the heating is concentrated at the few segments on which the brushes rest. As soon, however, as a certain speed is acquired, the heating is distributed over all the segments and the conditions are ameliorated. This source of sparking is, then, most serious at the moment of starting. This difficulty has been minimized by operating at a lower frequency than 60 Hz and by the employment of leads of high resistance connecting the winding to the commutator segments.

The simple single-phase series motor has therefore two major faults, low power factor and poor commutation at low speeds, confining its use to fractional horsepower and very high speed applications.

8.7.3 Single-phase compensated series motor

In all except the smallest sizes, it is usual to employ a compensating winding on the stator, in series with the armature and so arranged that its mmf as nearly as possible counteracts the armature mmf. A commutating winding is also frequently used, which somewhat overcompensates the armature reaction along the interpolar, or commutating-zone, axis and so provides a voltage to aid the current reversal, just as in a dc motor. By these means, the flux along the brush axis is reduced to a small fraction of its uncompensated value, and the power factor of the motor is greatly improved. Further improvement of the power factor is secured by using a smaller air gap and correspondingly fewer field ampere-turns than in an uncompensated motor, thus reducing the reactive voltage in the series field to a minimum.

8.7.4 Universal motors

Small series motors up to about ½-hp rating are commonly designed to operate on either direct current or alternating current and so are called *universal motors*. Universal motors may be either compensated or uncompensated, the latter type being used for the higher speeds and smaller ratings only. Owing to the reactance voltage drop, which is present on alternating current but absent on direct current, the motor speed is somewhat lower for the same load ac operation, especially at high loads. On alternating current, however, the increased

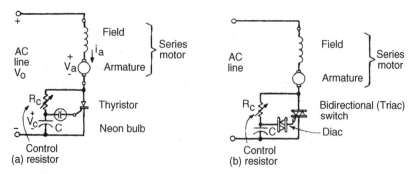

Figure 8.8 (a) Half-wave series universal motor circuit; (b) full-wave univer-sal motor circuit, and the holding torque angle curves. Stepper motors are available with holding torques up to 400 oz · in.

saturation of the field magnetic circuit at the crest of the sine wave of current may materially reduce the flux below the dc value, and this tends to raise the ac speed. It is possible, therefore, to design small universal motors to have approximately the same speed-torque per-formance over the operating range, for all frequencies from 0 to 60 Hz. On a typical compensated-type ¼-hp motor, rated at 3400 r/min, the 60-Hz speed may be within 2% of the dc speed at full-load torque but 15% or more lower at twice normal torque, while on an uncom-pensated motor the speed drop will be materially greater.

The commutation on alternating current is much poorer than on direct current, owing to the current induced in the short-circuited ar-mature coils, and this provides a definite limitation on their size and usefulness. If wide brushes are used, the short-circuited currents are excessive and the motor-starting torque is reduced, while if narrow brushes are used, there may be excessive brush chatter at high speeds, causing short brush life. Good design, therefore, requires careful pro-portioning of commutator and brush rigging to meet conflicting elec-trical, mechanical, and thermal requirements. Universal motors are generally used for vacuum cleaners, portable tools, food mixers, and similar small devices operating at maximum speeds of 3000 to 10,000 r/min.

The speed of the universal motor is controlled by means of a half-wave thyristor, or full-wave triac, as shown in Fig. 8.8.[3] The control device governs the half-wave average voltage applied to the motor as a function of the firing angle. The firing circuits are usually relatively simple. The speed is controlled by changing a resistance value, such as R_c, in Fig. 8.8. The characteristics of a universal (series) motor with half-wave control are shown in Fig. 8.9.

[3]A. Kusko; *Solid-State DC Motor Drives;* Cambridge, Mass., MIT Press, 1969.

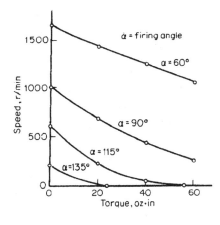

Figure 8.9 Measured speed-load torque characteristics of series motor and half-wave thyristor control.

8.8 Fractional-Horsepower-Motor Applications

8.8.1 Scope

A fractional-horsepower motor is defined by NEMA as either (1) a motor built in a frame with a NEMA two-digit frame number, or (2) a motor built in a frame smaller than the NEMA frame for a 1-hp, open-construction, 1700- to 1800-r/min, induction motor. The two-digit frame number is defined as 16D, where D is the height of the shaft centerline above the bottom of the mounting base. Fractional-horse-power motors include one-phase and three-phase induction and synchronous motors, one-phase universal motors, and dc motors. Ratings, with minor exceptions, are 1/20 to 1 hp, inclusive, Motors of smaller ratings are classified as subfractional or miniature.

8.8.2 Purpose

General-purpose motors are of open construction, rated at 60°C temperature rise by resistance over a 40°C ambient temperature. They are designed according to standard ratings with standard operating characteristics and mechanical construction for rise under usual service conditions without restriction as to a particular application or type of application. A definite-purpose motor is any motor designed according to standard ratings with standard operating characteristics or mechanical construction for use under service conditions other than the usual or for on a particular type of application.

8.8.3 Selection of type

The principal characteristics of fractional-horsepower motors are shown in Table 8.1. For any application, the motor selected should

TABLE 8.1 Characteristics of Fractional-Horsepower Motors

	Alternating				
	Single-phase				
	Split-phase types				
	General-purpose	High-torque	Two-speed, pole changing	Capacitor-start	Capacitor (1-value, or perm, split)
Schematic diagram of connections. Arrangements shown are typical or representative; most of the types illustrated have numerous other arrangements which are also used.	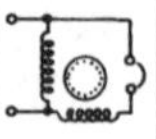			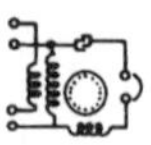	
Characteristic speed-torque curves. Ordinates are speed; 1 division = for all ac motors, 20% of syn. r/min; for universal motors, 1000 r/min; for dc motors, 20% of full-load rpm. Each abscissa division = 100% of full-load torque.			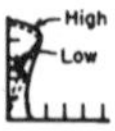 		
Rotor construction	Squirrel-cage	Squirrel-cage	Squirrel-cage	Squirrel-cage	Squirrel-cage
Built-in automatic starting mechanism	Centrifugal switch	Centrifugal switch	Centrifugal switch	Centrifugal switch	None required
Horsepower ratings commonly available	1/20–1/3	1/6–1/3	1/8–3/4	1/8–3/4	1/20–3/4
Usual rated full-load speeds (for 60-Hz ac motors; also dc motors)	3450, 1725, 1140, 865	1725	1725/1140 1725/865	3450, 1725, 1140, 865	1620, 1080, 820
Speed classification	Constant	Constant	Two-speed	Constant	Constant, or adjustable varying
Means used for speed control			Two-speed switch		2-speed switch or auto-trans-former
Comparative torques: Locked-rotor	Moderate	High	Moderate	Very high	Low
Comparative torques: Breakdown	Moderate	High	Moderate	High	Moderate
Radio interference, running	None	None	None	None	None
During acceleration	One click	One click	Two clicks	One click	None
Approximate comparative costs between type, for same horsepower rating: Below 1/20 hp .					
1/20–1/4 hp	100	75	210	125	140
1/2–3/4 hp	80	54	150	100	100–110
General remarks	For constant-speed operation, even under varying load conditions, where moderate torques are desirable or mandatory, this type is often used in preference to the more costly capacitor-start motor. Meets NEMA starting currents. Typical applications: blowers; centrifugal pumps; duplicating machines; refrigerators; oil burners; unit heaters.	High locked-rotor currents (in excess of NEMA) limit the use of this type on lighting circuits to applications where the motor starts only very infrequently, because of a tendency to cause flickering of the lights. Principal applications: washing and ironing machines; cellar-drainer pumps; tools for a home workshop.	Used where two definite speeds independent of load are required. Ratings above 1/4 hp usually made capacitor-start. Motor shown always starts on high-speed connection; transfer to low speed made by starting switch. Common applications: belted blowers for warm-air furnaces or for other purposes; attic ventilators; air conditioning apparatus.	A general-purpose motor suitable for most applications requiring constant speed under varying loads, high starting and running torques, high overload capacity. Also available as two-speed pole-changing motor above 1/4 hp. A few important applications are: refrigeration and air conditioning compressors; air compressors; stokers; gasoline pumps.	Primarily used for unit heaters, or for other shaft-mounted fans. Essentially a constant-speed motor, but by means of a two-speed switch, or by means of an autotransformer, other speeds can be obtained, *with fan loads*, of horsepower rating selected closely matches the fan load. Can also be made in intermittent ratings for plug-reversing service.

Standard motors are ordinarily designed to operate in ambient temperatures from 10 to 40°C. Variations in line voltage of plus or minus 10%, or variations in frequency of plus or minus 5% are allowable.

Locked-rotor currents for single-phase motors, except split-phase high-torque and synchronous types, usually do not exceed the following limits established by NEMA:

Rating, hp	Amperes at 115 volts	Amperes at 230 volts
1/6 and smaller	20	10
1/4	23	11½
1/3	31	15½
1/2	45	22½
3/4	61	30½

Fractional horsepower motors are built for across-the-line starting.

The standard direction of rotation is counterclockwise facing the end opposit the shaft extension.

Current Motors	1, 2, or 3 phase			Polyphase	Dc or ac (60 Hz or less), universal types		Direct current	
	Repulsion-start	Shaded-pole	Nonexcited synchronous (reluctance)	Squirrel-cage induction	Without governor	With governor	Shunt or compound	Series
			Stator winding may be: split-phase, capacitor-start, capacitor, polyphase					
	Drum-wound; commutator	Squirrel-cage	Cage, with cutouts	Squirrel-cage	Drum-wound; commutator	Drum-wound; commutator	Drum-wound; commutator	Drum-wound; commutator
	Short-circuit	None	Depends on stator winding	None	None	None	None	None
	$\frac{1}{8}-\frac{3}{4}$	$\frac{1}{2000}-\frac{1}{8}$	$\frac{1}{6000}-\frac{1}{3}$	$\frac{1}{8}-\frac{3}{4}$	$\frac{1}{150}-1$	$\frac{1}{50}-\frac{1}{20}$	$\frac{1}{20}-\frac{3}{4}$	$\frac{1}{25}-\frac{1}{20}$
	3450, 1725, 1140, 865	1450–3000	3600, 1800, 1200, 900	3450, 1725, 1140, 805	3000–11,000	2000–4000	3450, 1725, 865	900–2000
	Constant	Constant, or adjustable varying	Absolutely constant	Constant	Varying, or adjustable varying	Adjustable	Constant, or adjustable varying	Varying, or adjustable varying
		Choke or resistor			Choke or resistor	Adjustable governor	Armature resistance	Resistor
	Very high High	Low Low	Low Moderate	Very high Very high	Very high	Very high	Very high	Very high
	None Continuous	None None	None	None None	Continuous Continuous	Continuous Continuous	Continuous Continuous	Continuous Continuous
	128 100	100	200–200 275	165–195 100	75 105–175	110 140–160	175–225 120–140	185

A constant-speed motor suited to general-purpose applications requiring high starting torque, such as pumps and compressors. An associated type, the repulsion induction (buried cage) is used for door openers and other plug-reversing applications. Has been displaced for many applications by the capacitor-start motor.

For ratings below $\frac{1}{20}$ hp, this is a general-purpose motor. For fan applications, speed control is affected by use of a series choke or resistor. Applications: fans, unit heaters, humidifiers, hair driers, damper controllers.

Cutouts in rotor result in synchronous-speed characteristics. Curve shown is for split-phase stator. Pull-in ability is affected by inertia of connected load. Used for teleprinters, facsimile-picture transmitters, graphic instruments, etc. Clocks and timing devices usually use shaded-pole hysteresis motors rated at a few millionths of a horsepower.

Companion motor to capacitor-start motor with comparable torques and generally suited to same applications if polyphase power is available. Inherently plug-reversible and suitable for door openers, hoists, etc. High-frequency motors used for high-speed applications, as for woodworking machinery, rayon spinning, and portable tools.

Light weight for a given output, high speeds, varying-speed and universal characteristics make this type very popular for hand tools of all kinds, vacuum cleaners, etc. Ratings above ¾ hp usually compensated. Some speed control can be effected by a resistor or by use of a tapped field. Used with reduction gear for slower speed applications.

By means of a centrifugal governor, a constant-speed motor having the advantages of the universal motor is obtained. Governor may be single-speed or adjustable even while running. Speed is independent of applied voltage. Used in typewriters, calculating machines, food mixers, motion-picture cameras and projectors, etc.

A constant-speed companion motor for the capacitor-start or split-phase motor for use where only d-c power is available. For unit-heater service, armature resistance is used to obtain speed control. Not usually designed for field control.

Principally used as the d-c companion motor to the shaded-pole motor for fan applications. Used in these small ratings in place of shunt motors to avoid using extremely small wire.

meet the application and power supply requirements at the least cost. Numerous trade-offs are possible, for example, speed changing infinite steps compared to continuously adjustable-speed solid-state drives.

8.8.4 Ratings

Standard voltage and frequency ratings for ac motors are listed in Table 8.2 and for dc motors in Table 8.3.

8.8.5 Service conditions

General-purpose motors are designed to operate under the usual service conditions of 0° to 40°C ambient temperature, altitude to 3300 ft (1000 M), and installation on rigid mounting surfaces where there is no interference with the ventilation. Some general-purpose, definite-purpose, and special-purpose motors can operate under one or more unusual service conditions, which include exposure to dust, lint, fumes, radiation, steam, fungus, shock; operation where voltage, frequency, waveform, and form factor deviate from standards; and over-speed, overtemperature, and excess altitude operation. The manufac-

TABLE 8.2   Voltage Ratings of AC Fractional-Horsepower Motors

Motors	Frequency, Hz	Voltage
Single-phase	60	115, 230
	50	110, 220
Three-phase	60	115, 200
		230, (460)
	50	220, 380
Universal	60*	115, 230

*Can operate from dc to 60 Hz.

TABLE 8.3   Voltage Ratings of DC Fractional-Horsepower Motors

Primary power source	Rating, hp	Armature voltage	Field voltage
Los-ripple dc	$\frac{1}{20}$ to 1	115, 230	115, 230
1-phase rectifier	$\frac{1}{20}$ to $\frac{1}{2}$	75	50, 100
		90	50, 100
		150	100
	$\frac{3}{4}$ to 1	90	50, 100
		180	100, 200
3-phase rectifier	$\frac{1}{4}$ to 1	240	100, 150
			240

turer should be consulted for operation under unusual service conditions.

8.8.6 Thermal protection

Many single-phase motors are now available with a built-in thermal protector which affords complete protection from burnout due to any type of overload, even a stalled rotor. Most such devices are automatic-resetting, but some are manual-resetting. Motors that are protected usually are marked externally in some way to indicate the fact.

8.8.7 Reversibility

In general, standard motors of the kind listed in the table can be arranged by the user to start from rest in either direction of rotation. There are exceptions, however, Shaded-pole motors, unless of a special design, can be operated in only one direction of rotation. Small dc and universal motors often have the brushes set off neutral, preventing satisfactory operation in the reverse direction. Single-phase motors which use a starting switch ordinarily cannot be reversed while running at normal operating speeds, because the starting winding, which determines the direction of rotation, is then open-circuited. By use of special relays this limitation of split-phase and capacitor-start motors can be overcome when necessary. Such motors are built for small hoists. High-torque intermittent-duty permanent split capacitor motors; repulsion-induction (buried-cage) motors; and split-series dc or universal motors are often built for plug-reversing service. Standard polyphase induction motors can be reversed while running, as can the smaller ratings of dc motors; such applications should preferably be taken up with the motor manufacturer.

8.8.8 Mechanical features

Rigid and rubber-mounted motors are commonly available. Sleeve and ball bearings are both standard. Sleeve-bearing motors are designed for operation with the shaft horizontal, but ball-bearing motors can be operated with the shaft in any position. For operation with the shaft vertical, sleeve-bearing motors may require a special design. Rubber mounting is widely used for quiet operation, because all single-phase motors have an inherent double-frequency torque pulsation. An effective and common arrangement uses rubber rings concentric with the shaft and so arranged as to provide appreciable freedom of torsional movement but little other freedom. Sometimes the driven member picks up the double-frequency torque pulsation and amplifies it to an objectable noise, for example, a fan with large blades mounted rigidly on the shaft. The cure for this difficulty is an elastic coupling between the shaft and the driven member; no amount of elastic suspension of the stator can help. Standard motors are generally open and of drip-proof construction. Splashproof and totally enclosed motors are easily available.

8.8.9   Inputs of small single-phase motors

See Table 8.4. Full-load torque, in terms of horsepower and rated speed, is

$$\text{Full-load torque, oz} \cdot \text{ft} = \frac{84,000 \times \text{hp}}{\text{r/min}}$$

8.8.10   Application tests

The primary object of any application test is to determine the power requirements of the appliance or device under various significant operating conditions. A convenient way of doing this is to use a motor of approximately the right horsepower rating and of predetermined efficiency at various outputs. Watts input are carefully measured under each condition. From the watts input observed (never use current as a measure of load except for dc motors) and the known efficiency, the load is readily determined. Care should be taken in measuring the watts input to correct for the meter losses.

A second, and equally important, object of the test is to determine the actual locked-rotor and pull-up torques required by the appliance. The locked-rotor and pull-up torques of the test motor should be known or measured at rated voltage and frequency. (Locked-rotor

TABLE 8.4   Approximate Starting- and Full-Load Current for Single-Phase 115-V Motors

Rating, hp	Max. locked-rotor current, A		3450 r/min		1725 r/min		1140 r/min		865 r/min	
	Des. O	Des. N	A	W	A	W	A	W	A	W
⅛	50	20	2.9	207	2.7	176	3.9	207	5.4	245
⅛	50	20	3.2	254	3.0	214	4.3	254	6.0	296
¼	50	26	4.2	352	3.9	301	5.6	352	8.1	414
⅓	50	31	5.3	460	4.9	395	7.0	460	9.8	540
½	50	45	7.4	678	6.9	574	9.8	678	—	—
¾	—	61	10.6	981	9.9	835	—	—	—	—
1	—	80	13.3	1260	—	—	—	—	—	—

NOTE: A = amperes; W = watts; Des. = Designation.

torque often varies with slight changes in rotor position.) Using a transformer or induction regulator to obtain a variable voltage (do not use a resistance or choke for this purpose), measure the minimum voltage at which the motor will start the appliance and also the minimum voltage at which it will pull it up through switch-operating speed. Assuming that the pull-up and lock-rotor torques each vary as the square of the applied voltage, it is then a simple matter to determine the actual locked-rotor and pull-up torques required by the device. After a motor has been selected, it should be determined whether or not it can operate the device at 10% above and below normal rated voltage of the motor or over a wider range of voltage, if desired. If exceptional load conditions may occasionally be encountered, use of a motor equipped with inherent-overheating protection is often desirable.

8.8.11 Definite-purpose motors

For a number of important applications, involving large quantities of motors, NEMA has developed standards to meet these special requirements effectively and economically. Motors built to these standards are usually more readily obtainable and economical than special motors tailored to one application. Highlights and distinguishing features are given in Table 8.5. More details can be obtained in NEMA Standards.

8.8.12 Small synchronous motors

Small synchronous motors in the 1.5- to 25-W range for timing, tape drives, small fans, and record players are available as brushless dc motors or as hysteresis motors. The brushless dc motor consists of a permanent-magnet field, two-phase, synchronous motor driven by transistors from a dc source. The transistors are switched from a Hall-device signal which senses the rotor position. A regulator maintains constant speed.

Shaded-pole hysteresis motors, which operate at synchronous speed, are essentially the same as shaded-pole induction motors except that they use rotors of hardened-steel rings of a material having high hysteresis loss. Large quantities of such motors are built for clocks and timing devices. Clock motors have an input of 1.5 to 2 W and an output of a few millionths of a horsepower. Large motors with inputs up to 15 W are built for heavier duty applications. Rotor speeds are commonly 450, 600, and 3600 r/min. Most of these motors are furnished with built-in reduction gears to give output speeds of 60 r/min to 1 r/month.

TABLE 8.5   NEMA Standards for Definite-Purpose Motors

Application	Principal types	Distinguishing features
Universal motor	Universal: Salient-pole and distributed field	Dimensional standards: common practices utilizing parts
Hermetic motors	Split-phase, capacitor-start, polyphase	Parts only for hermetic refrigeration condensing units
Belt-drive refrigeration compressors	Capacitor-start, Repulsion-start polyphase	Open; sleeve bearings, extended rear oiler; automatic-reset thermal overload protection
Jet-pump motors	Split-phase, capacitor-start, repulsion-start polyphase	3450 r/min; ball bearings, open; machined back end shield; automatic-reset overload protection
Motors for shaft-mounted fans and blowers	Split-phase, permanent-split capacitor, polyphase	Enclosed; horizontal; sleeve bearings; vertical, ball bearings; extended through bolts; capacitors on front end shield
Shaded-pole motors for shaft-mounted fans and blowers	Shaded-pole; two-speed, three-speed	Open or totally enclosed; sleeve bearings; high slips
Belted fans and blowers	Split-phase, capacitor-start, repulsion-start; two-speed split-phase and capacitor-start	Open; sleeve bearings; resilient mounting; automatic-reset overload protection; extended rear oiler
Stoker motors	Capacitor-start; repulsion start; polyphase	Totally enclosed recommended; automatic reset overload protection
Motors for cellar drainers and sump pumps	Split-phase	Vertical, dripproof, 50°C; two ball bearings, or one ball, one sleeve; mounts on support pipe; built-in float-operated line switch; overload protection
Gasoline-dispensing pumps	Capacitor-start, repulsion-start polyphase	Explosionproof; sleeve bearing, built-in line switch and capacitor; voltage-selector switch on single-phase
Oil-burner motors	Split-phase	Enclosed, face-mounted, round-frame; manual-reset overload protection; two line leads

TABLE 8.5 (Continued)

Application	Principal types	Distinguishing features
Motors for home-laundry equipment	Split-phase	Low-cost, high starting current; open, 50°C; round-frame with ungrounded mounting rings; shaft extension with flat and hole for coupling
Motors for coolant pumps	Split-phase, capacitor-start, repulsion-start, polyphase	3450- and 1725-r/min; totally enclosed; ball bearings;, machined back end shield
Submersible motors for deep-well pumps	Split-phase, capacitor, polyphase	3450 r/min; designed for operation totally submerged in water not over 25°C (77°F); use external relay for starting

Reluctance motors, both self-starting and manual-starting types, are available for similar applications. Another type used is the synchronous-inductor motor, which is essentially an inductor alternator used as a motor; field excitation is furnished by a permanent magnet.

8.9 High Torque Motors

8.9.1 Introduction

Of course all motors are intended to produce torque, so the notion of "high torque" motor is perhaps a bit odd. What is meant by this term is a description of motors in which torque density (torque per unit weight or torque per unit volume) is the important attribute and in which tip speed or other speed-related effects are not important. Motors of this description would be used for applications such as direct drive of ships or wheel motors in automobiles.

This section focuses on generic features of machines and avoids detail as much as is possible. It includes a few comments on radial and axial gap machines. Two exotic machine types are described.

Some degree of caution is required here. Some first-order (partial) optimizations are carried out, assuming that all parameters, other than the one or two we are considering, are (or can be) held constant. This is not always strictly true.

8.9.2 Cylindrical rotor machine

In "standard" machine geometry, the stator is on the outside and consists of ordinary windings in slots and "back iron" to make up the magnetic circuit. Analysis of rotor outside machines would not differ from this by very much.

Torque produced by a cylindrical rotor is

$$T = 2\pi R^2 \ell \tau$$

where τ is the average shear stress produced by the electromagnetic interaction, R is the rotor radius and ℓ is the active length of the machine. Here, we assume that the radial flux density and the axial surface current density are ideally situated

$$B_r = \sqrt{2} B_{r0} \cos(p\theta - \omega t)$$

$$K_z = \sqrt{2} K_{z0} \cos(p\theta - \omega t)$$

Of course, these quantities may be viewed as the *active* components: in the stator winding, the total flux density may contain a component in quadrature with B_r or the surface current may contain a component in quadrature with K_z. In either case, it is the components of radial flux density and axial current density that are in phase with each other that produce torque.

Rotor volume is simply

$$V_R = \pi R^2 \ell$$

so that the torque per unit volume is

$$\frac{T}{V_R} = 2\tau$$

The total volume of a machine is substantially more than just rotor volume, so it is necessary to include radial space for the stator windings and back iron and axial space for the end windings

$$R_O = R + h_s + t_b$$

$$L = \ell + \Delta\ell$$

It may be appropriate to make a few first-order estimates of these quantities. The radial dimension of the current carrying region of the machine is

$$h_s = \frac{K_z}{J_a \lambda_s}$$

where J_a is current density in the stator slots and λ_s is azimuthal slot factor. There are a number of limits to both current density and total current. One of these is heating, both obvious in nature and difficult to cast in general terms.

8.9.2.1 Tooth flux. A second set of armature-current limits arises because stator currents produce magnetic fields, both across the gap and within the stator (slot leakage), and these fields add to the flux density that must be carried by the teeth and back iron of the magnetic circuit. To model this, assume that the part of the magnetic circuit that consists of slots and teeth is very highly permeable in the radial direction ($\mu \to \infty$), but has a much lower permeability in the azimuthal direction ($\mu_\theta \approx \mu_0/\lambda_s$). Then, in the ideal case the cross-slot flux density is

$$B_\theta = \mu_\theta \frac{r - R_i}{h_s} K_{z0} \cos p\theta$$

where the slot bottom radius is $R_i = R - h_s$. At the tooth tips, this is

$$B_\theta = \mu_\theta K_{z0} \cos p\theta$$

This would impose one limit on total surface current, since this azimuthal flux density and the radial flux density of interaction must be carried in the teeth. Since these fluxes are orthogonal, they add to form a peak flux density

$$|B_t|^2 = (\sqrt{2}B_{r0})^2 + (\sqrt{2}\mu_\theta K_{z0})^2$$

Since $\tau = B_{r0}K_{z0}$, the maximum value of shear stress indicated occurs when

$$B_{r0} = \mu_\theta K_{z0} = \frac{B_s}{2}$$

where B_s is the maximum working flux density in the teeth (ordinarily this would be saturation flux density). This would, in turn, indicate a maximum surface stress level of

$$\tau = \frac{B_s^2}{4\mu_\theta}$$

and since $\lambda_s \approx \frac{1}{2}$

$$\tau = \frac{B_s^2}{8\mu_\theta}$$

If saturation flux density is taken to be about 2 T, this works out to be about 400 kPa (about 58 PSI). Interestingly, the most highly stressed machines currently operated are also the largest, turbine generators, and they achieve on the order of 30 PSI.

For completeness, consider another limit which is unlikely to be reached: cross-slot leakage flux must be carried through the teeth and magnetic circuit. To estimate the load on the tooth roots, note that flux density has no divergence

$$\nabla \cdot \overline{B} = 0$$

this means that

$$\frac{\partial B_r}{\partial r} = -\frac{1}{r}\frac{\partial B_\theta}{\partial r} = -\mu_\theta \frac{1}{R}\frac{r - R_i}{ph_s} K_{z0} \sin p\theta$$

(Note this analysis assumes that the slots have uniform width so that μ_θ/r is constant and equivalent to its value at the radius R.) The magnetic flux density in the tooth region can be computed as

$$B_r(r) = B_{ri} + \int_{R_i}^{r} \frac{\mu_\theta}{R}\frac{r - R_i}{ph_s} K_{z0} \sin p\theta\, dr = B_{ri} + \frac{\mu_\theta}{2}\frac{K_{z0}}{p}\frac{h_s}{R} \sin p\theta$$

Assume that none of this flux crosses the air-gap. Then, assuming that all of the radial flux density is in the teeth, one can estimate the tooth root flux density

$$|B_{ri}| = \frac{\mu_\theta}{2\lambda_s}\frac{K_{z0}h_s}{pR} \sin p\theta$$

Note that, unless the machine has very deep slots, this flux density is unlikely to approach the saturation limit.

8.9.2.2 Back iron. Flux density in the core is easy to compute, and is

$$B_c = \frac{R}{pt_b} B_{r0}$$

so that, assuming that the core and teeth will be operating at about the same flux density (presumably saturation density), we can estimate a requirement for core depth

$$t_b = \frac{R}{p}(1 - \lambda_s)$$

8.9.2.3 End windings. The end windings are, of course, subject to specific construction details. They will add, however, some length to the motor which is of the form

$$\Delta\ell \approx \beta\frac{R}{p}$$

The coefficient β will depend on construction, but is on the order of one.

8.9.2.4 Volume torque density. The volume torque density can be expressed in a number of ways. Using overall volume, it is

$$\frac{T}{V_t} = 2\tau \left(\frac{R}{R + h_s + t_b}\right)^2 \left(\frac{\ell}{\ell + 2\Delta\ell}\right)$$

Perhaps more interesting, particularly when considering high torque density machines, which tend to have large values for p, is to consider the *active* volume. Suppose we have a machine in which the active part of the rotor has the same radial extent as the stator: the active area will be

$$A_A = \pi\big((R + h_s + t_b)^2 - (R - h_s - t_b)^2\big) = 4\pi R(h_s + t_b)$$

This indicates a torque per unit volume

$$\frac{T}{V_a} = \frac{\tau}{2}\frac{R}{h_s + t_b}\frac{\ell}{\ell + 2\Delta\ell}$$

A few immediate conclusions arise from this:

1. Torque density increases directly with rotor radius, pushing high torque machines to maximum practical radius.
2. Large pole numbers (p) reduce the impact of both back iron (t_b) and end-winding length ($\Delta\ell$).
3. The value of large pole numbers hits a point of diminishing return, when t_b becomes less than h_s.

8.9.3 Axial flux machines

These are often called "disk-type" machines because the active elements are shaped like disks. They are characterized by multipole flux in the axial (z) direction and by radial currents and they can be of any of the major classes. This analysis treats a machine-type which could be either synchronous or induction, but permanent magnet and even reluctance machines can be made in this format.

Shear stress in an axial gap machine is of the form

$$\tau = T_{r\theta} = K_r B_z$$

Now, since surface current has no divergence ($\nabla \cdot \overline{K}$), the radially directed current is

$$K_r(r) = K_{ri} \frac{R_i}{r}$$

But note also that, if the slots in the magnetic circuit are of constant width, the average operating magnetic flux density is also a function of radius. Noting $\sqrt{2}B_0$ as the *peak* flux density (in a tooth), the RMS flux density at some radius r is

$$B_{z1} = B_0 \frac{w_t(r)}{w_t(r) + w_s} = B_0(1 - \lambda_x(r))$$

The width of the teeth increase with radius

$$w_t = \frac{2\pi r}{N_s} - w_s$$

where N_s is the number of teeth and w_s is slot width. If we note the *slot* fraction at the inner radius as

$$w_s = \frac{2\pi R_i}{N_s} \lambda_{si}$$

then the tooth fraction is

$$w_t(r) = \frac{2\pi}{N_s} \left(1 - \frac{R_i}{r} \lambda_{si} \right)$$

The surface traction (shear stress) is then

$$\tau(r) = K_r B_z = K_0 B_0 \frac{\lambda_{si} R_i}{r} \left(1 - \frac{\lambda_{si} R_i}{r} \right)$$

where K_0 is *slot* surface current density and B_0 is RMS fundamental (in the teeth) flux density.

To get actual torque, integrate over the active radius range of the disk

$$T = \int_{R_i}^{R_o} 2\pi r^2 \tau(r) dr = 2\pi K_0 B_0 \left(\lambda_{si} R_i \frac{R_o^2 - R_i^2}{2} - \lambda_{si}^2 R_i^2 (R_o - R_i) \right)$$

It is possible to maximise this with respect to the radius ratio $x = R_i/R_o$ and the slot fraction at the inner radius λ_{si}. The process is standard (even if a bit naive) and results in $x = \frac{1}{3}$ and $\lambda_{si} = 1$. That is, the optimum value for inner radius is one-third of outer radius and the optimum value of slot fraction at the inner radius is unity. Thus, the slots occupy the whole of the inside circumference and one-third of the outer circumference.

This optimum is at best questionable, since it was derived with no variables other than the active disk in mind. Of course other things, including end-turn shape, machine cooling and structure will affect these choices, but the final machine will probably not be too far away from this. The result is a torque of

$$T = \frac{2}{9} 2\pi K_0 B_0 R_o^3$$

Note that this does not yet yield a volume. To find a volume, one must include the end turns and active length of the machine.

8.9.3.1 End turns. These are perhaps more properly called "edge turns". This is really the space required to connect windings at the outer periphery of the rotor (and stator). The total width of conductors that must be carried from one half-pole to the next half-pole (one-half of the width of all conductors in a pole) is

$$\Delta R = \frac{\pi}{2} \frac{R}{p} \lambda_{so}$$

This reflects the notion that the radial extent of the end windings is equal to the azimuthal width of all of the conductors coming out of one-half of a pole. Now, the slot fraction at the outer periphery is

$$\lambda_{so} = \lambda_{si} \frac{R_i}{R_o}$$

so that the fractional edge winding is

$$\frac{\Delta R}{R_o} = \frac{\pi}{2} \frac{R_i}{R_o} \frac{\lambda_{si}}{p}$$

The length of the machine will be affected by "end iron". The return path will be nearly the same as for a radial flux machine

$$t_b = \frac{R}{p}(1 - \lambda_{so})$$

One very interesting feature of axial flux machines is that they can have multiple interactions disk pairs with only one set of end return paths. That is, there can be a multiplicity of rotor and stator disks interacting with flux that threads its way from one end of the machine to the other. This is shown in Fig. 8.10 If there are n_R rotors, total machine length is

$$L = n_R(h_r + h_s) + 2t_b$$

where the axial length of air-gaps is ignored.

Machine volume is then

$$V_t = \pi R_o^2 \left(1 + \frac{\pi}{2} \frac{R_i}{R_o} \frac{\lambda_{si}}{p}\right)^2 \left(n_R(h_r + h_s) + 2\frac{R_o}{p}(1 - \lambda_{so})\right)$$

With n_R rotors, torque is

$$T = n_R \frac{2}{9} 2\pi \tau_0 R_o^3$$

so that torque per unit volume is, assuming stator and rotor length are the same ($h_r = h_s$)

$$\frac{T}{V_t} = \frac{4}{9} \frac{R}{2h_s} \frac{\tau_0}{\left(1 + \frac{\pi}{6p}\right)^2 \left(1 + \frac{R}{p} \frac{4}{3} \frac{1}{n_R}\right)}$$

Now, note that if p is large enough, the expressions for relative

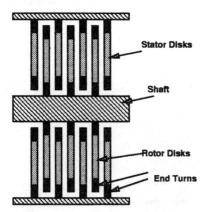

Stator Disks

Shaft

Rotor Disks

End Turns   Figure 8.10 Disk motor cartoon.

torque per unit volume for the two machine morphologies (radial and axial flux) are nearly the same (since $4/9$ and $1/2$ are fairly close in value). But note also that the axial flux machine uses nearly all of its volume (nearly all of it is active), whereas the radial flux machine has a large "hole" in the center, and the *total* volume of such a machine will be much larger than the *active* volume.

8.9.4 Step-motor like PM machines

This machine has a relatively high pole number and is built in such a way that its *active* radius is as large a fraction of its outer radius as possible. These features make this geometry suitable for relatively low-speed applications such as traction drives for buses and automobiles.

The machine is shown in cartoon format in Fig. 8.11. Its rotor is the outermost element, making this an "inside out" machine. A row of permanent magnets is mounted to the inside of an outer shell which forms the magnetic return path as well as the shaft. The magnets are alternately polarized.

The machine stator consists of a number of poles different (usually smaller) than the number of magnets. Wound on these poles, one coil per pole, is a multiple-phase armature winding. The required number of armature phases may be greater than three. Shown in the cartoon, for example, is a 12-pole (six-phase) stator winding and a 14-pole rotor. The electrical frequency is $\omega = p\Omega$, in this case $p = 7$.

The stator poles have concentrated windings around them, so the end turns are uncomplicated and short. The back iron of the rotor

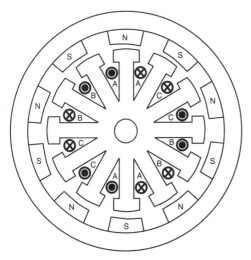

Figure 8.11 Step-motor-like PM machine.

must carry only one-half of the flux associated with one pole—a relatively small part of the total flux used by the machine.

Approach this machine at first by using a current sheet approach. To start, look at one pole with a phase winding around it. If the pole covers an angular extent θ_p (a physical, as opposed to electrical angle), it produces an MMF

$$F = \sum_{n=1}^{\infty} \frac{2}{n\pi} NI \sin n \frac{\theta_p}{2} \cos n\theta$$

(Note that this isolated pole also has a "DC" flux component which can be ignored because of what happens next.) Now, assume that there are N_p of these spaced about the circle and driven by appropriate currents. Using the index j to denote the progression of these, we find total MMF to be

$$F = \sum_{j=0}^{N_p-1} \sum_{n} \frac{2}{n\pi} NI_j \sin n \frac{\theta_p}{2} \cos n \left(\theta - \frac{2\pi j}{N_p} \right)$$

Assume the machine will be driven with currents which are periodic and which lag each other in time

$$I_j = \sum_{m \text{ odd}} I_m \cos m \left(\omega t - \frac{2\pi j}{N_p} \right)$$

Assembling all of this, the MMF distribution is

$$F = \sum_{n} \sum_{m} \frac{2}{n\pi} NI_m k_n \frac{N_p}{2} \{A_+ \cos(m\omega t - n\theta) + A_- \cos(m\omega t + n\theta)\}$$

where

$$I_m = \frac{4}{m\pi} \sin m \frac{\theta_{on}}{2}$$

$$k_n = \sin n \frac{\theta_p}{2}$$

$$A_+ = 1 \quad \text{if and only if} \quad m - n = kN_p$$

$$A_- = 1 \quad \text{if and only if} \quad m + n = kN_p$$

and in the last two expressions, k is an integer.

Of course, the interesting current component is the one with $n =$ the number of magnet pole pairs.

To get some idea of what is happening here, note that the total number of ampere turns could be written as either

$$2\pi r K_0 = 2 N_p N I_0$$

or

$$NI_0 = \frac{\pi r}{N_p} K_0$$

where I_0 is the maximum current amplitude and K_0 is the measure of the total current sheet magnitude. Thus

$$F = \sum_n \sum_m \frac{r}{n} K_0 f_m k_n \{A_+ \cos(m\omega t - n\theta) + A_- \cos(m\omega t + n\theta)\}$$

Equivalent surface current here is:

$$K_z = -\frac{1}{r}\frac{\partial F}{\partial \theta} = \sum_n \sum_m K_0 f_m k_n \{A_+ \sin(m\omega t - n\theta) - A_- \sin(m\omega t + n\theta)\}$$

This looks very much like the expression obtained with a conventional stator winding. This results in large current component at the right space and time order by picking angles that make f_m and k_n right at the right orders.

Suppose, for example, that the number of magnet pairs $n_0 = N_p + 1$. This would be a likely combination. Some pictures here have been calculated and drawn for $N_r - 10$, $n - 11$. If $\theta_p \ \pi/N_r$, an approximation that ignores the need for slot openings

$$k_{n_0} \approx \sin\left(\frac{N_p + 1}{N_r}\frac{\pi}{2}\right)$$

This is not too far from one if R is relatively large.

This produces a whole sequence of *synchronous* harmonics

$$k_{zs} = \sum_m K_0 f_m k_{(R+1)m} \sin m(\omega t - (N_p + 1)\theta)$$

8.9.5 Transverse flux machines

Particularly for low-speed machines, the use of short-pole pitch has advantages: it reduces required end-turn length and depth of back iron in the magnetic circuit. On the other hand, short pole-pitch machines have relatively large magnetizing MMF requirements because every pole must be excited. Further, low-speed machines are inefficient because conduction losses are normally associated with torque, while power is torque times speed.

The transverse flux machines described here are an attempt to take advantage of really short-pitch configuration to shorten further the current path in a machine, reducing the conduction loss. They do this by distorting the magnetic circuit (lengthening it considerably). They also use permanent magnets to provide for the relatively large MMF requirements.

Figure 8.12 shows two variations on the configuration. The simplest version is a single-phase machine (typically, multiple phases are arranged in different "rings". Current is carried in a single-circular coil, linking flux with a ring of permanent magnets. Engaging the tips of the yokes that go around the coil are pairs of permanent magnets driving flux around each yoke. The magnets labeled "S" are matched behind the coil with "N" magnets. The two magnets are linked by back-iron elements below the magnets. Current as labeled in the coil would be driving flux downward in the near legs of the yokes (as seen in this view), attracting the "S" poles and providing torque in the "clockwise" direction.

A two-phase version of the same machine could be made by using yokes both above and below, alternating with flux return elements. Two circular coils would drive currents in quadrature phase relationship. This second configuration might be re-drawn as shown in Fig. 8.13. This is a flattened-out version, justified by the fact that the ma-

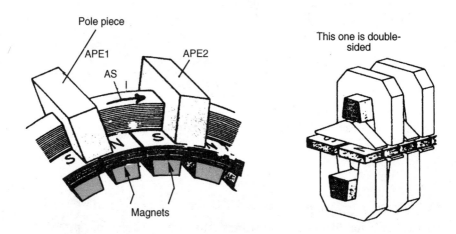

Figure 8.12 Transverse-flux machine geometry. *Source:* New Permanent Magnet Excited Synchronous Machine with High Efficiency at Low Speeds, H. Weh, H. Hoffman, and J. Landrath. *Proceedings: International Conference on Electric Machines,* 1988.)

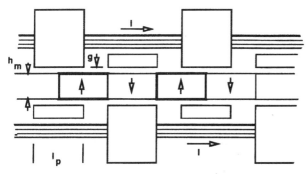

Figure 8.13   Two-phase transverse-flux machine.

chine it models will have a lot of poles and therefore relatively little curvature.

This one is most easily approached using an energy method. Magnetic co-energy is, for this system

$$W'_m = \int_{V_m} \int \overline{B} \cdot \overline{dM} + \int_{\text{coil}} \lambda dI$$

For this configuration, coil inductance does not depend on rotor position, so

$$W'_m = W'_{mo} + \lambda(x)I + \frac{1}{2} LI^2$$

Note that the first term, which relates to the co-energy of the magnets alone, will, in general, be a function of rotor position. This will produce "cogging" torque due to the permanent magnets interacting with the cores but it will have zero average value, so we can ignore it. Force is then

$$F^e = \frac{\partial W'_m}{\partial x} = \frac{\partial \lambda}{\partial x} I$$

Flux is a "triangle" wave with peak value (per yoke)

$$\lambda_0 = \mu_0 M_0 N_c \frac{h_m}{g} w \ell_p$$

where $B_r = \mu_0 M_0$ is magnet remnant flux density, N_c is the number of turns in the coil, w is width of one leg of the yoke and ℓ_p is yoke length.

This flux distribution is shown in Fig. 8.14. The slope of the triangle wave is the peak flux over engagement distance, so if I has proper sign

$$F^e = 2\mu_0 M_0 \frac{h_m}{g} wI$$

To get some idea of what kind of force density can be produced with this machine, see that with two phases, two-thirds of the surface is covered with active cores. Since the current and magnets aid in producing flux, a limiting value for current is

$$B_g = \mu_0 M_0 \frac{h_m}{g} + \mu_0 \frac{I}{2g}$$

Remember there are two sides to the yoke. Or I is limited to

$$I = 2 \left(g \frac{B_g}{\mu_0} - M_0 h_m \right)$$

Using a little shorthand

$$\alpha = \frac{\mu_0 M_0}{B_g} \qquad \xi = \frac{h_m}{g}$$

$$F^e = 4\mu_0 M_0 \frac{h_m w}{g} \left(g \frac{B_g}{\mu_0} - M_0 h_m \right) = 8\alpha\xi \frac{B_g^2}{2\mu_0} wg\,(1 - \alpha\xi)$$

Now, since spacing is one-half of core length and actual device width is $2w$, neglecting coil width, force density is

$$\tau = \frac{F^e}{\text{Area}} = \frac{8}{6} \frac{B_g^2}{2\pi_0} \frac{g}{\ell} \alpha\xi(1 - \alpha\xi)$$

This expression is still highly simplified since it assumes an arbitrary

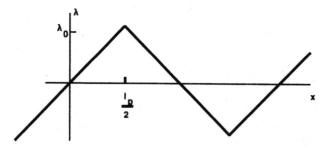

Figure 8.14 Flux linked by yoke.

value of g/ℓ. Note: it does not account for coil width, but does charge for both upper and lower surfaces of the rotor.

Suppose g/ℓ = ⅕ is possible. The maximum force density occurs when $\alpha\xi$ = ½. If core flux density $B_g \approx 1.8T$, then $\tau = \approx86$ kPa (about 12.5 PSI).

This might be carried even further by considering a flux concentrating version of this, which turns out to be somewhat similar to a PM biased variable reluctance machine.

Shown in Fig. 8.15 is a single-phase machine which has two coils, two sets of yokes, two rows of magnets and poles and problematical structure. The path of flux is complex: assume that the poles are lined up with the yokes, see that flux goes around one yoke, splits in the pole underneath and goes through two of the magnets and then around the two yokes underneath those magnets. In those yokes, of course, there is also flux from the adjacent top yokes, preserving symmetry.

Magnet MMF is

$$F_m = M_0 \ell_m$$

Magnet reluctance is

$$\mathcal{R}_m = \frac{\ell_m}{\mu_0 h_m w_p}$$

Gap reluctance (facing one magnet) is

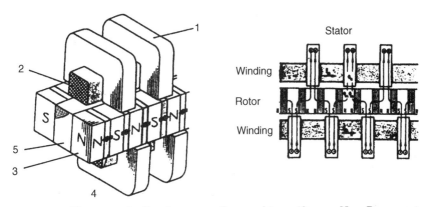

Figure 8.15 Flux-concentrating transverse-flux machines. (*Source:* New Permanent Magnet Excited Synchronous Machine with High Efficiency at Low Speeds, H. Weh, H. Hoffman, and J. Landrath. *Proceedings: International Conference on Electric Machines,* 1988.)

$$\mathcal{R}_g = \frac{2g}{\mu_0 \ell_p w_p}$$

where ℓ_m is magnet circumferential length, h_m is magnet radial height, w_p is pole (and magnet) axial width, ℓ_p is pole circumferential length, and g is physical gap dimension.

Flux per magnet is

$$\Phi_0 = \frac{F_m}{\mathcal{R}_m + \mathcal{R}_g} = \frac{\mu_0 h_m w_p M_0}{1 + \beta}$$

where

$$\beta = \frac{2g h_m}{\ell_p \ell_m}$$

Co-energy exchange over one-half cycle is

$$\Delta W'_m = 2\Phi_0 I$$

Note that this involves the same current I in both coils, since half of the magnet flux is in an upper yoke and half in a lower yoke. Then total force is

$$F_{em} = \frac{\Delta W'_m}{\ell_m + \ell_p}$$

since that is the "pitch" of the co-energy change.

It is a bit difficult to get a hard limit on the actual physical quantities here. Assume that current is limited by magnetic saturation. Saturation flux is

$$\Phi_s = B_s h_m w_p$$

and if we assume that $\Phi_0 = (1 - \alpha)\phi_s$, then

$$I = \alpha \Phi_x (\mathcal{R}_m + \mathcal{R}_g)$$

The co-energy change becomes

$$\Delta W'_m = 2\Phi_s^2 \alpha (1 - \alpha)(\mathcal{R}_m + \mathcal{R}_g)$$

With a little manipulation, this becomes

$$\Delta W'_m = 4 \frac{B_s^2}{2\mu_0} \alpha(1 - \alpha) h_m w_p \frac{\ell_m}{\ell_m + \ell_p} (1 + \beta)$$

That is the force associated with one magnet, which has total area (both sides, magnet plus associated poles) of

$$\text{Area} = 2w_p(\ell_m + \ell_p)$$

This implies a maximum developed shear of

$$\tau = \frac{B_s^2}{2\mu_0} \, 2\alpha(1 - \alpha) \left(\frac{h_m}{l_m + l + p} \right) \left(\frac{l_m}{\ell_m + \ell_p} \right) (1 + \beta)$$

There are no good rules nor limits for these dimensional parameters, but suppose $\ell_m/(\ell_m + \ell + p) = \frac{1}{3}$, $h_m/(\ell_m + \ell_p) = \frac{5}{3}$ and $\beta = \frac{1}{2}$ (which are not out of the ordinary). Then the saturation limit is reached when the peak force density is

$$\tau = \frac{5}{12} \frac{B_s^2}{2\mu_0}$$

This would correspond with nearly 540 kPa or 78 PSI.

The literature suggests somewhat lower limits—on the order of 135 kPa.

8.9.6 Suggestions for further reading on fractional-horsepower-motor applications

Miller, R.; *Fractional Horsepower Electric Motors;* New York, Macmillan, 1984.

Standards for Fractional Horsepower Motors; *NEMA Publ.* MG2, 1951, New York, National Electrical Manufacturers Association.

Veinott, Cyril G.; *Fractional Horsepower Electric Motors,* 2d ed.; New York, McGraw-Hill Book Company, 1948.

Veinott, C. G., and Martin, J. E.; *Fractional and Subfractional Horsepower Electric Motors;* New York, McGraw-Hill Book Company, 1986.

Motor Noise and Product Sound

R. Lyon

9.1 The Role of Sound In Product Acceptance

We often think of the sound of products in general, and of motors in particular, as "noise"—unpleasant and unwanted. People want to buy motors that are quiet, and designers and manufacturers seek to provide them. However, the goals for the sound of the motor, or the complete product can be elusive. Sounds identify products and inform us about how they are working. The sound of an automobile can conjure up an image of a sportscar of the 1950's and 1960's, motorcycle sounds are used to advertise and even represent a particular brand, and the soft sound of a room ceiling fan is said to establish a safe and comfortable home environment. The sound of a new car door closure reinforces our satisfaction with owning a quality product. Sound is so much more than noise.

Of course, unwanted noise is indeed a feature of product sound, and electric motors are often a major ingredient of that sound. The sound of a vacuum cleaner has a large component due to motor noise, a sound that is annoying and at the same time conveys a perception of greater cleaning power. The motor and fan in a room air cleaner or computer are doing a useful job, but over a long period of time, they can lead to annoyance and dissatisfaction with the product.

We also use product sounds to detect problems. A motor with squeaking brushes will quickly convey a message of poor quality. A projector that has a 120 Hz hum broadcasts cheap parts and construction. Motor sounds are both wanted and unwanted, and the first step for us is to be able to tell the difference.

9.2   Descriptions and Causes of Motor Sounds

A designer must be concerned with the physical causes for the various sources of motor vibration and sound, and their correlation with the subjective reaction to the sounds. One aspect of the subjective reaction is the words that people use to describe motor sounds. We do not expect the verbal descriptions to necessarily uniquely define the source of sound. Words like *whine, squeal, ticking, buzz,* and *hum* are used in the motor industry as well as by the general population. If there were a one-to-one correspondence between such descriptions and physical causes for the sound, our life would be somewhat easier, but the unfortunate fact is that sounds, due to different physical causes, may end up being described by the same word.

An example of this is the *whine* of a motor. Unless the motor is driving a gear (which can *whine* as well), *whine* often arises in a motor with commutator and brushes. It may be due to the fluctuating force between the brush and commutator, but it can also arise from electromagnetic force variations on the rotor and stator. In both cases, a resonance of the motor structure is excited by the force, leading to a perceptible increase in the sound as a certain speed is attained by the motor. Figures 9.1 and 9.2 show examples of *whine* caused by brush-commutator forces. Figure 9.1 shows a peak in noise due to a resonance between the motor and the stator, while Fig. 9.2 shows a *whine* as the brush forces excite a resonance of the stator shell.

9.3   Sound Quality (SQ) and Jury Testing

The complex pattern of annoyance and meaning as reflected in product acceptability is referred to as *sound quality*. In some cases, the relationship between acceptability and sound is fairly simple. A regulation or industry code may specify that product sound not exceed a certain number of decibels, either in sound pressure level L_p, or sound power level L_w (see below for the definition of these quantities). In such a case, meeting the requirement is confirmed by a physical measurement, which may be cumbersome, but is well defined. However, if the requirement is "customers should like the sound," then a physical specification or measurement will not be adequate.

People's preferences for perfumes, foods, or sounds can be determined from the technique of *jury testing*. Table 9.1 lists the major components of a jury test of product sounds. A group of respondents (jurors) is chosen based on the characteristics of the customers for that product (age, gender, socio-economic status, experience, etc.). A group of sounds, anticipated from variations in the product design, is gen-

1kHz A:AC/ 2V B:AC/ 0.2V INST 0/64 DUAL .5k
UNIT: dBV

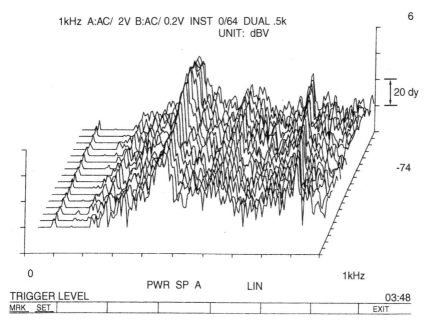

Figure 9.1 Waterfall spectrum shows amplified response as brush frequency coincides with motor dumbbell resonance at 420 Hz.

erated and recorded for later playback (the *stimulus set*). The jurors are asked to respond to these sounds in terms of various *perceived* product attributes, such as quality of construction, effectiveness in function, annoyance of sound, etc.

Commonly used response formats are *paired comparisons* (eg. "listen to the following two sounds and select the one that gives the impression of a higher quality product"), *fixed interval* testing ("judge the following sounds for *perceived quality* on a scale of 1 to 9), and *magnitude estimation*. Magnitude Estimation is less familiar but has advantages, particularly when the stimulus set (number of sounds to be compared) is large. It involves each respondent setting his/her own scale and responding accordingly. Then, at the end of the test, the respondent writes down his/her numerical values that correspond to major judgment categories (eg, the numbers corresponding to very high quality, moderately high quality, etc.) As a result, all juror responses can be rescaled to be consistent. A sample instruction set for magnitude estimation of *power* and *effectiveness* is shown in Exhibit A. The response form that corresponds is shown in Exhibit B.

The relationship between the subjective response and physical changes in the sounds from product components is obtained by subjecting the response data to *regression analysis* Regression analysis

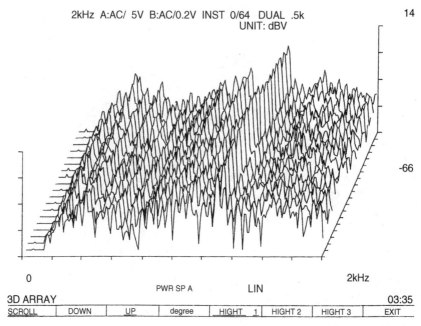

2kHz A:AC/ 5V B:AC/0.2V INST 0/64   DUAL .5k   14
UNIT: dBV

-66

0   2kHz
                    PWR SP A   LIN
3D ARRAY   03:35

| SCROLL | DOWN | UP | degree | HIGHT 1 | HIGHT 2 | HIGHT 3 | EXIT |

Figure 9.2 Waterfall spectrum shows amplified response as brush frequency coincides with shell resonance at 1300 Hz.

TABLE 9.1   Components of a Jury Test

1 Selection of jury members
  —what is the population (the cohort)?
  —are there special requirements (eg., hearing)?
2 Training the jury members
  —in scaling methods
    (magnitude estimation, fixed interval, paired comparisons)
  —calibration using standard attributes
    (eg., loudness)
3 Preparation of stimulus signals
  —component modifications
  —temporal/spectral modification
  —sequencing of stimuli
4 Design and conduct trials
  —form of presentation
    (headphones, speakers, mono, stereo)
  —scheduling, briefing, monitoring of listeners
  —assembling results
5 Data analysis
  scaling and normalizing data
  —statistical interpretation
  —presentation of results

Training in Magnitude Estimation: "The shapes you will see on the screen have different sizes. I want you to assign numbers to these shapes which indicate how large or small you think they are. Pick any whole positive number you want for the first shape. This number is going to be used to compare the size of the next shape you will see. Fill in the blank space on the sheet that is labelled with the same letter as that of the shape you are looking at. Write the number you have picked in that place.

Now look at the second shape on the screen. Your number for this shape should be assigned in proportion to how much larger or smaller you feel this one is. If it is the same size, your number should be the same. If it is twice as big, the number should be twice as big, etc. . . . ."

---

develops a functional (mathematical) relationship between the sound levels of various machine components and the psychological attribute. From these data, choices can then be made regarding modifications in the components, including all or some parts of the motor, which should lead to greater sound quality. The dependence of the attribute on a particular source component (motor brushes, airflow, gear noise) might be dropped from the function, either because the dependence is very weak (small value of coefficient in the regression function), or because a statistical test indicates low confidence in the derived value of the coefficient (jurors unable to make consistent judgments).

Figure 9.3 shows, for example, how the *acceptability* of a hypothetical vacuum cleaner is changed by alterations in the motor and airflow sounds. In this case, there is a combination of reductions in the component sounds (about 5 dB each) that is expected to optimize *acceptability*. In a study of the motor only, the sound sources might be brush-commutator interaction, cooling fan, bearings, imbalance and electromagnetic forces. Jury judgment categories might be *loudness, perceived power,* and *perceived quality*.

9.4 Noise Control (NC) vs. Design for SQ

The traditional approach to noise reduction as a part of product design has been "noise control," which we will contrast with "design for sound quality" as seen in Table 9.2. Noise control is the subject of a large number of texts and handbooks which emphasize add-on devices such as vibration isolators, damping materials, mufflers and enclosures, and related methods for reducing radiated sound. Because NC is add-on, it is generally after-the-fact, independent of the basic design, and readily costed independently. It is this after-the-fact character that allows one to speak of "dollars per dB." But if one can make such a

Exhibit B: Jury Response Form

Name:_____ ID# _____ | Group 1 |

Date: _____ Time:_____

1

RUN 5   BLK AF1

Power	Acceptability
1. _____	1. _____
2. _____	2. _____
3. _____	3. _____
4. _____	4. _____
5. _____	5. _____
6. _____	6. _____

2

RUN 5   BLK AF4

Power	Acceptability
1. _____	1. _____
2. _____	2. _____
3. _____	3. _____
4. _____	4. _____
5. _____	5. _____
6. _____	6. _____

3

RUN 5   BLK AF3

Power	Acceptability
1. _____	1. _____
2. _____	2. _____
3. _____	3. _____
4. _____	4. _____
5. _____	5. _____
6. _____	6. _____

4

RUN 5   BLK AF2

Power	Acceptability
1. _____	1. _____
2. _____	2. _____
3. _____	3. _____
4. _____	4. _____
5. _____	5. _____
6. _____	6. _____

5

RUN 5   BLK AF5

Power	Acceptability
1. _____	1. _____
2. _____	2. _____
3. _____	3. _____
4. _____	4. _____
5. _____	5. _____
6. _____	6. _____

*You have completed RUN 5.
Please turn to the next page.
Prepare to rate the power
and acceptability of the sounds
in RUN 6.*

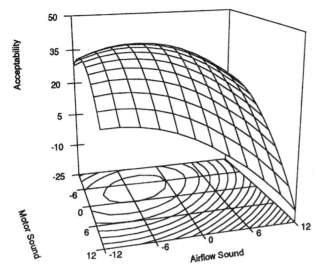

Figure 9.3 Subjective *acceptability* of vacuum cleaner sound as a function of motor and airflow noise.

TABLE 9.2   Comparison of Noise Control and Design for SQ

Noise Control	Design for Sound Quality
*Technology is in Place	*Principles are understood
*Implemented by Handbook	*Practitioners are Scarce
*Add-on after Product Design	*Integrated with Design Cycle
*Independent Design	*Highly Interactive
*Costing is Straightforward	*No Separate Costing
*May Reduce Reliability, Maintainability, Safety Utility	*May Enhance Function or Other Attributes

calculation, it indicates an after-the-fact approach, which is not desirable and adds expense, reduces maintainability and may also hinder maintenance.

From the manufacturer's viewpoint, add-on devices for NC have another disadvantage, in that they are a fixed cost for every produced unit. The cost of SQ design effort can be spread over all units, and the cost per unit therefore *decreases* as volume increases. Designing the reduced noise element into the product has more up-front cost, but may well cost less and result in a more competitive product in the long run.

Design for SQ integrates acoustical principles into the design process, along with materials choice, structural configuration, and lubrication. Because of the interactions among design parameters involved, an independent costing of design features for sound is not possible (nor is it for lubrication or structural design). The basic principles of relating mechanism forces and actions to the vibrations they induce and the sounds they radiate are known. These principles have to be applied differently though, if the noise source is electromagnetic as contrasted with bearing roughness or imbalance.

Still, incorporating acoustics into the basic design may result in enhanced performance in other areas. For example, perforating a motor cover to reduce its sound radiation may improve air cooling, and reduce material usage and weight. An example of this approach is presented below.

9.5 Measures of Sound and Vibration

Audible sound is a pressure fluctuation in the range of frequencies from about 50 to 15000 cycles per second (hertz or Hz). The conventional unit for pressure in acoustics is newtons per square meter (N/m^2 or pascals, Pa). It is common to express the pressure as a *level* in logarithmic units

$$L_p = 20 \log_{10} \frac{p}{p_{ref}} \; (dB).$$

The pressure p can, in fact, be any pressure, but in acoustics it is customary to use the measured or calculated root mean square (rms) pressure in the numerator. The reference pressure $p_{ref} = 2 \times 10^{-5}$ Pa, and a sound pressure level would be stated as X dB re 20 micro pascals (20 μPa), although the reference is usually not stated since it is universal in air (not so for vibration quantities). Typical values of sound pressure levels in the vicinity of a motor can range from 50 to 90 dB.

The total sound power radiated can also be expressed as a level in logarithmic units

$$L_w = 10 \log_{10} \frac{W}{W_{ref}} \; (dB)$$

where $W_{ref} = 10^{-12}$ watts (1 picowatt, or 1 pw)

is the accepted reference power in acoustics. Typical values of power level for the sound radiated by a motor range from 10^5 to 10^9 pw ($L_w = 50$ to 90 dB re 1 pw). Note that a rather high value of sound power level (90 dB) corresponds to a rather small amount of power (1 mw). A noisy motor may signal inefficiency, but not because of the power lost in sound energy!

The display of the way that sound pressure or power is distributed in frequency is called a spectrum. Typically, the contributions to power, or to the mean square (ms) pressure by different frequency components are grouped into frequency bands, which may be of fixed or proportional bandwidth as a function of frequency. Most "FFT" (fast fourier transform) analyzers calculate and present spectral data in constant (fixed) frequency bands, but this data can also be processed and presented in proportional bands.

The most common proportional (or constant percentage) band is the octave band. Each band is labelled by its geometric center frequency f_c and extends from $f_c/\sqrt{2}$ to $f_c\sqrt{2}$. The next band is centered at $2 f_c$, the next at $4 f_c$, etc. This frequency scale is presented in Fig. 9.4 using ISO/ANSI standard center frequencies. Octave bands are rather wide (about 71%), so if more frequency resolution is desired, one can present the data in third-octave bands, which extend from $f_c/\sqrt[6]{2}$ to $f_c\sqrt[6]{2}$, for a total bandwidth of $f_c\sqrt[3]{2}$. This corresponds to approximately a 23% bandwidth. The third-octave scale is also shown in Fig. 9.4, using the standardized band center frequencies. A radiated third-octave sound pressure spectrum is shown in Fig. 9.5.

When vibration data are to be graphed, the frequency spectra are typically presented using the same frequency bands as those for sound. As in the case for sound, vibration amplitudes are also expressed in logarithmic units, but the reference values are not very well standardized, and it is extremely important to indicate the reference quantity. For example, an acceleration level, calculated from $L_a = 20 \log_{10}(A/A_{ref})$, (dB), might use as a reference $A_{ref} = 1$ m/sec^2, 1 g (acceleration of gravity = 9.8 m/sec^2), or a variety of other values. That is why it is essential that vibration spectra presented in dB always indicate the reference quantity when absolute values of vibration amplitude are required. A sample third-octave spectrum of vibratory motor acceleration is shown in Fig. 9.6.

When motors encounter rapidly changing loads, or have changing operational conditions, the frequency spectra of vibrations or sound will also change. An example of this was shown in Figs. 9.1 and 9.2. Such presentations are called "waterfall" diagrams, and are possible because of the ability of modern analyzers to compute spectra for limited time windows very rapidly. Another form of such a presentation is the "spectrogram sometimes called a "sonogram" in underwater sound, or a "voiceprint" in speech analysis. An example of a spectrogram is presented in Fig. 9.7, showing intermittent "squeak" in a motor due to brush-commutator slip-stick (discussed below). In Fig. 9.7, the amplitude is represented by the changing darkness of the pattern. In other equivalent presentations, amplitudes are shown as color variations, or as contours as in a topographical map.

1/3 Octaves	Octaves
12.5	-
16	16
20	-
25	-
32	32
40	-
50	-
63	63
80	-
100	-
125	125
160	-
200	-
250	250
320	-
400	-
500	500
625	--
800	-
1000	1000
1250	-
1600	-
2000	2000
2500	-
3200	-
4000	4000
5000	-

Figure 9.4 Standardized center frequencies of octave and third-octave frequency bands.

9.6 Examples of Motor Sound Sources

Motor sound usually results from a force or motion that generates audible frequency energy, a structural response to that excitation (but not always), and the radiation of sound. We shall deal with these in sequence, first identifying some of the sources in this section, examples of response in the next, and then subsequently radiation mechanisms. It is useful to think of excitations as "dynamic" (specified force), or "kinematic" (specified motion), much as in electrical engi-

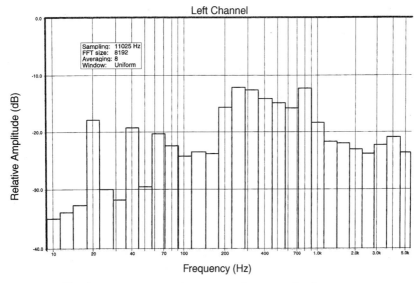

Figure 9.5 Third-octave spectrum of motor sound pressure level.

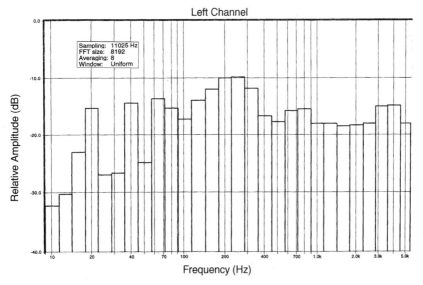

Figure 9.6 Third-octave spectrum of motor acceleration level.

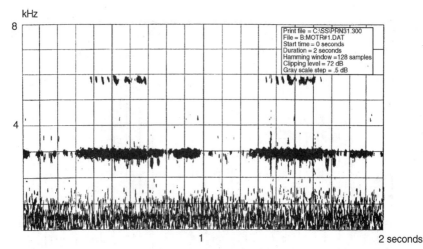

Figure 9.7 Spectrogram of a motor with squeaking brush-commutator interface.

neering where one thinks of sources as specified voltage or specified current, although as in electrical engineering, actual physical sources only approximate these idealized limits. Because the ways of ameliorating the excitation depend on whether it is dynamic or kinematic, such a division has practical consequences.

9.6.1 Imbalance

Mass imbalance in motors is usually a *rotating* imbalance, caused by an asymmetric distribution of mass about the axis of rotation. It can usually be resolved as two rotating lumped masses at two locations along that axis, and can be eliminated by "two-plane" or dynamic balancing. Most motor manufacturers carry out a single-plane or static balance of the rotor. This will leave an unresolved imbalance in the rotor, and vibration at the rotation frequency.

The method of balance usually involves adding bits of epoxy material to the rotor windings (if they exist), or grinding or drilling away bits of the rotor material. The latter method can be a source of noise itself, since it can be a source of additional "magnetic runout" (see below).

The effect of imbalance on the vibration spectrum is quite simply a line at the rotation frequency with some weak harmonics (20 to 40 dB down from the fundamental), as long as the imbalance does not excite other sources of vibration, such as rattling. The rotating imbalance in motors can be very significant at the rotation frequency, but at the harmonics or multiples of shaft rate, the amplitudes should die away quite rapidly—of the order of 20 dB per harmonic. If they do not, that is usually an indication of another source operating, such as magnetic runout.

9.6.2 Magnetic runout

Magnetic runout is a variation in the mutual (paired) force between the rotor and stator during a cycle of rotation. This may be caused by the rotor not being centered in the magnetic field, or any other variation in geometry, anisotropic material properties, or winding currents that cause flux linkage and the resulting forces to vary as the armature rotates. In general, these variations will have both spatial and temporal harmonics, leading to vibration harmonics at multiples of the shaft rate, with other frequencies associated with the temporal harmonics. Since the magnitude and phase of these forces are relatively unaffected by any vibrational response that they induce, we can regard them as a "dynamic" source of vibration.

An example of the sound spectrum of an inexpensive universal motor, typical of those used in consumer appliances, is shown in Fig. 9.8. The shaft rate is shown as f_{rot}, and several harmonics are clearly evident in the spectrum. Since this motor operates from 60 Hz power, the electromagnetic forces are temporally modulated by 120 Hz—the rate of variation in the magnetic forces. This creates "sidebands"

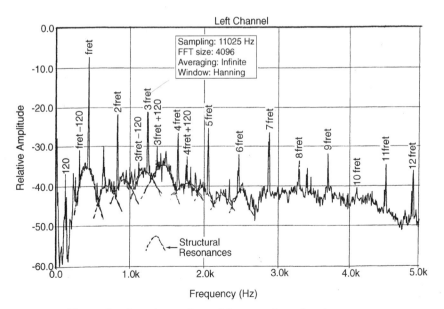

Figure 9.8 Narrow-band spectrum of sound from a universal motor.

around the harmonics of the rotation rate. We also see an underlying "foundation" to the spectrum, identified as structural resonances. We shall revisit this matter in the following section.

The paired electromagnetic forces that arise between stator and rotor can be quite complex, and have components in the radial, axial and circumferential directions. Analytical methods, supplemented by experiment, are useful in relating these forces to geometric and electrical design. Analytical methods include both fourier spectral calculations of the fields and currents, and finite element models for the magnetic circuits. Experimental methods use the vibrations induced by these forces, along with structural response data to "back out" the forces creating these vibrations.

9.6.3 Rotor slip

The rotating magnetic field in an induction motor revolves faster than the rotor, and the difference in these frequencies is the rotor slip frequency. Rotor slip results in a form of magnetic force variation in these motors. As the stator field moves through the rotor bars, induced currents produce forces that vary in time if the rotating field is not spatially uniform. These forces occur at the slip frequency and its harmonics. A particularly strong harmonic is found by multiplying the slip frequency times the number of rotor bars.

Although the slip frequency is so low in many induction motors that the rotor slip does not present a noise problem, it can produce audible sound in small split phase (such as shaded-pole) motors where the slip frequency is large. In some cases, the slip frequency can be about half the power supply frequency, and if there are many rotor bars, the vibration induced can generate audible sound. An example is shown in Fig. 9.9, which shows a spectrum of the vibration induced by this effect in a small shaded-pole fan motor.

9.6.4 Magnetostriction

Magnetostriction is a coupling of material strain to magnetic polarization that is common in ferromagnetic materials, and in certain other compounds as well (Terfenol D is a particular example). The laminations of a motor are particularly susceptible to this effect, and cause contraction and expansion of the magnet stack as the field varies in strength. If the field alternates, then the strain will have a double frequency component (typically 120 Hz). There will also be a vibration at the power line frequency if there is a strong bias field, as in a permanent magnet motor. Although magnetostriction will occur in a solid material, the vibrations can be significantly greater if the lamination stack is not rigid.

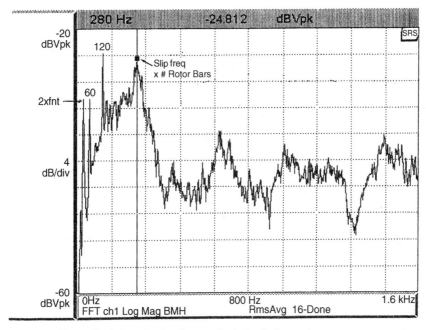

Figure 9.9   Rotor slip-induced noise from a shaded-pole fan motor.

This magnetostriction induced component can be seen in the spectrum of motor noise in Fig. 9.8. The frequency peak at 120 Hz is due to magnetostriction, but the motor itself does not have a structural resonance at such a low frequency and it is very small compared to the wavelength of sound (~2.9m) at 120 Hz. Therefore, any sound radiation at 120 Hz is very likely due to the vibrational response of the structure to which the motor is attached.

9.6.5 Brush noise

Universal and dc motors with brush and commutator elements have geometric and frictional variations at their interface that lead to dynamic forces, vibration, and noise. There is a "linear" version or component to this sound that is due to the geometric variations of the commutator profile that the brush follows. This source is kinematic in the sense discussed above. Of course, the brush may also bounce on the commutator. Both effects lead to brush motion that produces vibrations in the brush holder at mid and high frequencies. These vibrations may be enhanced by structural resonances in the motor shell or the armature (see the discussion of "whine" below).

The back end of the brush also radiates sound directly as it vibrates—generally above 4 to 5 kHz. This end of the brush acts like a small piston, and since it is small, it is a very inefficient radiator at lower frequencies, but as the sound wavelength becomes shorter (~7 cm at 5000 Hz), the brush is large enough to become a good radiator. An example of sound radiation computed from brush vibration, and compared with measured sound radiation from a motor is shown in Fig. 9.10. This type of sound is also referred to as "whine," but is not related to any resonance of the motor structure since it involves only brush motion.

9.6.6 Brush squeak

Under certain circumstances, a motor may squeak because of a brush-commutator interaction called "slip-stick"—a phenomenon associated with squeaking of brakes or fingernails on a chalk board. When slip-stick can couple into a structural resonance, the amplitudes of the vibration and the sound are greatly enhanced. When brakes squeal, the drum or disk supplies the structural resonance. Exhibit C describes the mechanism by which the sliding motion feeds energy into

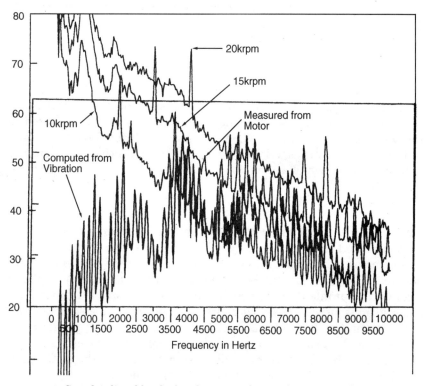

Figure 9.10 Sound radiated by the brush compared to total motor sound.

Exhibit C: Mechanism for Slip-Stick Oscillation

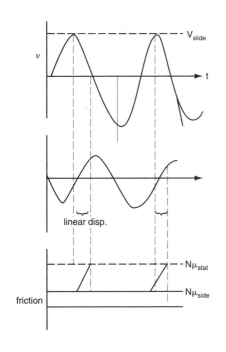

$$\Delta E = \frac{(\Delta\mu \cdot N)^2}{2K} \qquad \Delta\mu = \mu_{\text{sfd}} - \mu_{\text{slide}}$$

$$\Pi_{\text{in}} = f \cdot \Delta E = \frac{1}{4\pi} \frac{(N\Delta\mu)^2}{\omega M}$$

$$\Pi_{\text{diss}} = \omega\eta M\langle N^2\rangle = \omega\eta M \frac{V_{\text{slide}}^2}{2}$$

for oscillation $\Pi_{\text{in}} > \Pi_{\text{diss}}$ $(N\Delta\mu)^2 > 4\pi\omega^2 M^2 V_{\text{slide}}^2 \eta/2$

or $N\Delta\mu > \omega M V_{\text{slide}} \sqrt{2\pi\eta}$

There is a critical sliding velocity $V_{\text{crit}} = \dfrac{N\Delta\mu}{\omega M\sqrt{2\pi\eta}}$

vibration. The Exhibit shows a case where the static and sliding friction values are constant, but any situation where the frictional force decreases as sliding velocity increases can lead to oscillations that we might refer to as stick-slip.

The resonance that couples into squeak must cause vibration that is parallel to the direction of sliding. In a motor, this will typically involve a torsional mode of the rotor. Figure 9.7 shows the squeak of a motor at about 2.7 kHz, with a secondary response at about 4.1 kHz. Figure 9.11 shows torsional resonance frequencies for the rotor of this motor that coincide with the spectral peaks in the squeak sound.

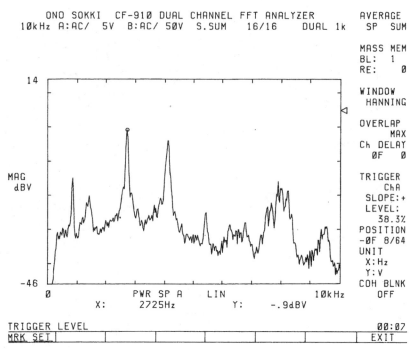

Figure 9.11 Impact response of motor armature showing strong torsional resonances at 2.7 and 4.1 KHz.

These resonances are excited by the sliding interaction between the brush and the commutator, as described in Exhibit C.

9.7 Structural Resonances and Amplification of Response

The structural response to the dynamic forces in a motor will be greater if the frequency of excitation is at or near a natural resonance of the motor structure. Usually this resonance is local to the motor itself, as described below. The higher vibration due to the resonance may radiate sound directly, or it may, in turn, cause the attached structure (base plate, frames, panels, etc.) to vibrate and radiate sound.

9.7.1 Motor whine

Motor whine is a highly tonal sound that results when periodic forces in the motor excite a resonance of the motor structure. These forces can result from various causes: brush-commutator interaction, magnetic forces on the rotor due to winding current or flux variations, and/or similar forces on the stator. It is because there are a number of possible causes that one must investigate whine very carefully to find a design solution when whine is a problem.

Although it is not necessary for a resonance of the structure to be excited in order for forces to generate sound, the sound will usually only be strong and intrusive when resonances are excited. We can distinguish three classes of resonances that are excited: "local" resonances of the stator, local resonances of the rotor, and mutual resonances in which both the rotor and stator participate. By a local resonance, we mean a vibration pattern that primarily involves only a portion of the overall structure. Such resonances are very common in all types of machinery.

A particular resonance of importance in this regard is the "dumbbell" resonance that involves both the rotor and the stator: the masses of these two elements "bouncing" against each other, while the rotor shaft, bearings, and bearing supports act as the springs. Typically, this resonance occurs at a few hundred hertz and may be referred to as "first critical." An example of the whine caused by this resonance is shown in Fig. 9.1, where the resonance occurs at about 420 Hz, and is excited at the commutation frequency (shaft rotation frequency × number of commutator bars) by the brush-commutator interaction forces.

Local resonances of the motor stator, including the magnet structure, the shell, brush plate, and end bells, can also participate in whine. Figure 9.2 shows whine that occurs at 1350 Hz due to a resonant mode of the stator shell. Again, this is excited by fluctuating forces between the brushes and the commutator. As the commutation frequency comes into coincidence with the resonance, the sound increases by more than 20 dB. Local resonances of the rotor are also excited by the forces mentioned, but because the rotor is more compact and massive, it does not vibrate with as large an amplitude. In addition, any sound that it radiates is enclosed by the motor shell, and the sound is blocked from getting out. It is therefore more difficult to find examples where rotor local resonances contribute to whine.

9.8 Nonlinear Vibrations and "Jumps" in Response

No excitation force or vibration in a motor is strictly a linear phenomenon, but much of the time the nonlinear effects are so weak that we can ignore them. There are cases, however, where the phenomenon depends strongly on a nonlinear characteristic of the response, and in such cases, we must deal with nonlinearities. One such example is a "jump" phenomenon in which the vibration (and sound) amplitude changes very rapidly, and in a non-reversible manner as the motor speed changes.

In a "dumbbell" resonance as described under *Motor Whine,* the spring comprises the rotor shaft bearings. Both hydraulic (oil-film journal) and roller element anti-friction bearings are nonlinear springs, and generally have a "hardening" characteristic as shown in Fig. 9.12. The frequency response of a mass-spring resonator with a hardening spring is shown in Fig. 9.13 for two degrees of nonlinearity. If the nonlinearity is strong enough, the response will drop off abruptly as the driving frequency is increased, as shown. As the driving frequency is decreased, the amplitude will abruptly increase, but at a frequency less than that of the downward jump.

An example of such behavior is presented here for a shop-type vacuum cleaner motor unit shown in Fig. 9.14. In this unit, the armature "hangs" from the motor housing, supported by a ball bearing. As the speed of the motor is slowly increased, the vibration generally increases, but abrupt downward transitions occur. As the motor speed is decreased, upward transitions occur at a succession of frequencies below those of the downward transitions as shown in Fig. 9.15. Although the motor armature hanging on the bearing represents a single dof resonator, multiple transitions occur because the harmonics of the running speed are able to excite the nonlinear resonance. In practice,

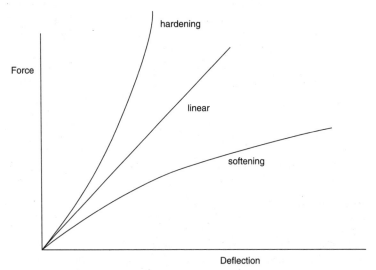

Figure 9.12 Force-deflection characteristics for linear, softening, and hardening springs.

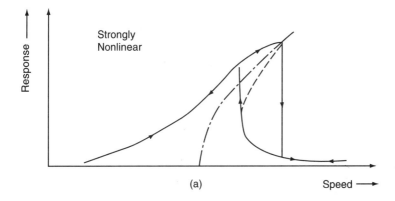

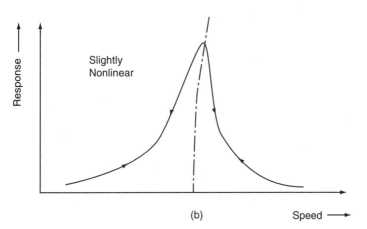

Figure 9.13 Frequency response curves for a hardening spring resonator.

only the transitions during motor slow-down can be heard because when power is applied, the motor speeds up too quickly for jumps to occur. If nonlinearity is the problem, then linearity may be the solution. In the motor shown in Fig. 9.14, a resilient piece of elastomer (an O-ring) is placed between the stationary outer race of the ball bearing and the motor housing bearing support to linearize the hardening ball bearing spring. The vibration as a function of motor speed with this modification in place is shown in Fig. 9.16.

This is only one of a number of potential nonlinear effects in motors, but in all cases, the message is "find out *why* the nonlinear effect is occurring, and find a way to reduce its influence."

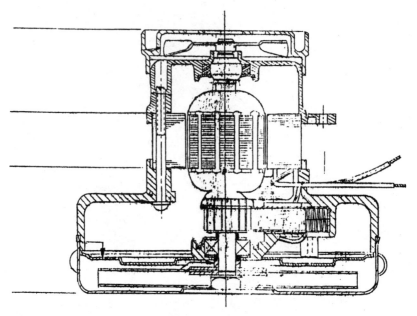

Figure 9.14 Sketch of a shop-type vacuum cleaner motor with armature supported by a ball bearing.

9.9   Reducing Sound Radiation by Decoupling from the Air

Sound radiation from a vibrating surface can be thought of as a product of vibrational response and coupling to the air. Mathematically, this can be expressed as follows

$$W_{rad} = \rho c A \langle v^2 \rangle_{x,t} \, \sigma_{rad}$$

where ρ is the density of air, c is the speed of sound (the product ρc is 407 mks units), A is the area of the radiating surface, $\langle v^2 \rangle_{x,t}$ is the space-time mean square (ms) velocity on the surface, and σ_{rad} is the "radiation efficiency" of the structure. The vibration of the structure's surface can be reduced by applying damping (which only affects the resonant amplification) and/or by applying a treatment to the surface that isolates (decouples) the vibrating surface from the air. Or, the decoupling can be accomplished by replacing a solid surface with a perforated surface, or its equivalent (perforated plate, expanded metal, truss structure, etc.).

9.9.1   Isolating and damping treatments

These treatments are usually "add-ons" applied to a structure to isolate the vibrations of the structure from the surrounding air. An il-

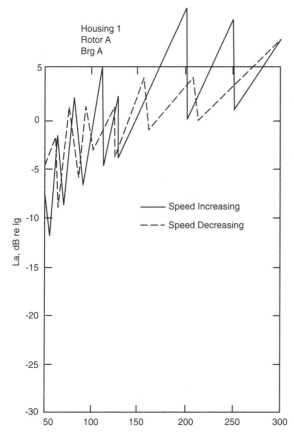

Figure 9.15 Vibration of shop-type vacuum cleaner vs. motor speed — shows jumps.

lustration of such a treatment is shown in Fig. 9.17. This particular design includes a "free layer" damping element which adheres directly to the structure. The thickness of a free layer damping treatment should be at least 50% of the structure thickness for plastic, and 90 to 200% of the structure thickness for steel or aluminum.

The layer of plastic foam and the outer septum form a mass-spring resonator. The septum will vibrate less than will the structure above the resonance frequency of this resonator. A typical transfer function relating septum to structure vibration for such a decoupler is shown in Fig. 9.18. One will usually realize a useful amount of sound reduction at frequencies more than an octave above the resonant frequency. The providers of these materials have charts that indicate the amount of reduction that can be achieved in practice.

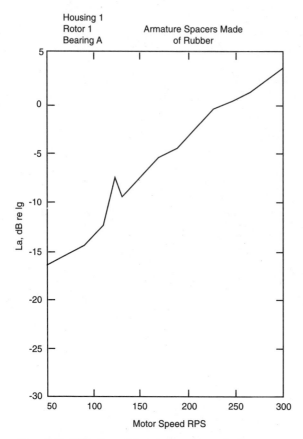

Figure 9.16 Vibration of shop-type vacuum cleaner vs. motor speed when bearing stiffness is linearized.

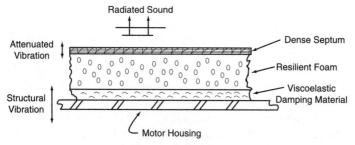

Figure 9.17 Sketch of a damping and isolation treatment for reducing radiation.

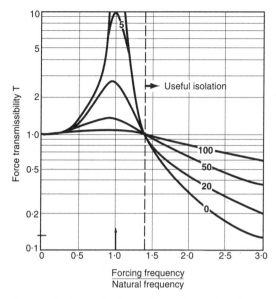

Figure 9.18 Theoretical vibration and sound radiation reduction by a foam–septum treatment.

9.9.2 Decoupling by perforation

If the solid surface of a motor is perforated, then its vibration amplitudes are not reduced (they may be increased slightly), but the ability of the structure to compress the surrounding air (and therefore generate sound) is reduced. This amounts to a reduction in σ_{rad}, the radiation efficiency. "Perforation" can be achieved by actual perforation of a solid sheet, or equivalently, by using expanded metal or a truss structure of some kind. Typically, a 9% or greater proportion of open area is able to achieve a useful amount of reduction.

Figure 9.19 sketches a brushless dc pwm motor of the type used in computer disk drives. The end bell of this unit that holds the stator coils was perforated as shown in Fig. 9.20. The result of this change on the radiated sound is shown in Fig. 9.21, and indicates more than 15 dB reduction in radiated sound due to the perforations.

Perforation of a motor housing can have other beneficial effects, such as improved cooling because the air can circulate better through the motor. But if oil, gases, or dust can pass through the openings and cause trouble, then perforation may not be possible. Therefore, perforation has to be regarded as a "target of opportunity," effective when it can be used, but it may not be available in many situations.

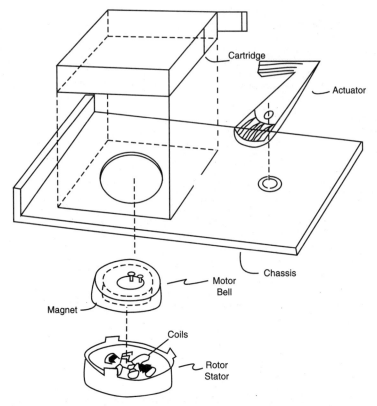

Figure 9.19   Sketch of brushless DC motor used in disk drives.

9.10   Product SQ as a Company-Wide Activity

Achieving product SQ on a consistent basis, like any product quality achievement, must result from a company-wide awareness and commitment. A diagram indicating the functional components of a product company is shown in Exhibit D. Below is a discussion of the role of each of these in a company-wide SQ program.

9.10.1   Marketing

Marketing has the job of looking "both-ways"—back to customer acceptance of products purchased, and forward to what customers will want to buy in the future. The comments customers make to dealers, correspondence with manufacturers, and service calls, etc. will indicate if there may be an SQ problem with the product. This understanding can be sharpened in focus group studies. If the study is properly designed, it can indicate where sound stands in the rank of important

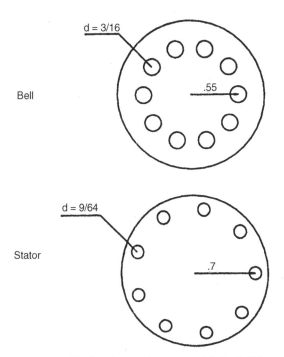

Figure 9.20 Perforation pattern on end bell of disk-drive motor to decouple vibrations from air.

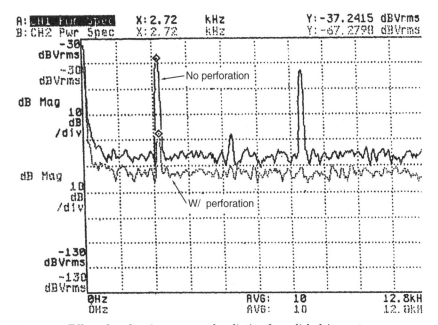

Figure 9.21 Effect of perforations on sound radiation from disk-drive motor.

Exhibit D: Functional Components of a Product Company

product attributes, and indicate also the kinds of words and descriptors customers apply to favorable and unfavorable sounds. Such information can be invaluable later on if a jury study is undertaken.

9.10.2 Product planning

In dealing with sound, product planners have often made the mistake of trying to give quantitative goals for SQ when they have no way of knowing if such a goal will really increase customer satisfaction. Unless a governmental or other regulatory requirement is to be met, or one wishes to counter an advertising claim by a competitor, giving a numerical value for a sound level to be achieved is not likely to be useful. "Our old model is 76 dB, so the new model should be 72 dB" is, unfortunately, Product Planning's too common response to a concern for noise. A statement such as "The new XP-9 motor needs to sound more powerful and less annoying than XP-8" may cause consternation to Engineering, but it says clearly what Product Planning wants. It should then be up to Engineering, perhaps with Product Planning's help, to decide how that goal is to be met with a new design.

9.10.3 Product design

Product design is the engineering function that converts the desires of Product Planning into a working prototype. The "soft-specs" of Product Planning must be converted into the "hard-specs" that the design must accomplish. One of the ways this is done is to run jury tests that will provide the kind of information that is shown in Fig. 9.3 in the case of a vacuum cleaner, but of course specialized to the product in question. A jury test can provide specific goals for component noise levels that should meet Product Planning requirements.

9.10.4 Manufacturing engineering

It would be very unusual for the design engineer's prototype to be a manufacturable product. As manufacturing engineers modify the design for production, it may occur that changes in materials or configuration, seemingly innocuous, will have large effects on sound radiation. A lightweight and stiff structure, desirable for weight and cost, may turn out to be a wonderful loudspeaker. Good communication between engineers in Design and Manufacturing can be critical in avoiding and catching these problems.

9.10.5 Manufacturing

Even well designed products can have SQ problems because of variations introduced in Manufacturing. Motor brushes with an incorrect spring setup force may squeak, or an armature with grinding cuts needed for balance may have too much magnetic runout. When problems occur, there is often a tendency to start changing parts, usually because the old parts did not exhibit the problem. This can reveal a design deficiency in that the design is not robust in providing stable behavior when variations in parameters occur. It is important that Manufacturing understand those features that are in the product for SQ, and not change those features without consultation.

9.10.6 Quality assurance

A design for good SQ will have certain features built in that can be subverted by the inevitable variations introduced in Manufacturing. Purchased components must meet specs that relate to acoustical performance. Manufacturing processes should not introduce variations that lead to unwanted sound. Whether the QA system is built around 90% testing or sample and audit procedures, processes need to be put in place to protect the final product from being compromised in its acoustical performance.

Servomechanical Power-Electronic Motor Drives

S. Leeb

A modern actively controlled motor drive is a system that combines a power-electronic circuit and a motor. The electronics in the drive system control the behavior of the mechanical shaft of the motor. For example, the electronics might be configured to provide the possibility of directly varying the shaft speed or torque based on a low-power, "signal level" command input. The earliest variable-speed drives avoided the need for (the unavailable) highly capable power electronic devices by cascading motors and generators, i.e., an M-G set. In an M-G set, a prime mover, such as a steam turbine or an induction motor running from a fixed frequency ac source, would be used to turn the shaft of a generator. The generator's terminal voltage could be controlled through a relatively low-power field or control winding. The generator's variable terminal voltage could then be used to alter the speed of a motor powered by the generator.

An M-G set involves a significant investment in rotating electric machines. For very high-power applications, cascades of electric machines, e.g., a "Ward-Leonard" style system[1], are still occasionally used or maintained. The advent of high-performance power-electronic components, however, has generally made it possible to develop electrical terminal drives that can directly control the flow of power through a motor from an electrical source to the machine shaft. Power-electronic circuits permit the construction of compact and highly efficient motor drive systems that are increasingly used in applications from the milliwatt to the megawatt range, including variable fan drives in ventilation systems, pumps and compressors, spindle controls in machine tools, and wheel or shaft drives in electric or hybrid vehicles.

A power-electronic circuit can control either the voltage or the current waveform applied to the terminals of a motor. It can control two basic features of this waveform: the average value or magnitude of the waveform, and the overall frequency content or shape of the waveform. This means that the circuit can drive the machine to control the shaft torque, speed, or position. For ac machines like the induction motor and the permanent-magnet synchronous machine ("brushless" dc motor), the power circuit also provides a specific waveform shape necessary to sustain the conversion of electrical energy into mechanical energy. In the case of the brushless dc machine, for example, the power-electronic drive effectively serves as an electrical commutator that replaces the action of the mechanical commutator in a conventional dc machine.

The next section explores the distinction between linear amplifiers and high efficiency switching power amplifiers. The following section examines the trade-offs associated with current versus voltage drives in the specific context of the dc motor. Models of dc motor servomechanisms with current and voltage drives will be developed that will expose fundamental design issues associated with all types of dc and ac motor drive systems. The dc motor is used to explore these issues because of the simplicity of the circuitry needed to create a basic dc motor drive. The final section explores useful power-electronic circuits for operating ac machines, either open loop or with active feedback control.

10.1 Power Converters: Linear vs. Switching

The purpose of a power converter is to process or "condition" electrical power, by interfacing a system with a given electrical specification (voltage and/or current), with another system with a different specification. Ideally, this conditioning is accomplished with the use of low-loss components including semiconductor switches, inductors, and capacitors. Efficiency as a design objective is what distinguishes a switching power supply from a linear regulator: a linear regulator is designed to process power in one direction, from a large reservoir of power to a small consumer, while in the process wasting a (possibly substantial) fraction of the power drawn from the source to accomplish regulation. The high efficiency of switching regulators (typically over 70%, with the state of the art approaching 85 to 95%) makes them desirable for powering high-density loads with high-power requirements and difficult thermal management problems.

The first part of this section examines linear regulators that could, for example, be used to drive a dc motor. The next subsection examines the canonical switching cell dc-dc converter, first presented by Landsman[5]. The canonical cell serves as a unifying circuit structure from which the common dc-to-dc high-frequency switching converter designs, i.e., the buck, boost, buck-boost, and boost-buck (Cúk) converters, may be derived[6].

10.1.1 Linear converters

Figure 10.1(a) shows an "open-loop," linear amplifier that uses an NPN bipolar-junction transistor configured as an emitter-follower to provide a controlled voltage across a load. The circuit is called a "linear amplifier" because the transistor is operated in the forward-active region, i.e., the collector current i_c is linearly related to the base current i_b by a constant, β

$$i_c = \beta i_b$$

In Fig. 10.1, the load is a resistor with value R_{load}, but this circuit could also be used to drive the terminals of a permanent-magnet dc motor, for example. The voltage V_{ref} is the command voltage that sets the voltage to be applied to the load. Assuming that the transistor is in the forward-active region (not saturated or cut-off), and also making the typical assumption that the forward drop across the base-to-emitter junction diode is abut 0.6 volts (optimistically low for a power transistor), Kirchoff's voltage law around the loop formed by V_{ref}, the base-emitter junction of the transistor, and the load resistor reveals that

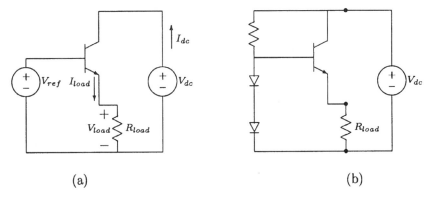

(a)   (b)

Figure 10.1 Linear voltage amplifier.

$$V_{\text{load}} = V_{\text{ref}} - 0.6$$

If the power transistor has a relatively high value of β (again, this is optimistic for a conventional power transistor; a Darlington combination of two transistors or a single Darlington power transistor would probably be essential in practice) then a negligible current in comparison with the load current flows out of the V_{ref} source. Hence, a low-power, signal-level voltage could be used to command the voltage across the load. This low-power voltage command might come from the digital-to-analog converter of a microcontroller, or from an operational amplifier circuit, for example. As long as we remember to account for the base-emitter voltage drop, the source V_{ref} essentially commands the load voltage. The bulk of the power provided to the load comes from the V_{dc} voltage source. The value of this source can fluctuate substantially without significantly affecting V_{load}, as long as the transistor stays in the forward-active region.

Assuming that the transistor β is large (100 or more), the base current is negligible compared with the load current, and Kirchoff's current law therefore reveals that

$$I_{\text{load}} \approx I_{\text{dc}}.$$

This approximation can be used to determine an upper bound on the efficiency of the amplifier. Efficiency is commonly defined as the ratio of the output power delivered to the load, to the input power:

$$\eta = \frac{P_{\text{out}}}{P_{\text{in}}} = \frac{V_{\text{load}} I_{\text{load}}}{V_{\text{dc}} I_{\text{dc}}}.$$

Recognizing that the load current and the input current are approximately equal

$$\eta \approx \frac{V_{\text{load}}}{V_{\text{dc}}} = \frac{V_{\text{out}}}{V_{\text{in}}}.$$

This is a general result for linear power amplifiers: the efficiency is bounded by the ratio of the output voltage over the input voltage. If, for example, the circuit is operating from an input source V_{dc} with a nominal voltage of 10 volts, while delivering five volts to the load (i.e., $V_{\text{ref}} \approx 5.6$ volts), the circuit will be 50% efficient. If it delivers 50 watts to the load (10 amps at five volts), it must draw about 100 watts from the input source. Where do the other 50 watts go? The power is dissipated as heat in the transistor! Hence, in this example, the case of the transistor would have to be connected to a heat sink adequate to dissipate 50 watts without allowing the temperature of the transistor to rise above its maximum specification.

In situations where the output voltage is fixed and known during the circuit design (i.e., a regulation, not a tracking application), and where the designer can pick the nominal value of an unregulated input voltage that is guaranteed not to fluctuate too much, it is possible to configure the linear amplifier so that its input voltage is only slightly above the output voltage. In this case, relatively high-efficiency operation can be achieved. However, this arrangement is unforgiving if the input voltage fluctuates substantially, or if a wide range of output voltage levels are needed. In general, therefore, most linear amplifiers operate with relatively poor efficiencies in comparison to switching power converters. This disappointing situation is common to all linear amplifiers. For example, a push-pull amplifier constructed from two transistors might be used to provide an ac voltage drive for a load. As long as the transistors in the push-pull stage are operated in the linear, forward-active region, the efficiency of the power stage will still be the ratio of the output and input voltages, and may be distressingly low from the standpoint of thermal management.

Nevertheless, for relatively low-power applications, i.e., around 100 watts, the thermal management of a linear amplifier does not pose insurmountable demands, and linear amplifiers are often used in many consumer applications, including stereo amplifiers and small motor drives. Both dc and ac linear drives are possible. Linear supplies may be especially valuable when electromagnetic interference (EMI) considerations are paramount or are perceived to be important, as in the case of audio amplifiers. They may serve as an effective way to generate a drive waveform with very low total harmonic distortion, for special machines that cannot tolerate high-harmonic components in the drive voltage (machine magnetic flux). They are also quick solutions for low-power (5 to 10 watts) regulation applications. For example, the popular 7800 series three-terminal voltage regulators essentially consist of the circuit shown in Fig. 10.1(b). The base of the power transistor is driven by a low-power voltage reference, shown in the figure as a stack of diodes. This could also be a zener diode, or another type of precision reference. Something like everything to the left of the dots in Fig. 10.1(b) is provided in the package of a three-terminal regulator, e.g., the LM7805 five-volt regulator[7]. The user provides the load and the input source V_{dc}.

The basic linear amplifier circuit can be modified to provide other improvements or operating modes. For example, a high-gain operational amplifier and a closed-loop feedback arrangement are used in Fig. 10.2 to ensure that the load voltage precisely follows the command voltage V_{ref}. The feedback loop minimizes the effect of the base-emitter voltage drop. This technique is used in many monolithic voltage regulators to ensure good output voltage regulation.

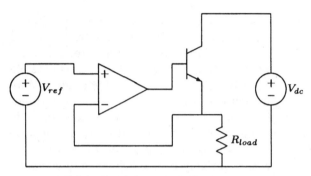

Figure 10.2 Voltage amplifier with feedback.

It is also possible to reconfigure the circuit as shown in Fig. 10.3 to provide a current-source amplifier. In this case, the transistor is arranged as a common-emitter amplifier. The command voltage V_{ref} sets the base current through the resistor R_b. The transistor draws a collector current through the load resistor that is β times the base current. Hence, in this circuit, the command voltage V_{ref} sets the load current, rather than the load voltage, with a gain that is a function of the value of the base resistor, R_b, and β.

10.1.2 DC-DC switching converters

A dc-dc switching regulator is a circuit that can provide a controlled or regulated dc output voltage from an unregulated dc input. It serves as an interface between two or more dc systems (hence the name dc-to-dc switching power supply), and can generally be designed to operate with high efficiency in comparison with a linear regulator. In the next section, for example, a "buck"-type switching regulator, or "chopper", will be used to provide a controllable voltage across the armature terminals of a dc motor, given an unregulated dc input voltage source.

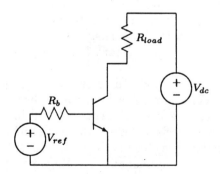

Figure 10.3 Current amplifier.

These converters might also be used to drive the field windings in ac or dc machines.

To get a better understanding of the distinction between different types of converters, consider how two dc systems might be interfaced. Start with a simple *two-port* model. This model is depicted in Fig. 10.4. The interface (the two-port box in Fig. 10.4) is presumed to contain no significant internal sources or sinks of power.

For now, no particular circuit topology is being derived here. Rather, the goal is to expose common features which relate several of the high-frequency, square-wave switching topologies. In this simple example, ignore any control or implementation issues. Also, assume that the terminal voltages and currents are constant with the polarities shown in Fig. 10.4. Under these conditions, the steady-state power sinked at one terminal must be sourced at the other terminal and vice versa. Notice that either port may source or sink power. Applying Kirchoff's current and voltage laws (KCL and KVL) reveals that the series switch must have average dc components of 50 volts across it and 15 amps through it. Similarly, the shunt switch must withstand an average of 150 dc volts across it and five amps through it.

Ideal switches provide a simple way to create consistent average waveforms in this configuration. They are particularly attractive because, ideally, when "off", they withstand a voltage without passing any current, and when "on", they pass a current with no voltage across the switch terminals. Hence, the product of voltage and current for the ideal switch is zero at all times and no power is dissipated.

To be consistent with KVL and KCL under the terminal conditions specified, the switches are operated to create voltage and current waveforms with appropriate average values. Specifically, during normal operation over a "switch period" of time T, the series switch turns on for a period of time $3T/4$. Then, this switch turns off and the shunt

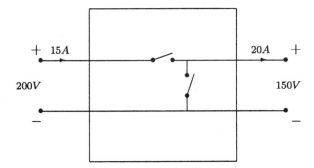

Figure 10.4 Simple dc-dc interface (Figure and example adapted from "Principles of Power Electrons"[6]).

switch turns on and remains on for a time $T/4$. At the end of this time, the shunt switch turns off and the cycle repeats itself. In the jargon of the trade, the series switch has a *duty cycle* of 75% and the shunt switch has a duty cycle of 25%. At no time are both switches on together, as this would result in a short circuit. Similarly, at no time are both switches open, as this would result in a cessation of power flow.

When the series switch is on, it carries a current of 20 amps to feed the 150 volt terminals. When the series switch is off, the shunt switch is on and the series switch is connected across the 200 volt input. Hence, the series switch withstands 200 volts. With the timing scheme described in the previous paragraph, we see that the current and voltage waveforms for the series switch are as shown in Figs. 10.5 and 10.6 respectively.

The average values of these waveforms are 15 amps and 50 volts respectively, as indicated in the figures (average variables are marked

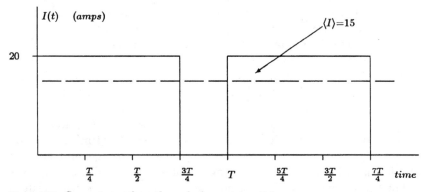

Figure 10.5 Current waveform through the series switch.

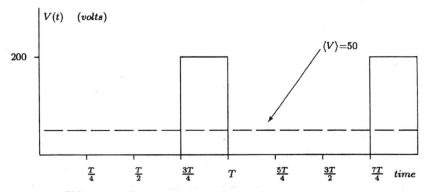

Figure 10.6 Voltage waveform across the series switch.

with ⟨⟩ symbols). A similar analysis of the shunt switch yields equally agreeable results. This simple switching scheme is apparently one option for interfacing two dc systems. Unfortunately, while the average values of the waveforms conform to their required values so that Kirchoff's laws are satisfied (on average), this scheme superimposes a substantial ac component to the terminal waveforms. Even though the waveforms in Figs 10.5 and 10.6 have the necessary average values, their instantaneous values *ripple*.

Filtering components must be added to smooth the waveforms. To maintain high efficiency, only inductors and capacitors are added. To help stiffen the bus voltage, a capacitor is added in parallel with the left terminals. An inductor is used to isolate the 150-volt terminal pair on the side with the shunt switch. Of course, the inductor will pass a continuous dc current, but will tend to block ac current. A revised version of the simple two-port interface is shown in Fig. 10.7.

This minimal topology "interface" is in fact the skeleton of four basic types of high-frequency dc-dc switching converters. This simple form, shown in Fig. 10.7, is so fundamental that it has been called the *canonical switching cell*[5]. It is often redrawn as shown in Fig. 10.8.[5,6]

As long as the switches are ideal (so that currents and voltage may flow or be blocked in any direction), power may flow bidirectionally through the circuit. By varying the interconnection of the designated

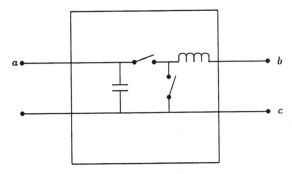

Figure 10.7 Simple topology for a dc-dc converter.

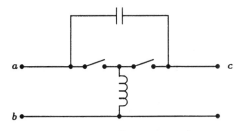

Figure 10.8 The canonical switching cell.

input and output terminals and the implementation of the ideal switches with real devices (which may or may not limit the flow of current or voltage in certain directions), all four basic high-frequency topologies may be derived. When the canonical cell is connected so as to allow power to flow directly from one terminal to another when one of the switches is closed, the buck or boost converters are formed. The type of converter formed depends on the direction of power flow; in the case of the buck converter, or chopper, power flows from a high-voltage source to a lower-voltage load when the controllable switch is closed. In the case of the boost converter, power flows from a low-voltage source to a load operating at a higher voltage when the diode is conducting. The classic buck and boost implementations of the canonical cell are shown in Figs. 10.9 and 10.10.

The braces in Figs. 10.9 and 10.10 indicate the canonical cell portions of the buck and boost converters. Note that actual implementations invariably make modifications to the canonical cell. For example, an extra output capacitor was added to make the buck topology shown in Fig. 10.9. For simplified discussions, the input capacitor is usually discarded from the buck configuration of the canonical cell, i.e., the voltage source is presumed to be "stiff," or to exhibit little internal impedance.

The buck-boost and boost-buck converters are implemented by configuring the canonical cell so that power never flows directly from one port terminal to another; energy is accumulated in one of the storage elements during part of the cycle and then is removed from the element to power the load on another part of the cycle. In the buck-boost

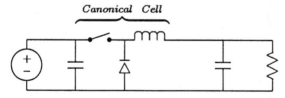

Figure 10.9   Buck converter.

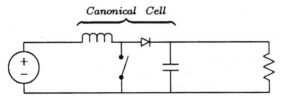

Figure 10.10   Boost converter.

topology, the most common of the "indirect" power supplies, the inductor is the intermediary energy-storage element. Also, the capacitance in the canonical cell is typically split between the input and output terminals. A simple buck-boost converter is shown in Fig. 10.11. A discussion of the less-common boost-buck converter is deferred to References 14, 16, and 22.

For now, all of the switching-power supply circuits have been drawn with a resistive load. Assuming that the switches, inductors, and capacitors in the power supplies are ideal, the resistive load is the only element in the circuits that can dissipate time-average power. This means that, in theory, any power taken from the input source is ultimately delivered to the load. In other words, the circuits are theoretically 100% efficient. Of course, the components would not be ideal in practice, but it is commonplace to achieve efficiencies in excess of 90% in actual implementations. The output voltage or current provided to the load can be actively controlled to a particular value or reference waveform by appropriately varying the duty cycle of the controllable switch. The next section examines a basic scheme for controlling the output voltage of a buck converter.

10.1.3 Buck converter with voltage control loop

To understand how a switching-power amplifier might be used in a servomechanical drive, consider the problem of making an output voltage that tracks a command reference using the buck converter shown in Fig. 10.12. This is identical to the task accomplished by the linear regulator shown in Fig. 10.2, whose output follows the command V_{ref}. Properly controlled, the buck converter should also be able to create

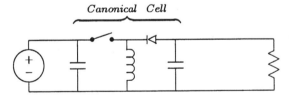

Canonical Cell

Figure 10.11 Buck-boost converter

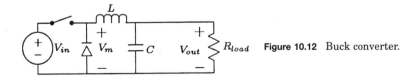

Figure 10.12 Buck converter.

an output voltage that follows a command reference, but with a greater efficiency over the full tracking range than was achieved with the linear amplifier.

Our goal is to control the output voltage of the converter. Given a particular input voltage source and a fixed switch frequency, our only possibility for controlling the output of the converter is to vary the duty cycle of the controllable switch. First, reconsider the open-loop behavior of the buck converter. Suppose, as in the previous section, that the controllable switch is operated with a fixed switch period and a duty cycle D that can be chosen. This approach is called pulse-width modulation (PWM), and is illustrated in the top plot in Fig. 10.13(a). The controllable switch is "on" for a time DT each cycle, and "off" for the remainder of the cycle. For now, focus on steady-state operation, and assume that the inductor is large enough that it is in continuous conduction—that is, there is always current in the inductor. When the switch is on, the input voltage will keep the diode reverse biased, i.e., "off". When the switch turns off, the inductor voltage will force the diode on in order to keep current flowing. Therefore, the instantaneous voltage across the diode, $V_m(t)$, looks like the bottom trace in Fig. 10.13(a).

When the load on the switches (e.g., the circuit driven by $V_m(t)$ in Fig. 10.12) is inherently low-pass in nature—that is, when the natural frequencies of the load are relatively slow compared with the switch frequency—the effect of a PWM drive is conveniently analyzed in terms of *average* variables. One definition for an average variable value in this context is

$$\langle w(t) \rangle = \frac{1}{T} \int_{t-T}^{t} w(\tau) d\tau \tag{10.1}$$

where $\langle w(t) \rangle$ indicates the average value of a variable $w(t)$ over the period T.

The waveform $V_m(t)$ can be thought of as the sum of two waveforms. One is the near-dc or average component $\langle V_m(t) \rangle = DV_{in}$. The remainder is $V_m(t) - \langle V_m(t) \rangle$, i.e., the "ac" component of the diode voltage waveform. If, for example, $D = 0.5$, then this ac component looks like a zero-centered square wave, with a maximum value of $V_{in}/2$ and a minimum value of $-V_{in}/2$.

In a well-designed buck converter, the values of the inductor and capacitor, L and C, will have been chosen so that, for a typical range of load values R, the transfer characteristic of the LRC low-pass filter formed by the "back-end" of the buck converter will look something like the magnitude Bode plot shown in Fig. 10.13(b). In other words, the filter will pass the low-frequency dc and near-dc components of

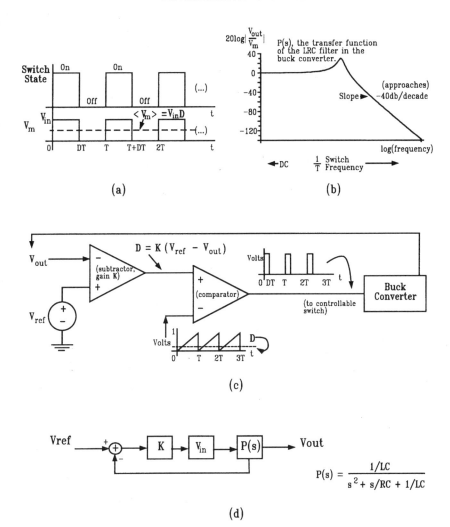

Figure 10.13 Buck converter: proportional control.

$V_m(t)$ straight through to the output. The values of L and C will have been selected, however, so that the ac components of $V_m(t)$ (generally at and above the switch frequency, $1/T$) will be severely attenuated by the low-pass filter. This means that $V_{out}(t)$ will look essentially like the near-dc $\langle V_m(t) \rangle$, with a small amount of AC ripple that passes through the second-order low-pass filter.

Setting a fixed duty cycle for the buck converter in Fig. 10.12 is therefore somewhat similar to setting V_{ref} in the linear amplifier shown in Fig. 10.1(a). Of course, the buck converter will generally operate with a higher efficiency than the linear regulator, unless special steps have been taken to limit the difference between the input and output voltage in the linear power supply. Notice that, contrary to the situation in the linear amplifier, the open-loop buck converter running with a fixed duty cycle provides absolutely no rejection or attenuation of variations at the output due to changes in the input source V_{in}. With a fixed value of D, the output voltage of the buck converter is essentially DV_{in}. If the input voltage changes, so will the output voltage.

This may not be of great concern if the buck converter is part of a servomechanical control system driving, for example, a dc motor. In this case, the buck converter and motor will be embedded in a closed loop attempting to control speed or some other mechanical quantity. Variations in the input voltage to the buck converter will be handled (by a well-designed control loop) as disturbances that get rejected.

If the buck converter is being used to drive a motor or some other mechanical or thermo-mechanical system such as a heating element, it may not be necessary to include the complete LC output filter shown in Fig. 10.12. Two motor-drive examples involving permanent-magnet dc motors are illustrated in Fig. 10.14. In Fig. 10.14(a), the low-pass output filter of the buck converter has been eliminated entirely. This circuit is sometimes called a "chopper". In this case, the dc motor will see an average voltage at its armature terminals equal to DV_{in}. It will also, of course, see a large ripple voltage around this average. However, if the motor is connected to a mechanical subsystem at its shaft that is substantially low-pass in character, e.g., a large inertia (fly-wheel) with some mechanical friction, the shaft speed will not respond to the rippling armature voltage if the switch frequency is high enough. It will only respond to the average value of the armature voltage set by the duty cycle command.

Taking advantage of a separation in time scales or time constants can be possible and valuable in other systems as well. Consider, for example, driving a resistive thermal heater to warm a large bath. Imagine that this load is suddenly driven with an input power of P watts by a linear, dc amplifier that provides "flat" waveforms of voltage and current. If the heater is warming a substantial mass, it might be

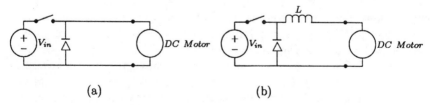

(a)   (b)

Figure 10.14 Chopper circuits.

several seconds or more before a measurable change in temperature is observed. Imagine that a switching amplifier is employed that provides $2P$ watts for half a switch period, and 0 watts for the other half, cycling periodically and rapidly in comparison with the time scale for response of the thermal mass. In this case, the net effect on the temperature of the mass would be essentially the same as if the amplifier delivered P watts continuously. The thermal system simply cannot follow the rapid changes in input power. Its "low-pass" character is incapable of following high-frequency variation in the input, averaging the effects of high- and low-input levels over time. Of course, this would not be the case if the switch period were not sufficiently short. A long switch period would provide significant time for the mass to "overheat" during the time when $2P$ watts were provided, and would also permit significant cooling of the mass during the fraction of the switch period when 0 watts were delivered. A practical example of this technique is switching with an inexpensive bimetallic strip to establish control for electric heating elements in residential kitchen stoves.

Even with a "low-pass" mechanical load, however, the situation in Fig. 10.14(a) may not be entirely acceptable, because the ripple voltage at the armature electrical terminals may cause a large ripple current to flow. This ripple current may cause large ripples in the shaft torque. If the mechanical system is slow, these ripples may result in only small oscillations of the speed. However, the large peak currents may exceed the switch current ratings in the chopper. They may also be responsible for unacceptable levels of acoustic noise form the motor windings, and could exceed the peak current specification of the armature electrical port. Generally, these problems show up most acutely in machines with relatively low armature impedances, e.g., high-performance *servodisk* motors. In such cases, one solution is to add sufficient inductance externally to the armature circuit to limit the voltage (current, torque) ripple, as shown in Fig. 10.14(b). External inductance must be added with care in the case of a dc motor in order to avoid creating unacceptable discharging in the mechanical commutator. Raising the switch frequency, if possible, can reduce the size of the impedance that must be added.

A simple switching circuit that might be employed as a bidirectional chopper is shown in Fig. 10.15. Notice that both switches can never be on at the same time, or a very low impedance path will form between the positive and negative 30-volt rails. The diodes provide "freewheeling" paths for current when the controllable switches are both off. These free-wheeling or "catch" diodes are essential if the load has non-zero inductance

Finally, if for some reason it is essential that the buck converter accurately follow an output voltage command reference and actively

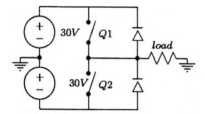

Figure 10.15 Bipolar chopper.

reject input voltage disturbances, it is possible to close a feedback loop that varies the duty cycle to achieve a desired output voltage waveform. An actively controlled converter could also serve as a component block in a larger seromechanism. One possibility for controlling the output voltage of the buck converter, shown for the case of a resistive load, is illustrated in Fig. 10.13(c). The duty cycle is computed by a *proportional* compensator as the error or difference between a command reference and the actual output voltage, times a gain K. This error computation could be accomplished, for example, by an operational amplifier configured as a subtractor with a gain of K. The duty cycle is compared with a sawtooth waveform by a comparator. The sawtooth has a period of T, a peak voltage of one volt, and a minimum voltage of zero volts. When the sawtooth waveform is below the value of the D output waveform computed by the subtractor, the output of the comparator is high. When the sawtooth rises above D, the comparator output goes low. Hence, the output of the comparator produces a PWM-drive waveform that could be used to directly drive a controllable switch (e.g., a transistor or MOSFET) in the buck converter.

The performance of the actively controlled buck converter can be analyzed with the feedback diagram shown in Fig. 10.13(d). This block diagram contains all of the functionality of Fig. 10.13(c). However, ripple voltage has been presumed to be small and is ignored in Fig. 10.13(d). That is, the block diagram describes only the relatively slowly varying dynamics (compared with the ripple frequency) of the output voltage. The difference between V_{ref} and V_{out} is computed by the subtractor, and fed to the proportional compensator with gain K to compute the duty cycle, D. The product of the duty cycle and the input voltage V_{in} produces $\langle V_m(t) \rangle$, which passes through the transfer function $P(s)$ of the LRC output filter of the buck converter shown in Fig. 10.12. Using Black's formula[8], the overall input-output transfer function for the actively controlled buck converter is

$$C(s) = \frac{V_{out}}{V_{ref}}(s) = \frac{\dfrac{KV_{in}}{LC}}{s^2 + \dfrac{s}{RC} + \dfrac{KV_{in} + 1}{LC}}$$

The second-order denominator of this transfer function has all positive coefficients as long as the circuit is loaded with a finite value of output resistance. Hence, the closed-loop system poles are guaranteed to be in the left half-s-plane (Routh criterion) and the system will be stable, in principle. For this proportional compensator, the steady-state error after a step change in the command output V_{ref} decreases as the gain K increases. However, the damping of the system also decreases as K increases. For larger values of K, the closed-loop system will therefore overshoot higher and oscillate more times than for lower gains. The proportional compensator is by no means the most sophisticated or flexible choice of series compensation for the buck converter, although it can produce reasonable steady-state performance. It is also possible to close a feedback loop around the buck converter that varies the duty cycle to ensure that a specific value of current is delivered to the load. That is, a current-source output could be made using the buck converter and an active control loop.

The next section will examine the effects of different power amplifier choices (current vs. voltage drive) in permanent-magnet dc motor velocity and position servo systems.

10.2 DC Motor Servomechanisms for Velocity and Position Control

Many electromechanical systems are tasked to provide precise control of a position or a velocity. A system could be required, for example, to *regulate* a position to a specific value or location, as in the case of a position controller for an antenna. It could also be required to accurately *track* a time-varying position or velocity reference, e.g., when following an aircraft with a radar dish, or serving as a drive motor in an electric vehicle. Direct-current machines are often used as actuators in such systems. Dc machines provide rotary motion and torque, and can also provide linear motion and force through clever mechanical arrangements such as lead-screw mechanisms.

Dc machines may be somewhat more expensive than comparably rated machines of other types (induction, stepping, and variable reluctance machines), and may also be more difficult to maintain. Because of limitations on the mechanical commutator, dc machines generally cannot be used at high altitudes or in a vacuum. However, from a power electronics standpoint, dc machines are relatively easy to control compared to many other motors, e.g., induction machines. Well-designed dc machines provide a smooth, nearly continuous motion with little "cogging," as would be found in stepper motors, for example. For these and other reasons, dc machines have been used in servomechanisms for over a century, and are still popular in many applications.

Because of the relative simplicity of the power electronics needed to drive a dc motor, this section reviews the basic issues of constructing velocity and position servomechanisms using the dc motor as an actuator. The following section will examine the power-electronic requirements for ac machines. However, the basic approach for controlling most ac machines is, to some extent, to first make them "look like" a commutator dc machine, at least in a mathematical sense. The basic approach to designing closed-loop controllers for position and velocity servos, therefore, is generally the same for both dc and also most ac machines (given an appropriate ac power amplifier).

10.2.1 Circuit analogue

This section will quickly review the basic wound-field dc-machine model which is described in more detail elsewhere in this text. The field winding in a dc machine is reasonably accurately modeled as a resistance in series with an inductance. The armature winding consists of wire coiled around a high-permeability rotor. The armature electrical port, therefore, has some electrical loss that could be modeled as a resistance in series with the back-electromotive force (back-EMF) source. It also has some inductance (typically small, in comparison to the field winding), which could be incorporated as an inductor in series with the resistor and back-EMF source. The mechanical shaft could incorporate elements to account for load torque, viscous damping, and other mechanical effects and components (e.g., a gear box). A reasonable circuit model of a dc machine is shown in Fig. 10.16. The voltage source labeled V_{bemf} in Fig. 10.16 is a speed-dependent voltage source that represents the back-EMF of the motor. The "source" is a transducer, therefore, connecting the mechanical subsystem (shaft of the motor) to the electrical subsystem (armature terminals). The torque produced by the shaft is a function of the field current and the armature current

$$\tau_m = GI_f I_a \tag{10.2}$$

The motor constant G is a function of the machine's construction (e.g.,

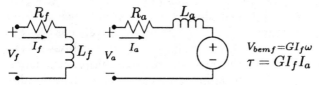

Figure 10.16 Circuit model.

number of armature turns, rotor radius, etc.). More complicated models accounting for nonlinear electrical effects like saturation and nonlinear mechanical effects like windage (air flow around the rotor) could be developed as needed. The model in Fig. 10.16 is often very satisfactory for control design, however.

Starting with the model summarized in Fig. 10.16, we can use basic circuit analysis techniques to develop a concise third-order dynamic (differential equation) model of the dc machine. The field circuit can be described by a first-order, linear, time-invariant differential equation relating the field current i_f to the field terminal voltage v_f

$$\frac{di_f}{dt} = \frac{1}{L_f} v_f - \frac{R_f}{L_f} i_f \tag{10.3}$$

The armature current i_a is described by a nonlinear differential equation that depends on both field current i_f and rotor speed ω

$$\frac{di_a}{dt} = \frac{1}{L_a} v_a - \frac{R_a}{L_a} i_a - \frac{Gi_f\omega}{L_a} \tag{10.4}$$

Finally, the mechanical subsystem is described by Newton's second law. The dc machine makes a machine torque τ_m described by Eq. 10.2. The motor shaft has inertia J and is connected to a load (motor operation) or prime mover (generator operation) that exerts a torque τ_l on the shaft. (This torque could be, and often is, a function of speed). For now, we ignore other possibly complicating details such as the presence of a gear box. Angular acceleration of the machine rotor results from differences between the two torques τ_m and τ_l

$$\frac{d\omega}{dt} = \frac{1}{J} (\tau_m - \tau_l) = \frac{1}{J} (Gi_f i_a - \tau_l) \tag{10.5}$$

One simple model for a common load torque is a friction τ_f that is linearly dependent on angular velocity, $\tau_f = \beta\omega$, where β is a constant.

In the case of a permanent-magnet (PM) dc machine, the field winding is replaced by a permanent magnet. This situation is functionally equivalent to driving the field circuit in Fig. 10.16 with a constant-current source of value I_f sufficient to create a comparable air-gap magnetic field to that produced by the magnet. The PM dc machine is accurately modeled by a second-order model consisting of Eqs. 10.4 and 10.5, with $i_f = I_f$. Block diagrams of a PM dc machine appropriate for control design are shown in Fig. 10.17. Laplace transforms of Eqs. 10.4 and 10.5 have been employed in the block diagrams, and the

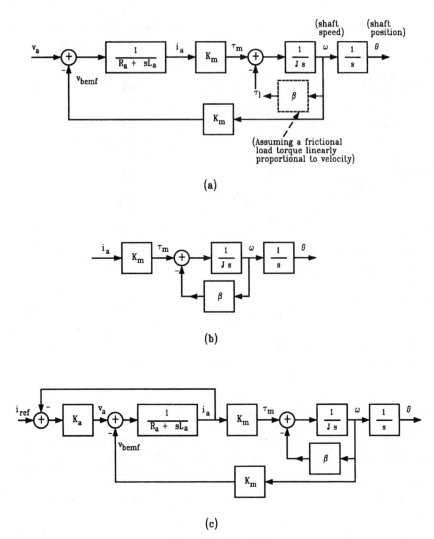

Figure 10.17 Permanent magnet dc machine block diagrams.

system is assumed to start from initial rest condition. The constant K_m equals GI_f, the product of the motor constant and the dc "field current" that represents the effect of the permanent magnet.

The first diagram in Fig. 10.17(a) shows how an ideal **voltage**-source drive on the armature would affect the shaft speed and position. A practical implementation of such a power amplifier could be made with a linear amplifier like the emitter-follower, or a switching amplifier like the chopper. Notice that, particularly in a high-quality machine with low armature resistance and inductance, a fixed armature voltage will approximately set the level of the steady-state back-EMF. Since the back-EMF depends on the shaft speed, setting the armature voltage to a particular value in a PM dc motor approximately sets the shaft *speed*. The next block diagram in Fig. 10.17(b) shows how an ideal **current** source driving the armature terminals produces torque, speed, and position at the machine's mechanical shaft. A current-source amplifier, in theory, directly commands armature current and therefore shaft *torque*. The details of the armature electrical circuit, i.e., armature resistance and inductance and motor back-EMF, are negligible to the extent that the current-source amplifier has sufficient voltage compliance to command current in the armature winding. The block diagram for the ideal current-source drive, therefore, is simpler than for the ideal voltage-source drive in Fig. 10.17(a).

In practice, it is not possible to drive the machine with an ideal current source. The power amplifier's ability to command current in a winding will always be limited by its ability to command an instantaneous terminal voltage necessary to set the current. This terminal voltage will not be infinite in magnitude or negligible in rise time, and the current command, therefore, cannot be arbitrary. This can be seen in any of the amplifiers—linear or switching—that have been examined in previous sections. Consider, for instance, the common-emitter amplifier shown in Fig. 10.3. Imagine that the load resistor is replaced by a winding that consists of some inductance and some resistance. In response to a step change in the base current, i.e., a sudden change in the command voltage V_{ref}, the best that the transistor can do is saturate, i.e., turn on "fully". The rise time of the current will then be governed by the L-R time constant of the load winding.

A more realistic model for a current-source drive, therefore, is shown in Fig. 10.17(c). Here, the current amplifier accepts a command labeled i_{ref}, which is compared to the actual armature current i_a. A proportional compensator with gain K_a drives an armature voltage v_a based on the error between the requested and actual current. This arrangement does not directly represent any of the circuits from previous sections, but it could be created with either the linear or switching amplifiers. It also does not account for nonlinear effects like saturation in the amplifier. However, the block diagram in Fig. 10.17(c) will serve to bring out some of the "real-life" issues that arise with a practical current-source amplifier.

10.2.2 Velocity servo–current-source drive

Because it is analytically easier to understand, let us begin by examining how to make a closed-loop system to control motor velocity using a current-source drive. A block diagram of a feedback loop for controlling motor speed with a current-source power amplifier is shown in Fig. 10.18(a). The gain of the tachometer is presumed to be one, but any constant gain or transfer function could be added if appropriate. The block with transfer function $Gc(s)$ represents a series compensator chosen to yield closed-loop stability and performance. This block includes the ideal current-source power amplifier. This section will consider several possible compensation options. First we consider a proportional compensator, i.e., $Gc(s)$ is a constant gain.

A circuit schematic of a unidirectional demonstration system based on the block diagram in Fig. 10.18(a) is shown in Fig. 10.19. The current-source amplifier is implemented with two NPN bipolar-junction transistors connected in a Darlington configuration. This arrangement

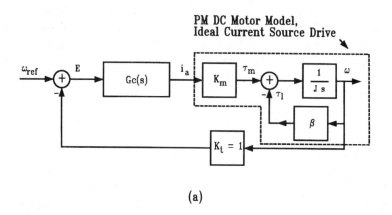

(a)

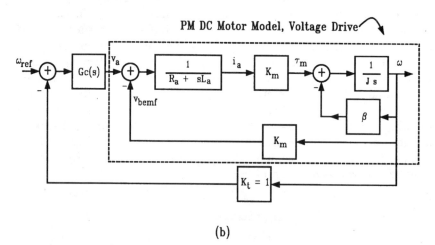

(b)

Figure 10.18 Velocity controllers with current-source and voltage-source drives.

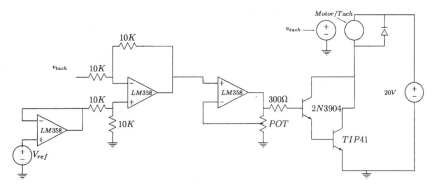

Figure 10.19   Unidirectional velocity servo: circuit schematic.

increases the overall current gain of the amplifier, minimizing the loading of the LM358 operational amplifier driving the transistors. The Darlington-connected transistors are used in a common-emitter arrangement with a "power" level (20 Volt) dc voltage source for the motor's armature circuit. A flyback or "catch" diode provides a free-wheeling path for the armature current in the event that the control loop is deactivated suddenly while the machine is running. Operational amplifiers implement the remaining functions in the feedback loop. An LM358 configured as a non-inverting amplifier provides a variable, proportional compensation gain. The error signal between the tachometer voltage v_{tach} and the speed reference V_{ref} is computed by an LM358 arranged as a subtractor. The voltage reference, which might be made by a potentiometer connected across the power supply rails, is buffered by an op-amp follower. The tachometer feedback signal v_{tach} might also be buffered by a follower if necessary (not shown).

The closed-loop transfer function for the velocity servo loop can be found by applying Black's formula[8] twice to the inner and outer loops shown in Fig. 10.18(a). For a proportional compensator, $Gc(s) = K$, a constant that includes the op-amp and transistor amplifier gains in Fig. 10.19. In this case, the transfer function relating motor speed to commanded speed is

$$\frac{\omega}{\omega_{\text{ref}}}(s) = \frac{\dfrac{KK_m}{\beta}}{\dfrac{J}{\beta}s + \left(1 + \dfrac{KK_m}{\beta}\right)} \tag{10.6}$$

Ideally, for a constant speed reference, this transfer function would approach unity in steady-state (zero frequency). That is, for dc excitations, the output should track the input or reference command perfectly. Unfortunately, for the proportional controller with a finite proportional gain, there is always a steady-state error or difference between the commanded and actual speeds.

In principle, the compensation gain could be increased indefinitely to make this error arbitrarily small. A typical root-locus of the closed-loop system pole locations for increasing positive gain, shown in Fig. 10.20, indicates that the system should remain stable as the gain is increased. Furthermore, a Bode plot of the magnitude and phase of the closed-loop transfer function, Eq. 10.6, indicates not only that the low-frequency behavior of the transfer function will approach unity as the compensation gain increases, but also that the closed-loop tracking bandwidth will increase. In practice, however, increasing the compensation gain indefinitely will most probably become seriously detrimental after a certain point. Higher closed-loop bandwidths demand excessive peak power requirements from the power amplifier. Also, as the gain is increased, we become more likely to "find" unmodeled poles. That is, the comforting stability argument made by the root-locus diagram is unlikely to be true when the effects of other, unmodeled system poles (e.g., poles from the tachometer and power amplifier) are included.

We could instead try an integral compensator, $Gc(s) = K/s$. The integral compensator provides a frequency-dependent gain that is, in

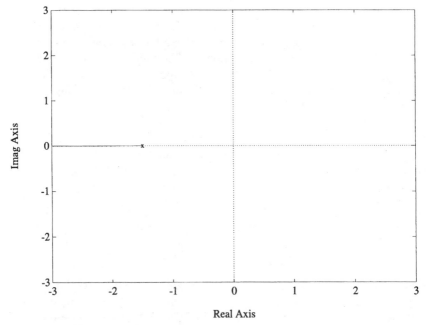

Figure 10.20 Root-locus: proportional compensation.

principle, infinite at zero frequency. In this case, the closed-loop transfer function is

$$\frac{\omega}{\omega_{\text{ref}}}(s) = \frac{\dfrac{KK_m}{\beta\alpha}}{s^2 + s\,\dfrac{1}{\alpha} + \dfrac{KK_m}{\beta\alpha}} \tag{10.7}$$

where

$$\alpha = \frac{J}{\beta}$$

Notice that the steady-state value of the closed-loop system under integral compensation is unity, i.e., the system tracks dc commands perfectly, regardless of the value of the gain K. The Routh criterion[8] can be used to show that the poles of Eq. 10.6 are always in the left half-s-plane, i.e., the system is always stable in principle. A root-locus of the system poles for positive integral gain K is shown in Fig. 10.21. With integral compensation, there is little advantage to excessively high values of the gain K. After a certain point, the closed-loop band-

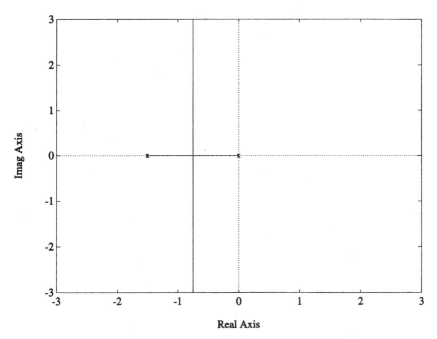

Figure 10.21   Root-locus: integral compensation.

width of the system essentially stops increasing, and the transient response of the system becomes more and more oscillatory.

There is a wide range of options for compensating this servo system to achieve desired performance; see References 10, 11, and 12 for a more complete discussion of the possibilities.

10.2.3 Velocity servo–voltage-source drive

Figure 10.18(b) shows a block diagram of a closed-loop velocity drive that employs a voltage-source amplifier to drive the motor. Because the amplifier is a voltage drive, the block diagram now shows explicit dependence on the armature impedance and the back-EMF. The loop with voltage drive will be shown to have an additional pole in comparison to a similarly compensated loop with current drive. This pole arises from the armature electrical subsystem.

The series compensator $Gc(s)$ includes the voltage amplifier. For now, assume that the system is compensated with a porportional gain, so that $Gc(s) = K$, a constant that includes the gain of the voltage-power amplifier. Employing Black's formula repeatedly on the block diagram in Fig. 10.18(b), the closed-loop transfer function can be shown to be

$$\frac{\omega}{\omega_{\text{ref}}}(s) = \frac{\dfrac{KK_m}{R_a\beta}}{(\tau_e s + 1)(\alpha s + 1) + \dfrac{K_m^2}{R_a\beta} + \dfrac{KK_m}{R_a\beta}} \tag{10.8}$$

The time constant $\tau_e = L_a/R_a$ is sometimes called the electrical time constant of the machine. In relatively small machines, this time constant may be small compared with the mechanical time constant α. If the electrical time constant is negligible (near 0), then the voltage velocity loop with proportional compensation is essentially first-order, and has a root-locus similar to that one shown in Fig. 10.20. If the electrical time constant is *not* ignored, then the root-locus will look something like the one shown in Fig. 10.22.

Whether or not the electrical time constant is ignored, the proportional compensator always leaves some steady-state error in response to a step input. To eliminate this steady-state error, we might again consider using an integral compensator, $Gc(s) = K/s$, where K is a gain to be chosen. Ignoring the electrical pole leads to a system that exhibits zero steady-state error in response to a step input, and which has a root-locus similar to the one shown in Fig. 10.21.

If we do not ignore the electrical pole, an integral compensator still leads to a closed-loop system with zero steady-state error in response to a step input. However, this system, with three poles (one from the

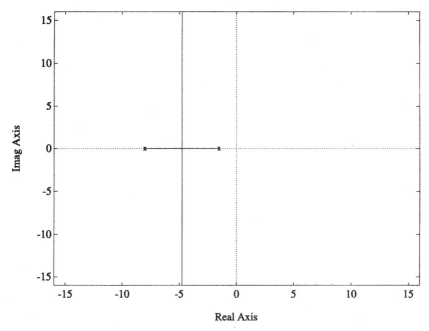

Figure 10.22   Root-locus: voltage loop with significant τ_e

integral compensator at the origin, one from the mechanical subsystem, and one from the electrical subsystem) will have a root-locus like the one shown in Fig. 10.23. For modest gains, the dominant poles (closest to the origin) of the system shown in Fig. 10.23 behave much like the dominant poles shown in Fig. 10.21. For higher gains, disaster ensues! If too high a gain K is selected, the system in Fig. 10.23 will be unstable. Even if the electrical pole is ignored during the design process, it or some other unmodeled pole is very likely to be present in that actual system. Great care must always be taken, therefore, when selecting compensator structures and gains. It is perhaps a too common practice to ignore the supposedly fast electrical poles in favor of the mechanical poles when designing a servomechanical drive system. Especially in very large machines, ac or dc, the electrical poles can actually be slower than the mechanical poles. Ignoring these poles in this case will almost certainly lead to an unstable servo system.

10.2.4 Velocity servo–practical current-source drive

Figure 10.17(c) shows a block diagram of a PM dc machine with a practical current-source drive, i.e., one with some dynamic limitation on the terminal voltage compliance. With some effort, the block dia-

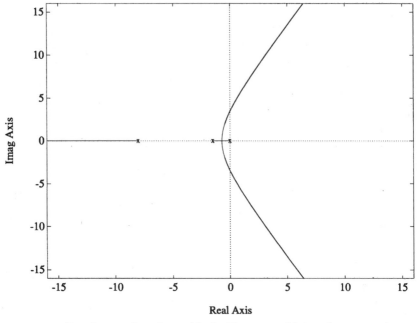

Figure 10.23 Root-locus: voltage loop with significant τ_e and integral compensation.

gram in Fig. 10.17(c) can be manipulated to reveal that the open-loop transfer function relating the output speed to the input current command is

$$
\frac{\omega}{i_{\text{ref}}}(s) = \frac{\dfrac{K_a K_m}{R_a \beta}}{\left(\tau_e s + 1 + \dfrac{K_a}{R_a}\right)(\alpha s + 1) + \dfrac{K_m^2}{R_a \beta}}
\tag{10.9}
$$

This transfer function is second-order. Closing a proportional loop around this system to regulate speed would result in a system with finite steady-state error in response to a step, and with a root-locus similar to the one shown in Fig. 10.21. Notice, however, the effect of the "minor loop" that attempts to force the actual armature current to follow the reference command i_{ref}. The electrical pole begins substantially deeper in the left half-s-plane. That is, as long as it does not saturate, the current minor loop has effectively decreased the electrical time constant, making the system "look more first-order", as we would have expected from the prior analysis with an ideal current-source drive. Also note that an integral compensator wrapped around this system to regulate speed would result in zero steady-state error, and a root-locus similar to the one shown in Fig. 10.23. However, because the minor current loop starts the electrical pole deeper in the left half-s-plane, the system with integral compensation might be able to support a larger range of variation of the integral gain K before going unstable.

10.2.5 Position control loops

Notice from Fig. 10.17 that the difference between a dc motor model that describes a velocity output at the mechanical shaft and one that describes position is the absence or presence, respectively, of a final integration block in the diagram (to integrate velocity into position). In a block diagram with a series cascade of blocks, the order of the blocks is irrelevant in determining the through-transfer function. Hence, any closed-loop transfer function derived for a velocity servo loop with an integral compensator will be identical to one derived for a position servo loop with a proportional compensator. This means, for example, that the transfer function Eq. 10.7 describes not only a velocity servo loop with a current-source drive and integral compensator, but also a position servo loop with a current-source drive and proportional compensator. Conclusions drawn in the previous three sections for velocity loops with integral compensators all apply, therefore, to position loops with proportional compensators. The exploration of other compensation possibilities are left to the reader and References 10, 12.

10.3 Power Electronics for AC Drive Systems

In a dc machine, the mechanical commutator ensures that current flows in the machine windings in a manner that will produce useful torque, even when the rotor changes position and speed. From a modeling and control standpoint, the presence of the commutator significantly eases the problem of designing appropriate power-electronic amplifiers for driving dc machines. Essentially, the problem becomes one of developing circuitry that can create flexible levels of voltage or current, without too much concern for the specific waveshape: either a linear amplifier or a chopper would generally produce adequate results, for example. Only the relatively slowly varying, average values of the terminal waveforms prove to be of concern in a well-designed system. In an ac machine, there is no mechanical commutator, and the electrical excitation of the stator must be appropriate to ensure sustained torque production. For a controllable motor drive, this generally means that the drive electronics must be capable of producing ac waveforms with controllable frequency and amplitude.

Switching power-electronic drives for ac machines are often (but not always) constructed as inverters, which operate from a dc input voltage and which produce a controlled ac output voltage waveform or waveforms. A dc bus that serves as the input voltage to the inverter can be created by rectifying a fixed-frequency ac utility service, for example. Figure 10.24 shows a full-wave rectifier set operating from a three-phase utility connection. A dc output voltage with relatively low ripple is produced across the capacitor. If necessary, the level of the dc output voltage can be controlled by replacing the diodes with controllable devices, such as silicon-controlled rectifiers. Controlling the firing angle of these devices permits control of the magnitude of the output voltage. Of course, a dc bus can be created in other ways. For a single-phase utility connection, either a single-phase, full-wave rectifier or a firing angle-controlled rectifier might be used. In an electric-vehicle drive system, the dc bus would come from a battery rack, and no rectification would be required.

An inverter uses the dc bus to create ac waveforms with controllable frequency and amplitude. Figure 10.25 shows a typical inverter configuration driving a balanced, three-phase, inductive load. This load could be the wye-connected stator of an induction or PM synchronous motor, for example. There are a variety of schemes for operating the switches Q1–Q6 to produce desired ac waveforms. Typically, the ultimate goal of an inverter drive for an AC machine is to make the machine appear from an electrical port to be a current-controlled torque source, just like the PM dc machine. Two approaches for operating the switches in a three-phase inverter will be discussed briefly here; others may be found in Reference 13 and especially in Reference 14. The goal of this section is to reveal how useful ac waveforms can be pro-

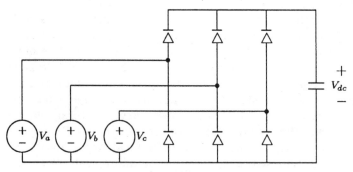

Figure 10.24 Three-phase rectifier.

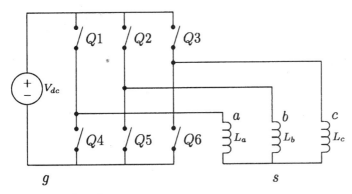

Figure 10.25 Inverter.

duced by an inverter given a DC bus. The waveform analysis presented is summarized from the excellent discussion in Reference 13.

In a "six-step", continuous current inverter, the controllable switches are operated as shown in the top six traces in Fig. 10.26. One leg or phase of the inverter consists of a top and bottom switch stacked together, e.g., Q1 and Q4. The stator connections a, b, and c can each be connected to either the top or the bottom of the dc input voltage by the switches. In any particular leg of the inverter, the two switches are never turned on at the same time in order to avoid shorting the input source. Also, in the continuous current inverter, one of the switches in each leg is always turned on to provide a path for phase current to flow. To emulate the behavior of a balanced, three-phase sinusoidal voltage source, the top or high-side switch in each leg is turned on 120 electrical degrees before the top switch in the next leg, and remains on for half of the electrical cycle.

When the high-side switch in a leg is on, the winding connected to that leg is connected to the top of the dc source. When the high-side switch is off, the low-side switch in the leg is on, and the winding is connected to the bottom of the dc source. The voltage between each stator terminal and point g, therefore, has a waveshape that looks like the switch state for the high-side switch in that leg. For example, the voltage V_{ag} has a waveshape like the Q1 switch state trace in Fig. 10.26.

This fact can be used to determine the line-to-line and line-to-neutral voltages seen by the load. For example, the line-to-line voltage V_{ab} will have a waveshape that looks like the difference of the Q1 and Q2 waveshapes. This line-to-line voltage is plotted in the seventh trace in Fig. 10.26. To determine the line-to-neutral voltage for Phase a, notice that, for stator terminal a, we can use Kirchoff's voltage law to discover that

Switch State

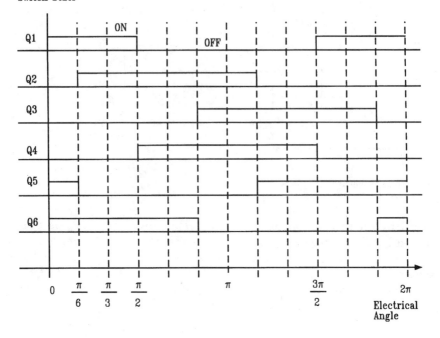

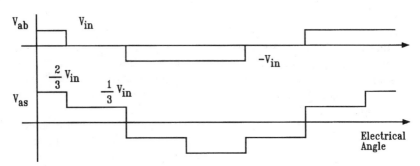

Figure 10.26 Six-step continuous inverter: switch states and voltage waveforms.

$$V_{ag} = V_{as} + V_{sg} \qquad (10.10)$$

and so on for the other two phases. Because the inverter and load are balanced and have three phases, we know that

$$V_{as} + V_{bs} + V_{cs} = 0$$

Therefore, the voltage

$$V_{sg} = \frac{1}{3}(V_{ag} + V_{bg} + V_{cg}) \tag{10.11}$$

Substituting Eq. 10.11 into Eq. 10.10 reveals that the Phase a line-to-neutral voltage is

$$V_{as} = \frac{2}{3}V_{ag} - \frac{1}{3}(V_{bg} - V_{cg}) \tag{10.12}$$

The waveshapes of V_{ag}, V_{bg} and V_{cg} are identical to the Q1, Q2, and Q3 wave traces in Fig. 10.26. Equation 10.12 and the traces Q1, Q2, and Q3 are used to produce the line-to-neutral voltage waveform V_{as} for phase a shown in the last trace in Fig. 10.26. The six-step inverter produces a line-to-neutral voltage that has a substantial sinusoidal component at the fundamental frequency, with some obvious harmonic distortion present at higher, odd harmonics of the fundamental.

The frequency of the output waveforms can, of course, be changed by varying the time allotted to complete one electrical cycle. In a PM synchronous machine or "brushless dc motor", the operation of the switches in the inverter is often "slaved" or synchronized to the rotor position, possibly by Hall-effect switches that sense the location of the rotor. This ensures that the AC waveform produced by the inverter will have a significant constant component when viewed in the rotor frame, as is necessary to sustain torque production. In a two-pole machine, for instance, the inverter would complete one electrical cycle for every revolution of the rotor. In essence, the inverter operates as an electrical commutator. The inverter can also be used to drive an induction machine. This drive could be "open-loop," i.e., the inverter can provide the induction motor with a fixed-frequency, balanced voltage set. It could also be synchronized to the position of the rotor, as would be essential in the implementation of a field-oriented controller.

In the case of the PM synchronous machine, once the inverter operation is synchronized to the rotor position, the machine essentially behaves like a conventional PM dc machine from the standpoint of the dc input to the inverter. Raising the inverter input voltage will increase the speed of the machine. The current flowing out of the dc input indicates the level of torque produced at the shaft of the machine. It is important, therefore, to be able to control the magnitude of the voltage or current applied to the machine. This can be done in at least two ways. The first approach would be to vary the level of the dc input voltage to the inverter. This might be done either to vary the machine terminal voltage directly, or perhaps to control the current injected into the machine with a minor loop. The second, pulse-width modulation approach uses the inverter switches to chop the voltage applied to the stator. The stator voltages can always be set to zero by turning on all three high-side, or all three low-side switches in the inverter (but never the high-side and low-side switches at the same time for a dc voltage input). The PWM switch frequency would be set significantly higher than the six-step electrical frequency. Varying the duty cycle will vary the average voltage applied to the stator terminals, again permitting voltage control or current control with a minor control loop.

10.4   References

1. A. E. Fitzgerald, C. Kinglsey, and S. Umans, *Electric Machinery*, McGraw-Hill, New York, 1983.
2. J. G. Kassakian and M. F. Schlecht, "High-Frequency High-Density Converters for Distributed Power Systems," *IEEE Proceedings–Special Issue on Power Electronics*, April 1988, pp. 362–376.
3. T. G. Wilson, "Life After the Schematic: The Impact of Circuit Operation on the Physical Realization of Electronic Power Supplies," *IEEE Proceedings–Special Issue on Power Electronics*, April 1988.
4. A. V. Oppenheim and R. W. Schafer, *Digital Signal Processing*, Prentice-Hall, New Jersey, 1975, pp. 204–206.
5. E. Landsman, "A Unifying Variation of Switching DC-DC Converter Topologies," *IEEE PESC Record*, 1979.
6. J. G. Kassakian, M. F. Schlecht, and G. C. Verghese, *Principles of Power Electrons*, Addison-Wesley, Massachusetts, 1991.
7. "Linear Databook," National Semiconductor.
8. William M. Siebert, *Circuits, Signals, and Systems*, McGraw-Hill, New York, 1986.
9. "(UC3842) Current Mode PWM Controller," *Linear Integrated Circuits Databook*, Unitrode Corporation, 1987 pp. 3-107–3-112.
10. J. K. Roberge, *Operational Amplifiers*, John Wiley and Sons, New York, 1975.
11. J. Van de Vegte, *Feedback Control Systems*, Prentice Hall, New Jersey, 1990.
12. G. S. Brown and D. P. Campbell, *Principles of Servomechanisms*, John Wiley and Sons, New York, 1948.
13. P. C. Krause, *Analysis of Electric Machinery*, McGraw-Hill, New York, 1986.
14. B. K. Bose, ed., *Adjustable Speed AC Drive Systems*, IEEE Press, New Jersey, 1980.

Index

ABOUT THE AUTHORS

H. WAYNE BEATY (Tulsa, OK) is Editor of the *Standard Handbook for Electrical Engineers* and the former Managing Editor of *Electric Light & Power* magazine and Senior Editor of *Electrical World* magazine.

JAMES L. KIRTLEY, JR. (Cambridge, MA) is Professor (on leave) of Electrical Engineering at Massachusetts Institute of Technology and Vice President at Satcon Technology Corp.

NIRMAL K. GHAI (Lexington, KY) is Manager of Generator Engineering at Magnetek, Inc. and former Engineering Manager in Westinghouse Electric Corp.'s Heavy Industrial Motor Group.

STEVEN B. LEEB (Cambridge, MA) is Associate Professor at Massachusetts Institute of Technology in the Department of Electrical Engineering and Computer Science.

RICHARD H. LYON (Cambridge, MA) is President of RHLyon Corp and Professor Emeritus of Mechanical Engineering at Massachusetts Institute of Technology.

It would seem apparent that one would want to make λ_s as large as possible, to permit high currents. The limit on this is that the magnetic teeth between the conductors must be able to carry the air-gap flux, and making them too narrow would cause them to saturate. The peak of the time fundamental magnetic field in the teeth is, for example

$$B_t = B_1 \frac{2\pi R}{N_s w_t}$$

where w_t is the width of a stator tooth

$$w_t = \frac{2\pi(R + h_d)}{N_s} - w_s$$

so that

$$B_t \approx \frac{B_1}{1 - \lambda_s}$$

6.5.1 Resistance

Winding resistance may be estimated as the length of the stator conductor divided by its area and its conductivity. The length of the stator conductor is

$$l_c = 2lN_a f_e$$

where the "end winding factor" f_e is used to take into account the extra length of the end turns (which is usually *not* negligible). The *area* of each turn of wire is, for an air-gap winding

$$A_w = \frac{\theta_{we}}{2} \frac{R_{wo}^2 - R_{wi}^2}{N_a} \lambda_w$$

where λ_w, the "packing factor" relates the area of conductor to the total area of the winding. The resistance is then just

$$R_a = \frac{4lN_a^2}{\theta_{we}(R_{wo}^2 - R_{wi}^2)\lambda_w \sigma}$$

and, of course, σ is the conductivity of the conductor.

For windings in slots, the expression is almost the same, simply substituting the total slot area

$$R_a = \frac{2qlN_a^2}{N_s h_s w_s \lambda_w \sigma}$$

$$B_r = \frac{\mu_0 NI}{g}$$

the space fundamental of that will be

$$B_1 = \frac{\mu_0 NI}{g} \frac{4}{\pi} \left(1 - \sin \frac{p\theta_t}{2} \right)$$

where θ_t is the angular width taken out of the pole by the magnets. So that the expression for quadrature-axis inductance is

$$L_q = \frac{q}{2} \frac{4}{\pi} \frac{\mu_0 N_a^2 R l k_w^2}{p^2 g} \left(1 - \sin \frac{p\theta_t}{2} \right)$$

6.5   Current Rating and Resistance

The last part of machine rating is its current capability. This is heavily influenced by cooling methods, for the principal limit on current is the heating produced by resistive dissipation. Generally, it is possible to do first-order design estimates by assuming a current density that can be handled by a particular cooling scheme. Then, in an air-gap winding

$$N_a I_a = (R_{wo}^2 - R_{wi}^2) \frac{\theta_{we}}{2} J_a$$

and note that, usually, the armature fills the azimuthal space in the machine

$$2q\,\theta_{we} = 2\pi$$

For a winding in slots, nearly the same thing is true: if the rectangular slot model holds true

$$2qN_a I_a = N_s h_s w_s J_s$$

where J_s denotes *slot* current density. Characterize the total slot area by a "space factor" λ_s which is the ratio between total slot area and the annulus occupied by the slots. For the rectangular slot model

$$\lambda_s = \frac{N_s h_s w_s}{\pi(R_{wo}^2 - R_{wi}^2)}$$

where $R_{wi} = R + h_d$ and $R_{wo} = R_{wi} + h_s$ in a normal, stator-outside winding. In this case, $J_a = J_s \lambda_s$ and the two types of machines can be evaluated in the same way.

THE STUDY OF INCUNABULA

THE STUDY OF

INCUNABULA

BY KONRAD HAEBLER

TRANSLATED FROM THE GERMAN

BY LUCY EUGENIA OSBORNE

WITH A FOREWORD

BY ALFRED W. POLLARD

NEW YORK

THE GROLIER CLUB

1933

Reprinted with the Permission of the Original Publisher
KRAUS REPRINT CORPORATION
New York
1967

FOREWORD

I FEEL it a very great honour to be asked to contribute a foreword to this most helpful treatise by Dr. Haebler on the *Study of Incunabula*. Dr. Haebler and I have been friends now for some thirty-five years and I have always had the highest respect and admiration for his fine qualities as a man and a scholar. The illustrated monograph which he wrote for the Bibliographical Society in 1896 on *Early Printing in Spain and Portugal* has been superseded by his own more exhaustive works on the same subject, but at the time it was written it was a notable piece of pioneering work and it was a great privilege to the Bibliographical Society, then only three or four years old, to be permitted to print it. With other monographs contributed in those early days by M. Claudin and Dr. Lippmann it helped to give the Society an international character which I regret that it has since hardly maintained. Dr. Haebler came to my help again after the death of Robert Proctor when I was struggling, by order of the Trustees of the British Museum, with organising a full-dress catalogue of the fifteenth-century books in their possession. His *Typenrepertorium der Wiegendrucke,* with its selection of different forms of the letter M as a help to the differentiation of gothic types, was a real standby to me when I was trying to follow in Proctor's footsteps and to test and develop his work without any share in his extraordinary natural gift for carrying in his head the forms of hundreds of founts of types, so as to recognise any example of them at a glance. In the specimen sheets of types devised as illustrations to the British Museum catalogue I seldom selected any example of a gothic fount which did not contain an M and I think these specimen sheets are the core of the whole work.

It is a great pleasure to me to note that in this *Study of Incunabula,* Dr. Haebler, now that he is no longer working with the materials for the *Gesamtkatalog der Wiegendrucke* at his elbow, has found the Museum Catalogue useful in providing him with examples of several of the points which he notes, *e.g.,* the occasional use of half-sheets of the larger size paper *(carta regalis)* in small folios, to the confusion of the rule that in folios the wire marks are always perpendicular, and instances of early books containing a table of rubrics, of wrong imposition, and of the troubles which arose when copy had to be divided between two compositors. It is good that the British Museum has been thus able to pay back a little of its debt, and all the better because the wealth of experience at Dr. Haebler's command is so great that it is an honour to anyone from whom he borrows.

It is this wealth of experience which lies behind every

section of Dr. Haebler's book, from its opening section on the earlier writers of incunabula down to its final page, which makes it of such surpassing interest. When I began early one afternoon to read the type-script sent me I went on reading it for the rest of the day and felt at the end of my reading that I had seldom received so much new information on so many of the points in which my own smaller experience had interested me. Histories of printing almost inevitably have been mainly written at second hand. Dr. Haebler's book abounds with discoveries of the kind which present themselves in the course of that leaf-by-leaf examination of the earliest books to which alone they give up their secrets. As he insists, the true incunabula are the books produced before the business of printing was mechanized by what for those days was large-scale production. The pioneers had to devise means of overcoming their difficulties as they encountered them. In the early days of book-building, difficulties were many and for this reason the leaf-by-leaf examination of incunabula is the best training which any bibliographer can have for enabling him to get at the secret of unusual features in books not merely of the fifteenth century only, but of subsequent periods as well. Dr. Haebler's book will teach anyone who uses it what sort of difficulties he may expect and what is their most probable solution. With its aid the student will be able to walk more boldly and feel greater certainty in his results. Whether the vast *Gesamtkatalog der Wiegendrucke* will provoke the new epoch of research Dr. Haebler seems to expect or prove rather crushing in its immensity remains to be seen. But with this introduction by Dr. Haebler in his hand the solitary student of fifteenth-century books will have an extraordinary help in the prosecution of any studies which he may have the pluck to undertake.

A. W. Pollard.

Wimbledon, April 21, 1931

PREFACE

WHOEVER deals for the first time with such a comprehensive subject as that of incunabula must make up his mind that he will not fulfill all expectations. On one point he will be considered to have placed too much, on another, too little emphasis; nor will either the limits which he has set for his subject nor the way in which he has treated it meet with universal approval. Naturally, therefore, I publish this book with some diffidence, for I foresee sharp criticism.

The *Study* has a long preliminary history. After my call to the Königliche Bibliothek in Berlin I gave its candidates for library service a series of lectures on the preliminaries of the study of incunabula. This was a new departure there, and Privy Councilor Schwenke made the request that I write a book which should serve as an introduction to this study. I felt obliged to decline, however, for the survey in behalf of the *Gesamtkatalog* was not at that time completed. When this point was reached, I was no longer in office. During an opportune stay in Berlin in the spring of 1923 I was, however, asked to repeat my series of lectures, and Privy Councilor Milkau was kind enough to delegate a stenographer to attend, so that the lectures might be made accessible to a wider circle of those interested; but I was dissatisfied with the stenographic report, for I had given the lectures extemporaneously, without preparation, and they needed revision. Since several copies of the report which circulated among my colleagues were so kindly received in spite of its faults, however, I felt under obligation to go into the subject at last more thoroughly and give it more careful treatment.

For this, the material for the *Gesamtkatalog* was of course no longer at my disposal. Instead I could make use of the British Museum catalogue of incunabula which was then far advanced, and to that I owe not only a great many illustrations of facts which I had already noticed, but also a number of suggestions as to the consideration and treatment of individual cases which I had not at first intended to include within the sphere of my observations.

I am especially indebted also to my associates on the Commission for the *Gesamtkatalog,* the Director, Dr. H. Schmidt, Darmstadt, Department Director Dr. E. Freys, Munich, Professor E. Voulliéme, and Dr. E. Crous, Berlin, and particularly the Chief Librarian, Professor Leuze, Stuttgart, who brought to my attention a number of peculiarities in Stuttgart incunabula. Nor do I feel less under obligation to Dr. Wiegand (Bremer Press) in Munich, for in the case of some of the earliest incunabula I have ventured to make use of certain conclusions which are the result of his keen observation.

PREFACE

This, then, is the origin of the *Study,* and I am hopeful that it will not altogether fail of its purpose, that of giving preliminary information in this special branch of bibliography.

K. Haebler.

Stuttgart, Spring, 1925

TRANSLATOR'S NOTE

This book is translated from the German edition of 1925, but embodies also certain revisions of the text made by the author in 1932.

CONTENTS

CONTENTS

THE STUDY OF INCUNABULA

THE STUDY

I. MEANING AND EXTENT

THE study of incunabula has to do with the earliest examples of printing. It cannot well be given independent rank, for it is in reality a part of bibliography, yet since its concern is with the work of that important period in which the printed book first came into being and gradually took definite shape, its claim to consideration is greater than that of bibliography in general; hence although it is but one part of bibliography, it justly demands such treatment as will ensure a realization of its own absolute individuality.

The term "incunabula" refers simply to the productions of a time so early that the art in question, no matter what, was still in its infancy; hence the German designation "Wiegendrucke." It is wrong to suppose the name derived from any kind of cradle-shaped tool in use in the earliest period of printing. The name has an entirely general significance, meaning only the products of an early time, and may therefore be used at will for the earliest expression of any art or other practice. A well-known French work bears for example the title *Deux cent incunables de la gravure,* and similar instances are known elsewhere as well. A reference to incunabula as such, however, is now generally understood as an allusion to the earliest examples of the art of printing.

[1]

Incunabula are productions in which we can follow the development of the book and in which the printer expressed himself in his work as an independent, creative master. As soon as this artistic phase had fully developed, it was succeeded by more intensive concern with the mechanical reproductive process, the germ of which had naturally lain dormant from the first in the printing art. The development of this process took a most varied course in various places and in different countries, so that a different time-limit would be the rule for the incunabula of each country and each town, if one were to go only by their own beginnings of the work. Indeed, the limits of the incunabula period itself have been very variously set by different scholars. An early catalogue of incunabula of the Universitätsbibliothek in Leipzig, in which the cradle-books of Italian origin predominate, ends as early as 1480. Another catalogue allows all printing to rank as incunabula which appeared before the year 1500, that is, to the year 1499. On the other hand, numerous catalogues, especially some of collections of modest size, have even added one or more decades of the sixteenth century to the period of early printing. Some end it, from a German point of view, with the year of the Reformation; others, as for example, Panzer, have arbitrarily taken a still later year as the closing one; but no one has ventured to come down further than to the year 1550, universally accepted as the end of the first century of the art of printing.

To-day, the end of the year 1500 is quite generally

accepted as the close of the incunabula period. It agrees with the usage prevailing from early times to the present, and is easily justifiable as an average date. The year 1500 is a very fitting limit for the early printing of Germany in particular, and as the home of the invention of the "black art" this country may well lay claim to special consideration. The mechanical side of the printing process had begun to gain a footing in some places in Germany even before the year 1500. The prominent workshops in which that was the case, those of Anton Koberger in Nuremberg, Johann Grüninger in Strassburg, Peter Drach in Speyer, nevertheless had so undeniably gotten their start in the incunabula period, and so, notwithstanding their attention to the commercial aspect, still retained so much feeling for the scholarly and artistic conception of printing that it would be wrong to deny their books the characteristics of true incunabula. In any case, most of the smaller printing-shops which appeared in Germany up to the year 1500 bore quite unmistakable marks of the incunabula period. As far as Germany is concerned, we might feel in favor of extending the limit of this period still further, to the Reformation, for that was the first event to mark a general and obvious change in the field of printing. To do this, however, would be unfortunate, for the years 1501-17 in Germany were a time in which printing played a considerably diminished rôle as much in comparison with the preceding as with the following period.

In Italy we find a somewhat different situation. Here

one must make a distinction between the printing of Venice, which began as early as the seventies to assume the leading rôle in the printing art in general, and the workshops of the rest of Italy, for their existence developed under essentially different circumstances. The incunabula period, as we have herein characterized its essentials, ended in Venice about 1480, for not in vain was Venice in the second half of the fifteenth century the indisputable centre of European commerce. The industrial activity usual there, especially in the form of trade-partnerships, seized upon printing at a comparatively early time, divested it of its individual artistic and scholarly character and substituted for it a plain, commercial, industrial aspect. Venetian book production from 1480 to 1500 had the true incunabula character only to a very limited degree, and students of incunabula might without material disadvantage eliminate the greater part of it from their researches. Nevertheless the particularly progressive appearance of books printed in Venice affords an interesting chapter in the development of the art, for although it deviates more and more from the specific character of the incunabula period, it is of unquestioned interest in the general history of the printed book. The peculiar development of Venice reached but little beyond the boundaries of the Republic. To be sure, a few cities of upper Italy were drawn more or less within its sphere, but general political conditions in Italy soon brought Venetian influence to an end. In the other Italian states conditions were almost the same as in

Germany. In several places in these states also there developed a transition to more modern, professional methods; but an overwhelming majority of the printing-shops in Italy preserved throughout the century the established artistic and scholarly traditions. In fact, the introduction of italic by Aldus Manutius in the year 1501 plainly strengthens our conception of the end of the true incunabula period as the year 1500.

In France, too, a distinction should be made between the printing of Paris and that of other cities. The limit for Paris was a considerably later one, however, than for Venice; yet about the middle of the nineties bookmaking even in Paris saw the introduction of methods of management which were contrary to the fundamental principles of the incunabula period. For example, here we see the ascendancy of the great publishing firms which took into their own hands the control of the field of book production and made the status of the printers that of hired craftsmen. That in so doing these publishers directly assumed control of scholarly and artistic aspects which had formerly been looked after by printers, did in a sense associate them with the customs of the days of early printing; yet printing itself shortly before the year 1500 lost the character which we have set up as an essential feature of the time of early printing, a development in which the provinces had no share. In Lyons, however, the publishers began early to play a prominent part in book production, although there it never came about that the printers were deprived of personal relations with their productions. Up to the beginning of the sixteenth century the early printed books of Lyons retained the characteristics of incunabula. The same was true on an even larger scale of the few other printing-shops which existed in France in the provinces before the year 1500, and for these the limit of the incunabula period could quite properly be assigned to one or two decades later.

The situation was similar in other countries into which printing found its way before the end of the fifteenth century. Up to that time, however, Spain and England entirely lacked a centre in which, as had happened for example in Venice or Paris, the art should have had a chance to develop earlier than was possible in the provinces. Notwithstanding all their presses, in comparison with Italy and France, they remained provincial, their entire output persistently retaining the characteristics of incunabula even down to the sixteenth century. In the north and east we find printing only sporadic, often as offshoots of the workshops which we have already met with in the older centres of culture. For all these countries, then, the limit of the year 1500 is altogether too early; yet this is not very significant, relatively speaking, since the entire book production in these regions in comparison with that of Germany, Italy or France is of no great importance. For all the reasons here set forth, then, the year 1500 is almost universally recognized as the limit of the incunabula period.

[6]

2. HISTORY AND LITERATURE

The year 1640 should be considered the birth-year of the study of incunabula, for in it the two-hundredth anniversary of the invention of the art of printing was celebrated in a number of the cities of Germany, on which occasion allusion was made to the fact that it would be worth while to list the oldest specimens of printing. This wish was soon to bear fruit. In the year 1643 appeared the work of J. Saubertus, *Historia bibliothecae reipublicae Noribergensis ... Accessit ... appendix de inventore typographiae itemque catalogus librorum proximis ab inventione annis usque ad a. Chr.* 1500 *editorum.* Noribergae 1643. 8°., which listed in its appendix some eight hundred and twenty-five incunabula which were found in the Stadtbibliothek at Nuremberg, founded, as is well known, in 1430. Ten years later the work by Philippe Labbé followed, *Nova bibliotheca mss. librorum s. specimen antiquarum lectionum latinarum et graecarum.* Paris 1653. 4°., (not to be confused with the 1657 folio edition with the same title), to which on pp. 337ff. was added as supplement IX a catalogue of "primae editiones illae, quae ante centum et quinquaginta annos in ipsis paene typographiae incunabulis prodierunt." Here the word "incunabula" first was associated with printing, even though not yet as a name for early printed books themselves. Labbé's catalogue included 1289 copies from the resources of the Royal Library in Paris. The first independent catalogue of incunabula was that by Cornelis à Beughem, *Incunabula typographiae s. catalogus librorum scriptorumque proximis ab inventione typographiae annis usque ad annum Christi MD inclusive in quavis lingua editorum.* Amstelodami 1688. 8°. This was no longer the inventory of a single collection but a first attempt to comprehend the collected literature of the period of early printing, the author having brought together approximately three thousand titles.[1] Naturally this catalogue dealt almost exclusively with such books as gave sufficient information concerning their origin, and described these works in a bibliographical way, *i.e.,* exactly as was customary for all books appearing in later centuries. Compared with this, the book by Michael Maittaire, *Annales typographici ab artis inventae origine ad annum MD.* Hagae Comitum 1719. 4°., was a decided improvement. Maittaire, actuated by an idea of development, was the first to arrange the presses in chronological order according to the year in which they appeared. He too gave titles in modern style, but often printed in voluminous notes identifying information as to the book, in the original form, prefacing the chronological groups which he formed, with extensive disquisitions of a general nature. He then continued his *Annales* with two volumes and an appendix which were to carry the bibliography down to the year 1664,

[1]Cf. v. Rath. *Vorläufer des Gesamtkatalogs der Wiegendrucke.* In *Werden und Wirken (Festschrift für Karl W. Hiersemann,* Leipzig, 1924). pp. 288-305. Peddie, R. A. *Fifteenth-century books. A guide to their identification.* London, 1910-14.

but before he had finished the indexes to his work he felt it necessary to have volume 1 appear in a second edition (Editio nova. Tom. 1, 1.2. Amstelodami 1733. 4°.) and this, although it did not render the first entirely superfluous, was nevertheless primarily the work which ensured his book a greater significance. While, namely, the first edition listed only about 1760 incunabula, the second contained nearly 5600, of which more than four thousand were described by Maittaire himself or by trustworthy authorities, and fewer than one thousand were enumerated based solely on quotations from other books.

In spite of this extraordinary increase, Michael Denis in his *Supplementum annalium Michaelis Maittaire*. Wien 1789. 4°., could, once more, more than double the number of traceable incunabula. In the Imperial Hofbibliothek in Vienna, preëminent for its resources, he traced 6311 incunabula hitherto unknown, and was the first to take into account, on a large scale, such incunabula as were without any information as to printing origin. Maittaire had recorded one hundred and thirty-six copies without places indicated, and forty-six with these but without year of printing, but listed only eighty-two books as incunabula which were quite unidentified. On the other hand, Denis listed 2237 editions which he believed he could prove to be incunabula although they themselves gave no evidence of their origin. In his hands the number of traceable incunabula grew to nearly twelve thousand.

Later, Franc. Xav. Laire's *Index librorum ab inventa*

typographia ad a. 1500 *chronologice dispositus.* Tom. I.II. Senonis 1791. 8°., assumed the title of a general catalogue of incunabula. In reality it was, however, only the catalogue of a private collection which Cardinal Loménie de Brienne had gathered and which was dispersed again by sale. This collection nevertheless contained among its 1332 items almost all the valuable examples of the first period of printing. What ensures the book a claim to lasting importance in the history of the study of incunabula is not however the list of recorded works, but the way in which Laire treated them. Here for the first time were scrupulously noted the characteristic features of early printing, the method of setting, the introduction of signatures, register, etc., while the index at the end of the second volume contained in a nutshell a survey of most of the salient features concerning which the student of early printing must be informed in dealing with his subject. It was the forerunner of a handbook to be used in the study of incunabula, something which, by the way, though that was more than a hundred years ago, has not yet been written.

Almost simultaneously with Laire began the publication of Georg Wolfgang Panzer's *Annales typographici ab artis inventae origine ad annum MD.* Norimbergae 1793-1803. Vol. I-V; Suppl. in Vol. IX-XI. As regards the style of the *Annales,* Panzer took from Laire only somewhat full reproductions of the colophons; for the rest, his book is decidedly inferior to his predecessor's in the method of treatment of the individual book. It became however a work which has even to-day a leading place among the indispensable tools of a student of incunabula, for Panzer gathered the whole number of incunabula known up to that time, grouped them first by country and place of printing, was the first to arrange them in chronological order within this geographical limit, and finally, recorded for each individual book, where he found it, or from what source he derived his information. This last detail in particular so added to the value of his work that even as late as the year 1900 Konrad Burger did incunabula research a real service when he published a concordance to Panzer and Hain, with additions from later catalogues of separate collections (K. Burger: *Supplement zu Hain und Panzer. Beiträge zur Inkunabelbibliographie. Nummernconcordanz von Panzers lateinischen und deutschen Annalen und L. Hains Repertorium bibliographicum.* Leipzig 1908). Moreover, Panzer was able to enrich our knowledge of early printing by nearly four thousand titles; but the chief value of his work lies not so much in this as in his perception that early printing must be considered from the point of view of its typographical origin.

Ludwig Hain, on the other hand, in his *Repertorium bibliographicum,* 1,[I-II] 2,[I-II] Stuttgart & Paris, 1826-1838. 8°., returned, as to the form in which it was published, to the older, purely literary standpoint, listing the books alphabetically by authors' names or by key-words. There is, however, no doubt that Hain would have added to this an extensive index of places of printing and printers, if death had not come before the completion of his book. Here too Konrad Burger was of assistance in filling the breach, in having Ludwig Hain's *Repertorium biblio-graphicum. Register*. Leipzig, 1892. 8°., appear in the *Beihefte zum Zentralblatt für Bibliothekswesen*. But the lack of indexes is not the only disadvantage which the *Repertorium bibliographicum* sustained through its author's early death. In the last volume a large number of names, Vincentius Bellovacensis, Virgilius, Jacobus de Voragine and many more, often referred to by Hain in the earlier part, were entirely neglected, since the publisher, apparently wholly without any comprehension of their importance to the value of the work, made not the slightest effort to fill these lacks. In another direction also Hain's *Repertorium* is inferior to the work of his predecessors, for it omits entirely any information as to his sources. Even that is not to be quite unqualified, if the asterisk beside some numbers means that Hain had those editions in his hands in the Hof-und Staatsbibliothek in Munich. In any case he used those as well as a number of other German libraries with richer incunabula resources for his work. If this, then, in spite of the imperfections mentioned, has remained for almost a century the foundation work for the student of incunabula not only in Germany but in all other cultured countries, one must recognize what a distinguished performance it was, although it was the achievement of an amateur forced to work under the most unfavorable circumstances—he is said to have written the book in part while in prison for debt.

That the *Repertorium* has become the foundation for all work on early printing, is due to its method. Until its appearance bibliographers had given early printed books a title according to contents, which form was usually borrowed from later editions of the same works, for incunabula almost universally lack a real book title. At best they would add to it with more or less fulness and accuracy according to the colophon, the information as to place, printer and date of publication. Hain first made it a rule to set down faithfully line by line and letter by letter, for every incunabulum which he himself saw, the beginning and the end, the "Incipit" and the "Explicit." He was the first thus to make it possible to identify a copy beyond dispute. To this textual description he then added a series of notes as to the format, designating size by "f. (folio), 4°, 8°"; kind of type by "g. (goticis), r. (romanis), ch. (characteribus)"; form of setting by "2 col. (columnis)"; number of lines by "33-ll (lineis)"; presence of signature by "c. sign."; of leaf-numbering by "c. fol. num."; of woodcuts by "c. fig. xyl. (figuris xylographicis)"; and of a register by "reg.". These facts (collation), which in their abbreviated form never take more than one or two of his short lines, comprise practically everything necessary for the typographical characterization of an incunabulum and make possible an accurate checking, which, together with the textual description, must result in identification. With this checking and with scrupulous care, an error is practically out of the question. As to number, Hain in the *Repertorium* with his 16,299 items scarcely surpassed his predecessors, of whose work, without however acknowledging it, he made extensive use. This is true particularly of Panzer's *Annales;* but Hain did not take everything from this indiscriminately, and in spite of an almost identical number of copies recorded, he brought out a number of new facts.

Built on this foundation erected by Hain a large number of catalogues of separate collections of incunabula appeared in the next decades. To be sure, a limited number had been published even before the appearance of his *Repertorium.* The previously mentioned works of Saubert, Labbé, Laire, were catalogues of single collections, even though not so named. There was in fact no lack of such catalogues, for as early as 1712, Wilisch had published his *Incunabula scholae Annaebergensis,* in 1787-92 Seemiller catalogued the incunabula of Ingolstadt, in 1788-9 Braun those of SS. Ulrich and Afra in Augsburg, in 1791 Gras those of Neustift in the Tyrol, in 1794 Hupfauer those of Beuerberg. In addition, there appeared in 1814-15 the most important catalogue of a private collection, Dibdin's *Bibliotheca Spenceriana.* In the eighteenth century however, catalogues confined strictly to incunabula were less common than those more comprehensive in scope, dealing with a general range of books rare from a bibliographical standpoint, belonging to institutions. In Italy an anonymous compiler had as early as 1681 published one of the Barberiniana, 1761-88 Audiffredi one of the Casanatensis in Rome, 1793 Fossi one of the Magliabech-iana in Florence. In Germany in 1699 Krantz followed their example for Breslau, 1712 Schurtzfleisch for Weimar, 1743-8 Götze for Dresden, 1746 Mylius for Jena, 1780 Reuss for Tübingen, 1786-91 Murr for Nuremberg-Altdorf and still later than Hain, 1849-52, Schoenemann for Wolfenbüttel. All these works included incunabula, treated with more or less attention to detail.

At the same time, first attempts were being made to go thoroughly into the history of printing in certain places or that done by certain printers. Audiffredi led off with his *Editiones Romanae saeculi XV* (1783), which was followed (1794) by his *Editiones Italicae*. At the same time (1791) Affo followed with his *Tipografia parmense* and Bandini with his *Iuntarum typographia.* In Germany Baur did a like service for Speyer (1764), as did Büttinghausen for Oppenheim (1763), Gesner for Lübeck (1782) and Zapf for Augsburg (1786).

The real period for the publication of catalogues of incunabula of special collections began however after Hain had laid the foundation for a new method of cataloguing. Of the great number of such publications only a few of the most important will be noted here. Holtrop made the beginning in 1856 with the *Catalogus . . . bibliotheca regia Hagana,* and there followed him, 1865, Ennen for the Stadtbibliothek at Cologne, 1866 Bodemann for the Königliche Bibliothek in Hannover; in Italy

1875-86 Pennino for Palermo, 1889 Caronti for Bologna; and in France, 1878, Desbarreaux-Bernard for Toulouse. Then Mlle. Pellechet's work began in France, when she compiled the catalogues of Dijon (1886), Versailles (1889), Lyons (1893), and Colmar (1895), and expedited the printing of that of Ste. Geneviève (Daunou, 1892). In addition, the catalogues of Besançon (Castan) which appeared 1893, and the Mazarine (Marais-Dufresne), deserve mention among the great numbers of French catalogues. In the north and east also the interest in incunabula was growing, as witness the works of Hellebrant (Un. Bibl., Budapest, 1886), of Bølling (Kgl. Bibl., Copenhagen), of Horváth (Un. Museum, 1895) and of Wislocki (Un. Bibl., Cracow, 1900). In Switzerland the catalogues of Gisi (Solothurn, 1886) and Scherrer (St. Gallen, 1880), and in Austria those of Schachinger (Melk, 1901), Schubert (Olmütz, 1901) and Hübl (Schotten Stift, Wien, 1904) deserve mention. In Germany at about the turn of the century a number of scholars straightway came forward as specialists in the realm of incunabula cataloguing, chief among them, Voulliéme, whom we must thank not only for the catalogues of Bonn (1894), Berlin (1906) and Trier (1910) but for his most excellent piece of research, the *Buchdruck Kölns bis zum Ende des XV Jahrhunderts* (1903); Ernst, who catalogued the incunabula of the library of Hildesheim and the special collection of the Kästner-Museum in Hannover; and finally Günther, who brought together the Leipzig resources (1909).

[16]

Many of these catalogues were merely indexes which contented themselves with appending the Hain numbers to the brief titles. Often however they catalogued all the copies in the same detailed way in which Hain had described them, and that in time resulted unfortunately, since the catalogues kept on being filled with descriptions of the same less rare books which were found in almost every fairly distinguished library of incunabula. Therefore in time a change came about, compilers contenting themselves with a short description of well known incunabula or with a simple reference to Hain, and with giving a detailed description only of the incunabula which either were wholly omitted by Hain or of which there could not be found a full description made from his own observation.

No less important than incunabula catalogues for the study of incunabula are the works which have for their subject the investigation of the printing of single workshops, single towns or whole territories. In this field the Italians especially showed very keen activity. By the year 1745 I. A. Saxius had already added to the *Scriptores Mediolanenses* of Argelati a list of the incunabula printed in Milan. In 1761 D. M. Manni published his *Della prima promulgazione de'libri in Firenze,* and in 1777 Gir. Baruffaldi his treatise *Della tipografia ferrarese dall'anno* 1471 *al* 1500. By the year 1800 one can count at least ten more similar monographs by Italian students of early printing, and even down to very recent times Italian biblio-

philes have been especially active in furthering the study of
the earliest history of native printing, to this end having
their rich archives searched through for information concerning the early printers and their products. Germany can
cite by comparison for the early period, only the work of
G. W. Zapf, *Augsburgs Buchdruckergeschichte* (Augsburg, 1786-91) and in France before the year 1800 a few
monographs appeared on the introduction of printing into
Paris. The wealth of literature concerning the local history
of early printing may be found listed in the notes of my
Typenrepertorium der Wiegendrucke. That such locally
restricted research was especially adapted to bring to light
unknown material on early printing is immediately apparent, and in it was laid the foundation for a national
bibliography of incunabula.

It was in Italy that a first attempt in this direction was
made, when in 1783 Audiffredi published his *Catalogus
historico-criticus Romanarum editionum saeculi XV,* and
followed it 1794 by his *Editiones Italicae saec. XV,* works
which even to-day are of significance for the study of incunabula. It was owing to their incentive that R. D. Cabal-
lero in his book, published in Rome 1793, *De prima typographiae Hispanicae aetate,* undertook to give a comprehensive list of all incunabula printed in Spain, an attempt
which was to be repeated on a larger scale in the year
1804, by Franc. Mendez in his *Tipografía española.*

It was a long time before these attempts were imitated.
When L. Ennen in 1865 in his *Katalog der Inkunabeln*

in der Stadtbibliothek zu Köln, in which Part I comprised only the examples from Cologne presses, or when J. W. Holtrop in the *Catalogus . . . bibliotheca regia Hagana* took the books of the Low Countries and described them with especial fulness, each work was limited to the resources of but one collection, however rich in its own field. F. A. G. Campbell, therefore, was the first to create a truly national bibliography of incunabula in his *Annales de la typographie néerlandaise du XVe siècle* (La Haye, 1874), which with its numerous supplements gives an exhaustive survey of the activity of the early printers of the Low Countries. In 1889 he was followed by G. E. Klemming, who however extended his Swedish bibliography from the beginning to 1530 because of the very small number of Swedish incunabula. In my *Bibliografía ibérica del siglo XV* (Leipzig 1903) I made a new attempt to bring together Spanish incunabula, but was obliged even as soon as 1917 to follow it by a supplementary second part, so numerous were the additions which came to light upon the publication of the book, a thing which frequently happens in such a case. Since then England has also acquired her national catalogue of incunabula, published likewise in the year 1917, in the work of E. G. Duff, *Fifteenth-century English books,* which appeared as Vol. XVIII of the Illustrated Monographs of the Bibliographical Society.

In time, however, the material contained in such separate publications attained such volume that the need of a new comprehensive catalogue became urgently felt. Once more it was Konrad Burger who in 1892 came forward with the plan to bring Hain up to date or at least to complete it by a general index. Both these suggestions fell on fruitful ground. Gottfried Zedler in the year 1900 published a general index of all incunabula which were to be found in the libraries of Hesse-Nassau. Similar plans were followed also in other German provinces. The idea of a general catalogue was adopted on an imposing scale in France, where an enthusiastic amateur, Mlle. Marie Pellechet, who had edited catalogues of single collections of incunabula (see p. 16), approached the Ministry of Public Instruction with a plan to list the entire number of incunabula in the public libraries of the country, in a general descriptive catalogue. Permission was given her in the year 1897 to publish the first volume (A-Biblia) of the *Catalogue général des incunables des bibliothèques publiques de France,* and her generosity assured the state of the means wherewith the work could be carried further by a competent compiler. After her death M. Louis Polain was able to bring out two more volumes (Biel-Gregorius, Paris 1915 ff.) and the completion of this very important undertaking, the joint achievement of public and private initiative, may soon be expected.

The French model for a national general catalogue of early printing was very widely imitated. Italy as well as Spain, Portugal and Sweden enthusiastically received the plan for a national listing of incunabula. Up to this time however, only Sweden has made a more serious start towards its realization, a start there, also, due only to the personal effort of the author of the idea. Isak Collijn, at the time General Director of Swedish libraries, published in several volumes the catalogues of the incunabula in almost all Swedish collections of any importance (Upsala 1907, Vesterås [1904] Linköping [1910], Stockholm 1914-16). In the other countries of the Old World, the idea is even now but a plan, the execution of which is still in the distance. On the other hand the librarians of the United States eagerly seized upon the idea and in the *Census of early printed books owned in America,* published by Winship 1919, carried it through in typical fashion. The work of bringing Hain's *Repertorium Bibliographicum* by means of supplements, to the standards of to-day, has met with greater results. W. A. Copinger, President of the Bibliographical Society of London, had, like Burger, planned to supplement Hain's *Repertorium* because of the quantity of new material available on the subject, and when they ultimately met, there came out 1895-1902 the *Supplement to Hain's Repertorium Bibliographicum* which appeared under Copinger's name and to which Burger, under the title *The printers and publishers of the XVth century,* added an index, itself a valuable contribution.

Next to Hain's *Repertorium,* Copinger's *Supplement* is the most necessary book for the student of incunabula. It lists in the first part several thousand additions to the descriptions of such incunabula as Hain had instanced, but of which he was unable to give exhaustive details. The two volumes of the second part give more or less full information concerning 6619 incunabula known for the first time after the appearance of Hain's *Repertorium*. To be sure, only a very insignificant percentage of these books has been described by the authors as seen by them, Copinger's *Supplement* almost always taking over descriptions without comment, as they were given in separate catalogues or else in bibliographical guides. As the catalogues from which the information was taken differed greatly in value, the same is naturally true also of the information in the *Supplement,* all the more as the adoption of it was made with a grave lack of discrimination. This disadvantage is obviated to some extent, as the *Supplement* gives the source for every description, a proceeding by which a regrettable omission in Hain is somewhat repaired; yet as regards its method of dealing with incunabula the *Supplement* is by no means an improvement, while because of its shortcomings in critical faculty it is in fact inferior to the *Repertorium*.

The same reproach as to a lack in critical information must be brought against a second attempt at a supplement to Hain's *Repertorium*. The *Appendices ad Hainii-Copingeri Repertorium bibliographicum,* Fasc. I-VI, Indices & Suppl., by Dietrich Reichling (Munich 1905-11 and Münster 1914) are far superior to Copinger's *Supplement* since, at least in Part 2, they list only such incunabula as the author himself had opportunity to examine. These are principally Italian copies, and (Part 7) those in Swiss libraries, of which new items are described according to the method of Hain's *Repertorium*. Furthermore, the library owning each item is accurately given and the signatures of the volumes are noted. So far as it is a question of copies with printer given (of the 2143 incunabula noted for the first time by Reichling seven hundred and eighty-three are more or less fully assigned), we have here an extraordinarily valuable addition to our knowledge of incunabula. But the descriptions of undated copies must be used with great caution, as in the decisions regarding them very serious errors have often crept in.

In comparison with these voluminous supplements to Hain's *Repertorium,* neither the *Aggiunte e Correzioni all'Index di K. Burger* by Tammaro de Marinis (Florence 1904) nor the *Nachträge zu Hains Repertorium bibliographicum und seinen Fortsetzungen von der Kommission für den Gesamtkatalog der Wiegendrucke* (Leipzig 1910), is of importance. The latter was nothing more than an experiment for the future *Gesamtkatalog* and the former contains only a few more than one hundred items.

Although Hain often took pains to determine by comparison the origin of such copies as lacked the printer's name, his interest was, generally speaking, directed rather toward the literary than the technical aspect of incunabula, and as his method has been followed for two generations a similar interest in the literary versus the technical aspect is to be noted also in the work of his successors; and we observe in the case of certain promising undertakings consid-

erably earlier than the publication of Hain's *Repertorium,*
that they failed to bear fruit.

In 1787 Gerhoh Steigenberger, canon regular of Pol-
ling, published a brief work entitled: *Literarischkritische
Abhandlung über die zwo allerälteste gedruckte deutsche
Bibeln,* these Bibles being preserved in the Kurfürstliche
Bibliothek in Munich. This essay is accompanied by four
plates on which type-alphabets and specimen passages of
printing by Mentelin, Eggestein and Fyner are shown,
and in the preface the author explains that these plates
are by Johann Baptist Bernhart, "of especial skill in this
field, and collector of the alphabets of the first printers."
Johann Baptist Bernhart[1] in 1782 entered the service of the
Munich Hofbibliothek in which he remained until his
death in 1821. He had by his own choice devoted himself
to early printing, and as a proof of this taste, left behind
him a magnificent collection of drawings of old types and
initials, from which one realizes that he purposed to collect type- and initial-alphabets as examples of the stock
of the earliest printers. He had been a friend not only of
Steigenberger but also of Placidus Braun, and it is for
that reason very probable that the eleven copper plates
with seventy-five alphabets of types of early printers which
Braun had used in his *Notitia historico-litteraria de libris
ab artis typographicae inventione usque ad annum MD
impressis in bibliotheca liberi ac imperialis monasterii ad*

---

[1] Cf. E. Freys. *Joh. Bapt. Bernharts Gesammelte Schriften.* In *Wiegendrucke und Handschriften. Festgabe K. Haebler gewidmet.* (Leipzig
1919) pp. 145-74.

SS. Udalricum et Afram Augustae extantibus (Augsburg 1788-9. 4°), are to be attributed to Bernhart or that at least they owe their origin to his inspiration. Franz Gras made a third attempt in this direction in the years 1789-91 when he published a *Verzeichnis typographischer Denkmäler, welche sich in der Bibliothek des regulierten Chorherren- stiftes des heiligen Augustin zu Neustift in Tyrol befinden.* This book has fourteen plates on which Gras reproduced no fewer than one hundred and twenty-one alphabets of early printers, although, to be sure, in a form conveying only a very imperfect impression of the originals.

These publications record the earliest efforts to further the technical study of incunabula, endeavors which, it is astonishing to realize, remained for more than fifty years without a sequel. Because of the exclusively literary re- searches of the Hain school they were completely forgot- ten, and for that reason there is absolutely no connection between them and the next step in the same direction.

In the year 1857 J. W. Holtrop began the publication of his *Monuments typographiques des Pays-Bas au quin- zième siècle,* issued in twenty-four parts up to the year 1868. They were to form a companion volume and supple- ment to the author's *Catalogus . . . bibliotheca regia Ha- gana* (Hagae Comitum 1856) and thus aimed to illustrate the entire stock of the early printers of the Low Countries by giving specimens of their work. It is true, in doing this, the working-out of the type-alphabet was almost entirely neglected, but the examples are so numerous that they practically suffice for a knowledge of the peculiarities of the several types used in the Low Countries. Although to-day far surpassed as regards technical details (the plates were lithographed from tracings) Holtrop's work has the great merit of having almost completely attained its object, the illustration of the early printing of the Netherlands.

Thierry-Poux followed Holtrop's example with his *Premiers monuments de l'imprimerie en France au XVᵐᵉ siècle* (Paris 1890). Thierry-Poux had a great advantage over Holtrop in that his illustrations were made by a photographic process and thereby gave an essentially truer picture of the originals than Holtrop's lithographs. On the other hand the abundance of the material to be dealt with imposed on him much greater limitations than was the case with the early printing of the Low Countries. Thierry-Poux could no longer bring to light the entire stock of every printer but was obliged to limit himself to reproducing the first examples from each press. He thus showed fairly exhaustive material for the small printing-shops of the provincial cities but was able only imperfectly to do justice to the great and productive workshops of Lyons and Paris.

It was much easier for Gordon Duff, who, in his *Early English printing* (London 1896), could show in his forty plates not only a collection of types of the early English printers but could also even give specimens of the work of all those who had printed on the continent in the English language or for the English trade.

What Holtrop accomplished for the Low Countries, my *Tipografía ibérica* (La Haye-Leipzig 1901-02) was to do for the early printers of the Iberian Peninsula. The range of material was very much the same, but the difficulties which naturally arose in a foreigner's work did not allow the book to attain anything like such completeness. An additional difficulty was that for making reproductions, line-engraving was here used for the first time on a large scale, and that the originals were not as satisfactory as they should have been to meet the warranted requirements of this process. On that account the picture which the *Tipografía ibérica* gives of the early printing of Spain and Portugal is not to be considered complete, though, owing to the inaccessibility of this part of the incunabula output, it may be taken as presenting a good many points hitherto unknown and of scientific interest to the student.

The lacks which works previously mentioned had left in illustrating the accomplishments of the several countries in the printing art have in the process of time been almost completely filled. Anatole Claudin's monumental work, *Histoire de l'imprimerie en France au XVe et au XVIe siècle* (Paris 1900-20) did not, it is true, owing to its author's death, reach the extent at which he aimed, but in its four volumes it comprehends the early printing of Paris and Lyons with a fulness attained in no other work. The reproductions, too, excel those of almost all other books of similar character. Claudin was the first to carry out the fundamental principle of giving for each kind of type not only specimen text but also complete alphabets of the type used therein. The only fault which one can find with his book is that the author lacked the requisitely keen eye for fine distinctions in the shapes of the types, so that his specimen texts are often not wholly in agreement with his alphabets and the conclusions drawn therefrom.

In his *Fifteenth-century English books* (Oxford 1917), Gordon Duff again illustrated English early printing, with the same scope as in his *Early English printing,* but with fifty-three plates. And I myself endeavored, in the *Geschichte des Spanischen Frühdrucks in Stammbäumen* (Leipzig 1923) to fill the gaps which the *Tipografía ibérica* had left.

Besides these works on the history of early printing surveying it by individual countries, are some which are general in character, and others which aim at a still narrower limitation of their subject. It would be quite impossible to note here all the monographs devoted to single places of printing and single printers or individual collections, and equipped with illustrations of specimens of the most noteworthy productions. It is also impossible to note all the reproductions of incunabula which have ever appeared. Of older works which contain facsimiles of incunabula in greater numbers, Sotheby's *Typography of the fifteenth century* (London 1845) and Lippmann's *Druckschriften des XV bis XVIII Jahrhunderts* (Berlin 1884-7) at least should be mentioned. The most imposing undertakings in this direction were those begun by Konrad Burger and published by the Imperial printing-office, *Monumenta Germaniae et Italiae typographica* (Berlin 1892-1913), which strove to do nothing less than to illustrate by specimens, all the types used in the fifteenth century in Germany and Italy. The first publisher must finally have recognized that this would be an impossibility within the limits of a single work, and it was an advantage to the undertaking that Ernst Voulliéme upon taking over the winding up of the publication after Burger's death, confined himself chiefly to the German field, and instead of absolute fulness strove for the most obvious possible characterization of individual printing-shops.

Two English undertakings are of value to the student of incunabula. The Type Facsimile Society, founded in London, which for ten years distributed annually among its members some fifty photographic reproductions, aimed particularly to give an illustration of every product of the early presses, which the individual student could but seldom have an opportunity to examine in the original because of its great rarity. The Woolley photographs on the other hand are illustrations of incunabula which were in the private collection of George Dunn at Woolley Hall, and embrace, though apparently without a scientific viewpoint, the whole field of printing in the fifteenth century.

All these works based on the illustrative method, with the exception of Claudin's *Histoire* and my *Geschichte des Spanischen Frühdrucks,* have confined themselves to reproducing single pages of incunabula as type specimens. The Gesellschaft für Typenkunde des XV Jahrhunderts, founded by Isak Collijn, first undertook to give for every single type a specimen passage as well as a complete alphabet. The plan to make an exhaustive representation of the stock of every printing-shop has been carried out in the 1350 plates thus far published, though, it is true, these are concerned with only a few printing towns (Cologne, Lübeck, Magdeburg, Leipzig). It is to be hoped, however, that with a return of better business conditions the Gesellschaft will be able to resume its publications.

In the supplements of Copinger and Reichling, the Hain methods found their last effectiveness. In the meantime, however, a new era for the study of incunabula had dawned. In the year 1898 Robert Proctor published the first part of his *Index of the early printed books in the British Museum and in the Bodleian library of Oxford*. The idea of bringing together these two greatest public collections of incunabula was less happy in that Proctor was not in a position to deal even as exhaustively with the Oxford specimens as with those of the British Museum. Moreover it was not the incomparable richness of the London collection through which his publication gained its crowning importance, but rather the method followed in his *Index*. In the first place Proctor did not arrange his material as Hain and all his followers had done, alphabetically by authors, from a literary point of view, but technically, according to places and printers, arranging the places in the order in which they received the art, and listing the printers of each place according to the first appearance of each. Then too, Proctor abstained from any description of the text. He furnished only, so far as the book revealed it, the year and day of its publication and followed the very brief title with the information as to what types were used in its production. To this end, a short survey precedes each press, of the types which were used by it, and of these types the measure of twenty lines is almost always given, together with a brief description consisting chiefly of references to other related types.

Thereby an entirely new element was introduced into dealing with incunabula, placing the study of early printing on a wholly new basis. Apart from the unsuccessful beginning of J. B. Bernhart and his imitators, the technical identification of early printing which is of the utmost importance for the great number of undated books, had up to that time been left to the personal judgment of individual students. The custom had become established of calling the types of a book with no printer given, similar to those of some item of known origin and thereby of determining its assignment to a certain press. With these "similarities" individual choice was naturally allowed a wide scope, for there are scarcely any two men whose opinions as to similarity absolutely coincide. On the other hand, an English bibliophile, Henry Bradshaw, in a very obscure place (his *Memorandum II: A classified index to the XVth century books in the De Meyer collection sold at*

Ghent, November 1869, privately printed 1870), had set forth in a note a plan by which incunabula could be classified by an exact method according to type-forms, by genus, species, etc. In a later publication, by giving examples from Dutch printing-shops he gave a practical demonstration of his method, thereupon declaring that in dealing with incunabula the literary arrangement of Hain's *Repertorium* should no longer be the rule but rather the geographical-chronological order as used in Panzer's *Annales* and Holtrop's *Catalogus.* Although later Bradshaw's article found a wider circulation through its inclusion in his *Collected Papers* published in 1889, it was Proctor who first helped him to full recognition, for he not only acknowledged Bradshaw as his teacher, but undertook to use the Bradshaw method throughout the whole range of fifteenth-century printing, thus gaining for its originator a more general consideration and making the method a scientific achievement. That he modestly acknowledged what he owed to Bradshaw's initiative cannot detract from the value of his service.

From the scientific standpoint it is regrettable that Proctor did not develop his exact method abstractly, but applied it only to the case of a single collection. Moreover it did not spring from his brain in full perfection. It was an unfortunate blunder that he used as the basis of his main work a kind of type measurement, of the unsuitability of which he was himself convinced. He took the measure of twenty lines in such a way that he measured from the base of the first line, without taking descenders into consideration, to the upper edge of the twentieth line without regard to the ascenders. This measure falls short of the measure of twenty full lines on account of the number of ascenders and descenders and therefore remains indivisible, so that it could be read only of twenty full lines. It would therefore be useless for all display types of which there would be often but one line, or at most only a few lines, to be measured at one time. He saw this error himself, and in the supplements of which he published four for the years 1899-1902, he recommended the measure of twenty lines from the base of the first to the base of the twenty-first line. This measure is divisible even to the minutest degree, so that it can be used for any number of lines. A second disadvantage was his strict chronological arrangement and the corresponding notation of his types. In spite of the extraordinarily rich store of incunabula in the British Museum it still could not fail to happen that the dates of some which later came to light upset Proctor's chronological order. Therefore in the supplement he was betrayed into altering the notation of the corresponding types in every such case, so that often the type numbers of the supplement no longer indicate the same types as do those of the main work. Finally, Proctor retained to a far greater extent than his method allowed, the idea of "similarity" of type-forms. It was peculiarly unfortunate that because of the form of the catalogue he was obliged to characterize the type in the most concise possible way. For this reason he indicated all types only by the press number and their own; hence the comparisons lack clearness, and are really only fully useful to one who has the entire resources of the British Museum or a somewhat similar and extensive collection at his command. Moreover, in these comparisons as I have said, he relied to a considerable extent upon an idea of "similarity" which is in absolute contradiction to the method itself. The danger of this procedure strikes one at once if he takes up Proctor's catalogue of the German books of the sixteenth century. Here for the purpose of comparison he reproduced characteristic specimens of presses, and numbered the types by them. In this he relied entirely upon his individual opinion as to similarity without being able to give exact and convincing reasons for it. Using this volume on the sixteenth century as a test, we should therefore be more or less cautious in accepting the conclusions of the earlier volume.

This is no reason however for doubting the significance of Proctor's pioneer method. He could not, unfortunately, make up his mind to assist in the improvement of his system, and accordingly I was obliged to resolve without his help to apply it with exactitude in the *Typenrepertorium der Wiegendrucke*.

The *Typenrepertorium* was first of all to be nothing more than the freeing of the Proctor system from a single collection. The first volume is almost entirely restricted to carrying out *in abstracto* what Proctor on the basis of the British Museum resources had offered in a way which was not sufficiently practical. The most important innovation was the introduction of the M-form in addition to the measure, the diversity of which had already occurred to me before I knew Proctor's work. The following volumes were even more independent of Proctor since I was convinced that the other countries in the collections of the British Museum were by no means as exhaustively represented as was true of Germany. For the rest, the *Typenrepertorium* attempted only, depending upon the methods of Bradshaw and Proctor, to do away absolutely with the idea of similarity in type comparisons, and to substitute for it the consideration of distinguishing characteristics.

More recent study of incunabula has recognized the *Typenrepertorium* as a foundation for the determination of their origins. The *Catalogue of early printed books in the British Museum,* which when completed is certain to replace the Proctor *Index,* has, it is true, kept an independent kind of type notation; but it is constructed throughout on the foundation laid by the *Typenrepertorium.* In the *Catalogue,* however, the method is carried still further, so that often in what the *Typenrepertorium* considers a certain type, the *Catalogue* make a finer distinction. The *Typenrepertorium* was to serve as a basis for just such research, and it is only to be regretted that up to this time work in this direction has not been carried on to a greater extent. The *Catalogue* of the British Museum shows how modern incunabula research has settled its problems not only as regards the study of type but in various other directions as well. It does justice not only to the technical but also to the literary aspect of incunabula study and as a special catalogue of a single collection sets a standard not thus far attained elsewhere.

The plans for a new edition of Hain's *Repertorium* never quite dropped after Burger's stirring up of the subject in 1892. This point only was clear, that it could not be satisfactory simply to assemble with the old resources of Hain as Burger had intended, the abundant material which subsequent publications had brought to light, but that the new edition of Hain's work must be founded on a fresh survey of resources. Preliminary steps to this end were being taken in several places when the Prussian Minister of Public Instruction represented by his Excellency Friedrich Althoff, determined to make plans on a larger scale. In the winter of 1904-5 a Commission for the *Gesamtkatalog* was appointed, in which a number of noted German students of incunabula were called together with the task first of all of putting the project on a sound basis by a careful inventory of German resources. At the same time connections were formed with foreign countries, with the object of promoting similar efforts, as in the case of Switzerland, the northern countries, Austria, Italy and America, or, as in Spain, England and Belgium, with that of sending workers from Germany. In 1911 when the inventory in Germany was finished, the Commission set about producing monographs according to a uniform plan, from the material collected both in Germany and in foreign countries. At the Exhibit of the Arts of the Book in Leipzig, specimens were shown of all sections of the projected *Gesamtkatalog der Wiegendrucke,* descriptions, printers' index, subject index, and information as to copies, but the outbreak of the war prevented these from receiving the desired consideration and criticism. The Commission was able, it is true, to continue its work during the period of the war without serious interruption, but in spite of this it was not possible to carry out the original plans to the fullest extent. The inventory in Austria was definitely broken off long before its completion, the just-gained contact with Italy was lost, and the German government declared itself not only not in a position to grant the means to carry out the publication as in previous cases, with a German personnel, but saw no prospect even of money for the printing of the catalogue. The Commission had perforce therefore to be glad to find at least a private publisher for the work, to keep it from the fate of being covered with dust in a library's archives. Five volumes have appeared in the years from 1925 to 1932, and it is to be hoped that publication will continue at the same rate. The future must show whether this work, designed to replace Hain's *Repertorium,* will succeed in giving the study of incunabula an impulse similar to that which it received from Hain.

3. HISTORY OF EARLY PRINTING

The early history of the art of printing is not in itself a part of the study of incunabula, but an independent branch of knowledge concerned with the events of the period of early printing, while, properly speaking, the study of incunabula considers the methods of the earliest bookmaking. Both however often touch upon the same material and their fields so often meet that one cannot entirely separate them. In spite of this, I believe that I can limit myself here to a very brief sketch.

That Johann Gutenberg of Mainz was the inventor of the art of printing cannot be doubted, judging by the unanimous evidences of the fifteenth century. The imperfectly established legends of the sixteenth century cannot shake this conviction. When and where however Gutenberg made such progress in his endeavors that he could make practical use of them is not known with certainty. They must already have gone far in Strassburg, for in Gutenberg's lawsuit with the Dritzehn brothers in 1438 the goldsmith Dünne testified that he had earned a considerable sum of money by furnishing printing materials. The earliest production which can be dated with any certainty is the astronomical calendar for the year 1448 which may have originated in Mainz. It is very doubtful whether one may consider the various small books, which were produced with the types of the 36- and 42-line Bibles, as developments; and it is rather improbable that so large a work as the German *Sybillenbuch,* from which the fragment of the *Weltgericht* derives, belongs to an especially early period. The various differences in the types of the earliest small books rather reduce to the supposition that a number of makers shared in them. Gutenberg for a long time earned his living by giving lessons in his "secret" art to certain interested persons in return for pay. One may naturally assume that these pupils must have taken pains to turn their learning to practical account in order to defray the cost of training. From these came probably many of the surviving fragments and it is only natural that in different hands slight variations in the types should be noticeable.

The first great production of the new art was the 42-line Bible, which probably originated with Gutenberg and was the subject of his lawsuit with Johann Fust. According to that, it may have been finished approximately in the year 1455. The issue of the suit between Gutenberg and Fust is unknown to us, but in later times the type of the 42-line Bible was in the possession of Peter Schöffer who, although Gutenberg's pupil, had gone over to Fust's establishment; hence probably the stock of the 42-line Bible of Gutenberg had been taken from him and adjudged to Fust. The Humery documents prove that Gutenberg had procured new stock. Whether he really was the printer of the 1460 *Catholicon* cannot be proved; in any case he died in 1468 without having organized another printing establishment of greater productive power.

The spread of the art of printing had already begun

before the taking of Mainz by Adolf of Nassau in the year 1462, and the influence of this event has been too much over-estimated. The printing-house of Fust and Schöffer was in existence during and after the ecclesiastical controversy in Mainz, and it can be proved that a number of presses existed before 1462. It is accepted as a fact that the Indulgences of 1454-5 were from two different workshops. The 36-line Bible probably did not come from either Gutenberg, Pfister or Schöffer; but Pfister was printing after 1460 in Bamberg. From still another press came the Vienna calendar for 1462, which must have come out at latest, in the beginning of that year. Of the printers whom one must consider as pupils of Gutenberg and Schöffer, Berthold Ruppel in Basel, Philip Kefer in Nuremberg, Johann Neumeister in Foligno, Ulrich Zell in Cologne, not one appears to have begun his work in connection with the Mainz catastrophe. Only in the second half of the sixties did the art spread to Cologne, to Strassburg, to Basel. At the same time it pushed on over the Alps to Subiaco and Rome. The impulse towards the new art must have been very strong at about that time, for with the year 1470, especially in Italy, new printing establishments sprang up like mushrooms, and each of these presses had to have at least one master who was thoroughly familiar with all of the duties connected with printing.

Gutenberg had chosen the gothic missal letter as a model, and all the earliest German printers used gothic type until Adolf Rusch in Strassburg about 1464 first undertook to use the humanistic roman letter also. Probably this type served as a model for Conrad Sweynheym and Arnold Pannartz in Subiaco. In Rome the roman type at once attained absolute supremacy, and in the type of Nicolaus Jenson in Venice reached its highest perfection.

In the seventies the triumphal progress of the art of printing had made its way over the whole civilized world. They were almost without exception German masters who introduced it into foreign countries, but, surprisingly, leadership had even so soon gone over to Italy. Not only French, Spanish, Dutch and English printers boasted of printing with Venetian types, but Germany herself voluntarily recognized the supremacy of the Italian school of printing, although German or at least German-trained masters had laid the foundation of its fame.

As early as the years 1480-1500 the new art began to gather the fruits of the tree which had quickly sprung up in the seventies. Printing had overcome its early imperfections and begun to be conscious of its mission to the history of civilization. In addition to reproducing the intellectual legacy of the past, the new method of dissemination began to avail itself of contemporary literary achievements. Printing was now at the service of various endeavors in every conceivable department of civilization. In the great intellectual movement of the Reformation this development first came to fulfilment; yet one may almost believe that by the year 1500 printing had already attained the form in which it has come down to posterity. It is

this process of growth which lends the study of incunabula a special charm.

THE BOOK · *Make-up*

1. PAPER

THERE is no one precise mark by which one can recognize whether a book belongs to the incunabula period. This is obvious at once when we realize that a very large number of printers whose activity had begun in the time of early printing were still at work far into the sixteenth century, and therefore, naturally used the same stock and followed the same customs not only before but after the year 1500. An absolutely certain decision is possible, consequently, first of all only for those incunabula which themselves record something regarding their origin. They did not, or at least only very exceptionally, do that as do modern books, on a title-page. The great majority of them wholly lacked such a thing, but adopted the custom from the makers of manuscripts, of putting this information at the end of the book. As a compensation this last passage, the colophon, often contains a much greater amount of information than is usually found on a title-page. The colophon as a rule names, besides the title and author of the book, not only the place and date of its origin as well as the printer's name, but also often includes valuable literary notes regarding the composition of the book, its editor and corrector, who prepared the text and oversaw the printing, as well as the publisher and the patron who made possible the publication of the work.

[43]

PAPER

To be sure, there is an extraordinarily large number of incunabula, estimated as at least one-third the whole, which lack such a colophon. Among these there naturally occur a great many with features not absolutely established as those of incunabula. But since, as I said, a general distinctive mark is lacking, the study of incunabula has followed the ·ule of considering as such all books which could derive from the early printing period, so long as neither external nor internal evidence can be adduced that they can only have made their first appearance later than 1500. In this way many books which have been listed as incunabula by the bibliographers, Hain and Copinger not excepted, later had to be omitted from the list, since an accurate examination showed that in the text persons and events were noted which belonged only to the period after 1500, or that they were printed with stock of the use of which in the fifteenth century no evidence can be obtained. In spite of this, the newest incunabula catalogues probably still include many a book which has not an absolute right to a place there, and fresh research will keep on divesting of this character some books which now pass as incunabula. But until such a proof is forthcoming in individual cases it is unreservedly recommended to record as incunabula all books which give any impression of being products of the period of early printing.

There are of course a large number of characteristic features by which one may assume that he is dealing with an incunabulum. Even the paper on which incunabula were printed gives an essential clue in judging them. It is true that even in the early period the quality of the paper used for printing gradually deteriorated as time went on. Local reasons also occasionally determined the use of inferior paper in books which nevertheless belong without doubt to the early period. In general, however, early printing is marked by its appearance on very stout paper which is not really white, but of a greyish tint. When one comes upon incunabula printed on paper which, although in stoutness and firmness like that of early printing, is chalky-white in color, he may always assume that it is a copy which has been washed in recent times. By this term we mean that in order to improve its appearance or to remove traces of earlier neglect, so increasing its value, it has been treated with sundry chemicals. Earlier incunabula were in any case usually printed on a thick, somewhat yellowish paper of such superior quality that to-day after more than four hundred years, wherever it has not met with especially unfortunate treatment, its original color and consistency still endure.

Italy was the country which supplied most of the paper used in early printing. There, even before the invention of printing, paper-making had become very widespread, and many of the small Italian printing-towns of the seventies owed the early advent of printing within their walls to the fact that satisfactory paper was to be found in the place itself or in its immediate neighborhood. How closely related were paper-making and early printing may be seen from the fact that in many of the earlier printing contracts even the supply of pulp for paper-making is mentioned. Some paper was also already made outside of Italy by the middle of the fifteenth century and the great demand which followed the spread of printing soon brought about the establishment of more places for paper-making in Germany and France and even also in Spain.

In order to estimate correctly the significance of paper in the study of incunabula one must grasp the process itself, at least in broad outlines. The pulp of that period was scooped by hand out of the vats which held the liquid mass, by means of molds of thick wire net with a wooden frame. The wires running closely side by side were held in position by others at right angles further apart, and in the centre of one half of the frame was usually introduced, likewise by the aid of wires, a figure which in the finished sheet appeared as a watermark. The earlier kinds of paper in the period of early printing almost without exception had a watermark, yet the lack of one is not a sufficient reason for saying that a book on unmarked paper could not be an incunabulum. In the incunabula period paper without watermarks was now and then used, more often however in Germany than in Italy, but it appears almost invariably that these papers were of inferior quality.

The examination of watermarks was for some time thought to prove of great advantage in the study of early printing, but it has been shown that the results attained have fulfilled these expectations only in very small part.

In the first place an examination has not yet been able to explain the significance of watermarks with absolute certainty. In the beginning, at all events, they were indications of origin, the trademark, as it were, of the paper-mill from which they came. Then by degrees they seem to have assumed the character also of a mark of quality. It is unreasonable to assume that the enormous amount of paper used in incunabula with the Venetian mark of a balance in a circle or the Genoese mark of a hand and star must have come from a single paper manufacturer. From documentary sources concerning paper supply it may be concluded that without doubt certain watermarks at least were at the same time marks of quality, as for instance may be cited the very wide currency given some marks (the bull's head in particular, as well as the above named). Hence the conviction of their significance as indicative of origin is shaken, and is further affected by the method of their execution. When one realizes that the watermark resulted from the impression of a wire figure which was fixed in the network of the deckle, it follows directly that an absolute similarity in marks of a certain mill would be an impossibility, since every efficient mill would employ a large number of deckles and even the greatest care would be unable to keep individual figures made by hand out of such stubborn material as metallic wire, from varying from one another in minor ways.

Still another fact appears. Investigations as to the printing of the 42-line Bible have led to the surprising conclusion that even this printing of the earliest period was not done on one kind of paper alone, but that the individual gatherings of the book were made up in a very systematic way from various kinds of paper and then put together. This was apparently an isolated occurrence, the reasons for which are not at once discernible. In reality, however, paper with different watermarks was used in almost all very large examples of the period of early printing and even copies of the same book by no means agree with regard to the paper used in them. Hence it appears that hard and fast conclusions in forming estimates as to incunabula are not to be gained by examining watermarks, and that not even a wholesale investigation of these in incunabula would warrant the hope that the amount of information forthcoming would be proportionate to the enormous labor involved.

I do not mean to say however, that for this reason just given, the examination of watermarks may not lead in some cases to very valuable conclusions, as for example in a book of Erhard Ratdolt's which bears the impossible date "M.cccclxviij.vij. calen. Decembris." (Mataratius, *De componendis versibus*. Hain 10889). This book was printed on paper with the rare mark of a flying dove, which Ratdolt used apparently but once more, in a Publicius, *Artis oratoriae epitome*, which bears the date of November 30 (pridie calen. decembris) 1482. It may be considered as almost absolutely certain that the *Mataratius* appeared at approximately the same time. Although, then, a general consideration of watermarks holds out few prospects in the study of early printing, their examination in individual cases will assuredly show them to be of real use.

2. SIZE

The paper of the early printing period was mainly of two sizes, a large folio, "forma regalis," and a small folio, "forma mediana" or "communis." The sheet of "forma regalis" measured approximately 70 x 50 cm. and folded once gave the large folio size of incunabula as seen in the *Speculum* of Vincentius Bellovacensis, the *Biblia* of Adolf Rusch, and other books. The dimensions of "forma mediana" were subject to somewhat greater variations. An average size might be 50 x 30 cm. From this size the majority of early folios were made.

The size of incunabula is not indicated quite as is that of the modern book. Nowadays the gathering of a book is identical with the sheet, and the size can accordingly be simply determined from the number of leaves combined in a gathering. The early printer on the contrary, especially in his folio size, combined whatever number of leaves he pleased in a gathering, so that this number is useless in determining the size, which may be done by referring to the chain lines in the paper. In the deckle the lines close together run parallel to the long side of the rectangle while those widely spaced run parallel to the short side. In the sheet folded once, the folio size, the lines close together must run horizontally; in a twice folded sheet or quarto, vertically, and once more horizontally in the four times folded, or octavo. By this at the same time is determined the position of the watermark, which is to be found in folio size in the centre of the leaf; in quarto, in the fold of the leaf; in octavo, again approximately in the centre of the upper edge. It appears therefore that the size of incunabula is as independent of their bulk as of the number of leaves in a gathering. It is possible however for the inside of a book to vary its size, for occasionally, although not usually, early printers, more often in Italy than in Germany, used for the gathering of the usual or small folio the half sheet of the "forma regalis," so that in some gatherings or even in single leaves of a gathering, the chain lines were in quarto position while the other leaves and gatherings were printed in folio. The same mixtures occurred in quartos, some leaves of which were added in octavo.[1]

The quarto gatherings of the incunabula period were seldom made up from a twice folded sheet. Moreover in the earliest times especially, the number of leaves in a gathering was arbitrary. Gradually, then, the custom was formed of making up the quarto gathering from two sheets, eight leaves. In this combination however the vari-

[1] For folio and quarto, cf. *Eusebius*, Eggestein s.a. Hain 6708. B.M.C. I p. 73; *Decisiones*, Schöffer 1477. Hain 6047. B.M.C.I p. 33; *Hilarion*, Lignamine 1473. Cop. 2978. B.M.C.IV p. 31; *Caracciolus*, Wendelin of Speyer 1473. Hain 4430. B.M.C. V p. 163. For quarto and octavo, cf. *Psalterium*, Eggestein s.a. Hain 13512. B.M.C. I p. 74; *Modestus*, Schurener s.a. Hain 11441. B.M.C. IV p. 59; *Galeottus*, Lerouge 1476. Hain 7437. B.M.C. V p. 215.

ous printers proceeded differently. Some folded every sheet by itself and laid the folded sheets into each other; others put the sheets together and folded them all at once. The process used is sometimes to be recognized only by the signatures where those exist, or else by the arrangement of the leaves having watermarks. The make-up of the quarto gathering from a single, twice folded sheet, also a gathering of four leaves, as became quite universal in the printing of the Reformation period, occurred only very exceptionally in the incunabula period. If one comes upon a book with four-leaf gatherings he may always assume that it belongs to the sixteenth century.

For the octavo, which became adopted very gradually in the early period and was at first but little used, it was customary from the start to make up the gathering from a quadruple sheet. Then it came about that two sheets were placed together for a gathering, amounting to sixteen leaves.

The unusual sizes, 12°, 16°, 24°, occurred towards the end of the early period, particularly in liturgical books, breviaries and prayer books, but they so seldom occur and their make-up is subject to so few fixed rules that there is no need to describe them in detail.

On the other hand the make-up of the gatherings in the folio deserves a still more detailed discussion. Gutenberg made up the 42-line Bible in gatherings of ten leaves, a fact which is noteworthy in that it required an extraordinarily large stock of type. It was not, it is true, absolutely necessary to assemble the type for twenty complete pages, for a gathering of ten leaves. Instead, when the setting reached the fifth double sheet (pp. 9-12), the printer could begin to print the inner sheet of the gathering, so that its type could be distributed during the setting of leaves 7-10. Still, the large gatherings depended on an especially large stock of type, for the advantage just noted would have been attained just as well with smaller gatherings. Nevertheless, Gutenberg's own pupils and in fact the earliest printers in general kept at first in a marked degree to the gathering of ten leaves, and the expression "quinternio" came into such common use that even such gatherings as no longer consisted of ten leaves were still designated as quinternions. The rule was so general a one however that even quartos were made up in gatherings of ten leaves (five half folio sheets).

Among the gatherings of ten leaves in the earliest books, some gatherings of more or fewer than ten leaves occurred, because even from the first, large volumes in the haste of their making were divided into a larger or smaller number of parts on which work could be carried on simultaneously. The occurrence of irregular gatherings is explained also by the fact that most of the older printing-houses were by no means dependent on a single press but usually had a large number of them at work. Investigations have shown that the 42-line Bible was made up of four separately set parts and that probably these parts came from a like number of presses. The earliest records concerning the first printing-shop in Italy either expressly state or imply that the publishers of the books originating in the early seventies almost always allotted them to a large number of presses. Even so unpretentious an establishment as that of Johannes Reinhardi in Trevi, which so far as we know issued but one extensive work besides one minor item, had no fewer than four presses at work. From this evidence we could also conclude that wherever an irregularity is noted in the make-up of the gatherings in a large book, it was due to a setting in parts, and the number of such parts in any one book will determine for us the number of presses which produced it.

This is true however only for the earlier incunabula period. The gathering of ten leaves was displaced in time by the gatherings in eights, which also became the rule for the folio, so that the number of leaves to a gathering became the same, generally speaking, for all three ordinary sizes.

A great many printers however did not make up their productions in gatherings of like numbers but in regular alternations of different numbers of leaves, and this surprising occurrence has not even yet been conclusively explained. The very earliest example of this is Perottus, *Rudimenta grammaticae*, which Wendelinus de Wila finished September 20 (duodecimo Kalendas Octobrias) 1475, which has, besides a first gathering of ten leaves, thirteen gatherings of alternating eights and sixes. In later times this was the custom of the Strassburg printers especially, and in their folios are often found in regular alternation not only gatherings of eights and sixes but also even of sixes and fours.

3. REGISTER

Because of such diversity in the make-up of the gatherings it is easy to understand that a need arose very early for some arrangement for the correct assembling of the printed sheets of a book. The printers of the 42- and 36-line Bibles had already done something of the kind, though to be sure the "Tabula rubricarum" which was added to these books on special leaves (which naturally were not preserved in the majority of cases) is to be considered quite as much a guide to the rubricator for the completion by hand of the titles to the separate parts. Still it could serve likewise for the checking of the completeness and the correct assembling of the book. The "Tabula rubricarum" which is first met with in some of the earliest Basel and Strassburg books as a special leaf,[1] very soon became an integral part of the book and was the forerunner of the modern table of contents.

The word "registrum" in incunabula had two meanings. In the books of German origin it was often used in the sense in which we use it to-day, that of a table of contents, which was usually known as "Tabula" or "Index." In the books which came from Italy, France or Spain the word however in most cases had a special meaning. It

[1] Cf. *Alphonsus de Spina,* Mentelin. Hain 872. B.M.C. I p. 55; *Bonaventura,* Eggestein. Hain 3535. B.M.C. I p. 70; *Biblia,* Ruppel. Hain 3045. B.M.C. III p. 714.

comprised a summary, usually made up from their first words, of the gatherings and sheets of the book, and was to enable the printer, bookbinder or owner to check the volume's completeness and correctness of make-up.

This register was an Italian device. The first to make use of it was the Roman printer of the *Epistolae Hieronymi* (Hain 8550), which was published without the printer's name but is attributed to Sixtus Riessinger, and was printed at latest by 1470 if not by 1469. To be sure in this case the register was still undeveloped and did not bear that name. The printer however under the title "Inchoationes quinternorum" gave a list of the first words of each of the eighty or so gatherings which made up the two volumes of the work. Under this name the register once more occurred, used by Adam Rot in Rome, who in 1474 supplied an edition of the *Consilia* of Petrus de Ancharano (Hain 945) with a similar "Inchoationes quinternorum." In reality however he gave not only the catchwords of the gatherings but, as in the normal register, those of the double sheets. Ulrich Hahn first gave the register this normal form in his Turrecremata, *Expositio Psalterii,* of October 4, 1470. He did not consider an explanation of the meaning of the register necessary, but other printers made up for the omission. Gerard Lisa in Treviso in the *Tesoro* of Brunetto Latini of December 16, 1474, did so at great length. His explanation ran as follows: "elle di sapere chel primo quinternio a nome a, el segundo a nome b, el terzo c: e cosi seguendo fino al ultimo quinternio, el quale ha nome o. Onde troverete qui da sotto la continuazzione della prima carta alla seconda, della segonda alla terza fin alla carta da mezzo di chiascun quinternio, e da puo si dice el cominciamento dell altro quinternio per ordine." Even in 1484 Georg Stuchs in Nuremberg in his *Missale Romanum* (Hain 11384) felt impelled to make a similar, briefer comment: "Nota primum verbum cuiuslibet folii usque ad medium quaterni." The usual form was to arrange the catchwords in as many vertical columns as the size of the book necessitated and to indicate the beginning of a new gathering by leaving a line blank. To the end of the incunabula period this remained not the only, but the most widely used form of register. Although the designation "registrum" was the rule, many other names also appeared. When they stopped printing in quinternions they wrote also "registrum quaternorum;" but they did not apply it exactly. In 1481 Georg Herolt in his Origenes, *Contra Celsum,* called his index "Registrum quinternorum," although of the thirty-two gatherings of the book only five were really quinternions; and even in 1495 Andreas Torresanus in Venice in his Paulus de Castro, *Super Digesto Veteri,* did a somewhat similar thing. Most printers solved the difficulty by naming the index "Registrum foliorum" or "chartarum." Many other names also came into use. Johannes de Colonia and Johannes Manthen wrote, 1475, in Alexander de Ales, *Super tertio sententiarum:* "Tabula chartarum secundum ordinem ponendarum," which Stephan Plannck, 1491, in his *Wie Rom gebauet ward,* in German, imitated thus: "die ordenung von den quatern vindestu nach der ordenung vnd vswisung des a.b.c." In 1477 Aloisius Siliprandus in the *Missale Romanum* called the register "Speculum presentis voluminis," in 1485 Anton Koberger in Paulus de Castro, *Consilia,* named it "Numerus et ordo quaternorum," and Andreas Torresanus, 1494, called it "Ordo chartarum," in a *Breviarium de camera.* In the *Super Digesto Veteri* of 1495 he called it "Signa quinternorum," and in 1486 Nicolaus Battibovis in Lucan's *Pharsalia* noted it as "Examen voluminis." Aldus Manutius had no designation for it in Greek, but Zacharias Kallierges in his *Etymologicum magnum* and elsewhere called it "Τῶν τετραδίων ἰθύτης."

Meanwhile other forms for the register had become accepted. Instead of the arrangement of catchwords in vertical columns usual at first, in which a line left blank meant the beginning of a new gathering, Georg Sachsel and Bartholomäus Golsch used a horizontal arrangement. In their Sabinus, *Paradoxa in Iuvenali* of August 9, 1474, they made this statement: "Sequitur tabula foliorum et quelibet riga continet unum quinternionem." But this arrangement was much less clear than the vertical, and it was less widely imitated. On the other hand, the register in columns was improved by designating the gatherings by the letters of the alphabet even before printed signatures in the books themselves came into general use. Eucharius Silber did this for example in his Pompilius,

Syllabae (Hain 13254), printed July, 1488. Especially in the case of the Roman printers did the register become so necessary a part of the book that we find one used by Stephan Plannck even in a book of only three gatherings (Petrus de Abano, *De venenis,* April 29, 1484, Hain 11 *et al.*). It was however a misconception of the real purpose of the register when Johannes Renensis in Viterbo in the *Miracoli della vergine* (Hain 11229), finished September 1, 1476, reproduced the first word of each page.

Which printer first conceived the idea of doing away with the enumeration of the catchwords after the introduction of printed signatures is difficult to determine. Printers were prepared for this development, for some, as for example Antonino Miscomini in Venice in his *Biblia Italica* of 1477 (Hain 3151), added to the signatures in the register how many leaves there were to a gathering (a, quinternio, etc.), and then added the catchwords of the double sheets. Only toward the middle of the eighties was the custom adopted of confining the characterization of the gathering to giving the number of leaves, omitting the catchwords. The first to do this was perhaps Hermann Liechtenstein who in the *Catholicon* of Balbus of September 24, 1483, gave a register by merely enumerating the signatures and adding to each, whether the gathering was quinternus, quaternus or ternus. This form became widespread among the Venetian printers, but only very few (*e.g.,* Simon de Luere) used it exclusively. The question of space seems to have been decisive in choosing one or the other form of register. If the printer had a blank page at his disposal he usually kept to the old form and enumerated the catchwords of the double sheets in vertical columns. If space was limited, he merely designated the signatures, with the size of the individual gatherings. In doing this the printers availed themselves of various forms. Even the "Inchoationes quinternorum," although to be sure without that name, came into use again. In a *Nonius Marcellus* . . . of June 8, 1492, Nicolaus de Ferrariis once more added the catchword of every gathering and indicated its size by a Roman numeral placed beside it. Usually it is true, signatures took the place of catchwords, and commonly instead of the number, the designation quaternus, ternus, duernus, etc. Less often was the form chosen which placed the signatures in a line side by side, while the numbers of the gatherings were placed in a second line above or below. In a *Doctrinale* of Alexander de Villadei of July 4, 1486, Petrus de Piasiis even used Arabic numerals, in which Bernardinus de Choris imitated him in his Nicolaus de Orbellis, *Super textu Petri Hispani* of November 7, 1489. Ultimately the printers went a step further and no longer entered the numbers of the gatherings singly but gave them only in a summary. Such a register was put in a few lines and after the row of signature letters had only the note: "Omnes sunt quaterni praeter xy qui sunt terni." This as well as the column form was good usage in Italian printing after about 1485.

The Italians introduced still another innovation in the

register. In a *Biblia Italica* of October 31, 1487, Johannes Rubeus in Venice gave a full register in five columns below which he noted: "In questa opera sono quinterni quarantacinque e charta una." This statement did not agree however with the proportions of the gatherings, which varied from six to twenty leaves to a gathering, but correctly gave the leaves as 452. Doubtless this had something to do with the custom which we often meet with in Italian publishers' accounts, of reckoning the price of the book by its quinternions, thirty to thirty-five of which were sold at a ducat at that time. This example had little following. Only one, Simon de Luere, after 1499 quite consistently gave after each register the size of the book, abbreviating the words to "qut." and "c".

An apparent difficulty occurred in the register in many Italian quartos. Most of their gatherings were in eights and were at first almost always treated in the register in the same way as folios. In other books, however, the register not only omitted, as always, the second half of the gathering, but also leaves of the first. On closer investigation it is plain that the register gave the catchwords of leaves one, three, five, etc. The reason for this is however quite simple. The quarto gatherings were made up of the same folio sheets as those of the folio, and for a gathering in eights, two of these were needed. If one laid these one upon another before folding, the signatures followed along on leaves one, two, etc. But if one folded every sheet by itself and laid the sheets so folded, into each other, then the sheet signatures stood on leaves one, three, etc. Roman, but especially Florentine, printers transferred this custom to the register, which thus took on a form less easy to understand. The regular use of the register by the Italians was accompanied by occasional carelessness and errors. That the prefatory matter, if it was made up of one gathering, was often disregarded should not be called absolutely an error, but the information concerning the size and make-up of the gatherings was frequently wrong. It was gross carelessness for a printer to use the register of his original text unchanged in a case in which it did not agree with his own edition. This happened for example in the second edition of the *Opuscula* of Philippus de Barberiis by Johannes Philippus de Lignamine in Rome (Hain 2455), and two Venetian printers, Georgius Arrivabene and Petrus de Quarengis, were also guilty of similar negligence.

The register was of Italian origin and was at first therefore used only in Italy. France and Spain soon adopted it from Italian sources and it quickly gained currency in both countries, although received with surprisingly little favor among German printers. It might be thought that this statement is not founded on fact, since a register is to be found in several earlier German books which, printed on a special leaf similar to a "Tabula rubricarum," was preserved in only a few copies. Laire notes one in an unknown Eggestein book, and according to Pellechet, Thomas Aquinas, *Summa,* pars tertia [Basileae, Michael Wenssler], contains one, but it is undeniable that the register was far from being as widely used in Germany as in Italy. When Friedrich Creussner, Nuremberg 1473, supplied a register for his Sixtus IV, *De sanguine Christi,* it was merely the result of his finding one in the original Rome edition, for elsewhere he almost never used one. The earliest independent register in German printing is that of the anonymous printer of Burgdorf who put one in his Jacobus de Clusa, *De animabus exutis,* 1475. Among the Swiss printers the register was on the whole more customary. In Germany the Nuremberg printers alone used it fairly often, but only in quite isolated cases did it occur elsewhere. In Cologne, so far as I know, it was not used at all.

4. SIGNATURES

Signatures served a purpose similar to that of the register. They were a German invention and their use is probably almost as old as the printing art itself. It is very evident that there was a need for a convenient method of assembling in orderly sequence, from the piles of loose printed sheets, the hundreds of copies of the same book. Albert Pfister used signatures in Bamberg, 1460-61, although this practice was probably not his invention but one learned by him from his teachers. These earliest signatures were naturally written by hand and were usually set so close to the right lower corner of the leaf that most of them were cut off when the book was bound.

The first to add signatures by printing them simultaneously with the text was Johann Koelhoff in Cologne.

SIGNATURES

Even the first specimens from his press (1472) contained signatures, which were placed directly under the right corner of the text on the double sheets so that the leaves of the first half of each gathering always had signatures while those of the second lacked them. This invention of Koelhoff's spread very rapidly in Germany. As early as the year 1472 the anonymous printer of the *Gesta Christi* in Speyer printed signatures in his books, and after 1474 they occurred, even though irregularly, in scattered books of German origin as well as in those of other countries. To be sure, manuscript signatures seem to have held their own with the printed ones somewhat longer in Italy than in Germany. Whether a step between the two, the addition of signatures by means of stamping them in by hand, can also count fleetingly as an intermediate stage, is doubtful. It appears that Mentelin in the *Speculum historiale,* Strassburg, 1473, marked his sheets in this way, but hand-stamped signatures in books from German workshops are rather rare and almost always belong to a period in which it had become customary elsewhere to print signatures and text at the same time.

Hand-stamped signatures are more often found in examples from Italian printing-shops. They seem to have been in general use in Milan especially, for a very long time, until the custom became established there of printing the signatures on the sheet at the same time as the text. Italy in general followed the German example somewhat tardily. Printed signatures in some books did occur after 1474 in scattered workshops of upper Italy but it was comparatively speaking a very long time before their consistent use became the rule in Italian printing-houses.

Sweynheym and Pannartz in Rome, in whose books manuscript signatures may be seen, may have used Arabic figures in marking the sheets of the *Lactantius* of 1470 (Hain 9808) instead of the usual letters of the alphabet. The matter is somewhat doubtful; but it is certain that the use of figures for marking gatherings became the custom especially in Venice at Franz Renner's press. The usual practice was to mark the gatherings with small letters, and the sheets of the gathering with Roman numerals, very often, however, with gothic letters. The alphabet then in use comprised twenty-three letters (a b c d e f g h i k l m n o p q r s t u x y z), but it often happened in many books that for r, s and u double forms ꝛ and r, s and ſ, u and v) were used. In the books which were made up in Latin countries by native printers the letter K was often omitted, while German printers and such foreigners as were of German training retained the use of this letter for marking gatherings, even in Italian, French and Spanish books in which there is no occasion for its use. If the alphabet used once did not suffice for this marking it was customary next to help it out with some marks of abbreviation (ꝛ, ꝰ, 9). If these also were not sufficient, a second alphabet, of capital letters, was generally used. Instead of this second alphabet Franz Renner in 1476 in a Carcanus, *Sermonarium* (Hain 4508) and a *Biblia Latina* (Hain 3063), was perhaps the first to use Arabic figures, which were occasionally, and sometimes exclusively, used instead of letters.

In works in more than one volume the separate parts were often signed in different ways, the first beginning with the lower-case alphabet, the second with the upper-case, the next with double letters or with combinations of upper- and lower-case. When the separate parts were very large, the signatures were made up of a number of letters, often five or six. The law books of Baptista de Tortis in Venice are especially notable examples of this.

In the large Venetian printing-shops where a great deal of presswork was going on simultaneously, a further step was taken in the period of early printing, *i.e.,* the marking of the sheets belonging to one press, by a sheet signature in addition to the regular one. Contrary to the modern usage of setting the sheet signature at the left corner of the leaf and only the regular signature at the right, the custom of the early period united signature and sheet signature at the lower right. The sheet signature did not however consist of a word printed in full but only of one taken from the title of the book, often, of a scarcely intelligible combination of letters. By this means, however, it was possible to distinguish easily between the leaves belonging to one book and those of another work marked with the same signatures.

A great many variations as much in position as in form occurred also among the regular varieties of signatures in incunabula.

In a number of books belonging without exception to

an early period the signatures were placed vertically, as in several books of Domenico de Lapi in Bologna (Hain 2296, 5180), in a *Rhetorica* of Cicero (Meiningen), the typographical origin of which cannot even now be established, and in books of Johannes Brito in Bruges, in which, strangely enough, his signatures were combined with leaf catchwords. Printed signatures in other books were placed not directly below the text, but, as was usual with written ones, so far out in the margin that they would be cut off by the binder. This was the case in the *Institutes* of Nicolaus Jenson in 1476 (Hain 9488), in books of Albrecht Kunne in Memmingen, and of Egidius van der Heerstraeten in Louvain. In a Bonaventura, *De Confessione* (Hain 3507), Keysere and Stoll in Paris placed the signatures at the lower left instead of in the right-hand corner. Guy Marchant once placed them in a Consobrinus, *De justitia commutativa,* 1494 (Copinger 1753a), in the centre at the bottom of the page, a procedure occasionally imitated by Jacob von Breda in Deventer. In a number of books in Germany and Italy the signatures were not placed below the last line of text but beside it, and Hans Sporer in Erfurt (Hain 4065), and the brothers Hist in Speyer (Hain 1167, 4117, etc.) inserted the signatures in the last line itself. In Nuremberg, the printer of the Rochus-Legende, and in Leipzig, Konrad Kachelofen (Hain 13514, Copinger 4944) placed them not on the recto of the leaf, but on the verso, in the position of catchwords. Again, in other books, as those of Paul Friedenperger in Venice (Copinger 1857), the signatures were at the right, above the text. Even Anton Koberger, who until 1481 generally printed without signatures, once did a somewhat similar thing in the *Quadragesimale* of Johann Gritsch 1479 (Hain 8066), where he marked the leaves at the top as if with signatures, with "Alphabetum primum— alphabetum XLVIII." As a parallel case may be instanced the special way in which Pieter van Os in Zwolle marked the *Sermones* of Bonaventura, 1479 (Hain 3512). In this book every verso bears a letter and the opposite recto a number; the letters run from a-z, and after all these have appeared in combination with the figure I, the alphabet begins again with figure II and this is repeated up to XIV. Akin to this is the peculiar way in which the gatherings in some missals were marked, some of which were printed in Lübeck, some in Stockholm, for northern dioceses. In them the signature was often printed on the outer margin in the centre, while at the same time a second kind of signature which did not at all correspond to the gathering was put at the top of the leaf. Nicolaus Laurentii made odd experiments in Florence with his signatures. In one of his books (Belcari, *Laude,* s.l.c.a. (Hain 2751), he so alternated signatures and leaf numbers that some of the leaves show only the former, others only the latter. At another time (Cherubini, *Regula vitae spiritualis* . . . 1482 (Hain 4935), he marked the eight-leaf gatherings with two consecutive letters, as "mn, op, qr, st, ux," on the versos of the first and second leaves of each gathering.

[67]

SIGNATURES

The boundary-line between signatures and leaf numbers can never be sharply drawn. Matthaeus Cerdonis in Padua was accustomed in a great many of his books to supply the first four leaves of each gathering with leaf numbers, leaving the second half unsigned, nor was he alone in this practice.

All kinds of variations from the rule occur in the form of signatures. Usually they were printed with the types of the text and made up of a combination of letters and Roman numerals. The printers of the Low Countries had a special liking for printing the signatures in another, larger kind of type than that used in the text. French printers also occasionally followed their example. In Germany only Konrad Kachelofen to my knowledge did anything similar.

Naturally in books printed in roman, roman lower-case letters were likewise generally used to sign the sheets of the gathering. But even the Italian printers used Arabic figures with the letters of the alphabet for signatures in books printed in roman. Still other forms of signing occurred sporadically. It once happened that a printer supplied his signature letters with a corresponding number of points instead of figures. Another (Heinrich Quentell in Turrecremata, *De potestate papae,* 1480) supplied the second and following leaf with two or three times the letter corresponding to the gathering. A leaf-numbering like that used by Matthaeus Cerdonis occurred when half the leaves of the gathering were supplied with the continuous letters of the alphabet, leaving the other half unsigned.

On the other hand it is not an irregularity to find signed in quartos and octavos only the leaves of a specific sheet. It is true, in a great number, perhaps even in the majority of small books, the signatures were so placed that here again the leaves of the first half of each gathering had signatures while the second half was unsigned. But it is entirely justifiable, considering the real purpose of signatures, if by the method of making up (see p. 51), only the first and second, or the first and third, in a quarto, and in an octavo generally only the first or the first and fifth leaves have signatures. Usually until the end of the incunabula period the signatures in the case of octavos were far oftener entirely omitted than in the other sizes.

5. LEAF-NUMBERING

In the discussion of signatures leaf-numbering has been mentioned occasionally. The custom of supplying the leaves of a book with continuous numbers was already so widespread in the period of the production of books in manuscript that it would not be at all surprising should we come upon it at a very early period in printing. That, however, strangely enough is not the case. Leaf numbers in early printing were for a long time supplied by hand and we find them printed only a few years before the time when printed signatures also came into fashion. While the use of both was scarcely unknown, it must be understood that in the earliest days, when the printer used leaf-numbering, the signatures were usually given up, and vice-versa.

In the study of incunabula, leaf-numbering is important only in its more unusual forms. As a rule, where leaf numbers existed at all, they were in Roman numerals in the type of the book, so that books in roman used roman and books in gothic, gothic characters for this purpose. Yet it is to be noted that German printers at home and, many times, abroad (*e.g.,* Paul Hurus in Saragossa) not uncommonly used roman type for numerals. It was a very general practice moreover to use larger type than that of the text.

Almost all the peculiarities of position which have been considered in the case of signatures recurred in the leaf-numbering of incunabula. These numbers were often so placed that the word "Folium" (in scattered instances one finds "Pagina" instead) occurred in the centre of the top margin of the verso, while the number had the corresponding position on the recto. It also happened that sometimes the "Tabula" assumed a leaf-numbering which in reality did not occur in the book. From the various workshops of almost all the countries concerned in early printing there occasionally issued books in which the leaf-numbering occupied the place usually held by signatures (to the right on the outer margin of the page), this happening many times even when the signatures were placed there also. The leaf number even occurred once at the bottom, in the centre. Even when it was put at the top of the leaf, its position was never always the same. In general, to be sure, it was either at the top centre, as often on the recto as the verso, or it held a position in the upper corner similar to that held below by the signature. Guy Marchant once did something very unusual in his *Transitus s. Hieronymi,* 1498 (Hain 8632), when he pushed the leaf number up to the right so far that it stood outside the limits of the type-page.

Leaf-numbering with Arabic figures gained currency in the beginning with the printers of Venice, first occurring there in the second half of the seventies in Franz Renner's liturgical books. It was very little used in Germany during the entire incunabula period. The earliest example of it is in a book of Helias Heliae in Beromünster (Mammotrectus 1470, Hain 10555) in which the first half the gathering is numbered with Arabic figures while the columns of each page are marked continuously with the letters of the alphabet. Such column-numbering occurs sporadically in a few other German books of the earlier period. Thus Johann Guldenschaff in Cologne in the *Lilium circa officium missae* of Bernardus de Parentinis (1484, Hain 12419), marked the columns of each page with the letters A-D, while Peter Drach in Speyer as early as 1478 (Leonardus de Utino, *Sermones de sanctis,* Hain 16135), supplied the columns of each gathering with numbers, the book also having leaf numbers on the rectos and versos of the leaves. Column-numbering did not gain currency in Germany to any great extent however On the other hand, it became very widespread in Italy, especially in the large folio law books in two columns which were intended for academic purposes. Here the numbering of columns usually took the place of leaf-numbering, yet examples occur even in Italy, with both kinds of numbering side by side.

Pagination on the other hand almost never occurred in the incunabula period. The only example which I can cite is a book of Aldus Manutius in Venice (Perottus, *Cornucopiae,* 1499 (Hain 12706), which exhibits besides, the also unique characteristic that even the lines of every page are numbered as in modern school-editions of the classics.

6. CATCHWORDS

Catchwords in the incunabula period served purposes similar to those of signatures and leaf numbers. By their use is meant the reproducing of the first word of a new page, at the foot of the preceding. They became quite widely used in books of the Reformation period, in these their task being to make it easy in reading, especially in reading aloud, for the eye to go smoothly from the bottom of one page to the top of the next. We can scarcely assume however that such a purpose was theirs in the period of early printing; for catchwords occur very exceptionally in incunabula, and only in the last years of the fifteenth century do we find occasional page catchwords. In general, on the contrary, catchwords were placed only at the end of the gatherings or like signatures at the end of the double sheets. They were evidently only a means of assuring the correct assembling of a volume from the separate sheets. It argues for the correctness of this interpretation that catchwords first appeared at the time when printers were trying to attain the same end by the aid of signatures, and it is evident also from the fact that signatures and catchwords were but seldom used together, for printers, influenced perhaps by their common source, decided on either the one or the other procedure.

Printed catchwords are found for the first time in the undated *Tacitus* (Hain 15218) which was printed in Venice by Wendelin of Speyer. It can scarcely however have been printed after the year 1471, since the printer designated it as "artis gloria prima suae." In 1472 in Bologna, Baldassare Azzoguidi followed his example in his Antoninus, *Medicina dell'anima* (Hain 1229), and the fact that up to the year 1480 catchwords occur in ten other Italian books, but only very scatteringly in specimens from German presses, makes it seem probable that this invention like that of the register was first used in Italy. Manuscript catchwords are found, it is true, in German books (Herpf. *Speculum aureum,* Mainz, Schöffer, Hain 8523), in the year 1474, but Johann Koelhoff, the inventor of signatures, was the first to print catchwords as well, as in his Thomas Aquinas, *Quaestiones de veritate* (Hain 1419) and his Fliscus, *Synonima* (Copinger 2532), both issued in the year 1475. The custom was never as widespread in German incunabula as in Italian, and we cannot therefore follow its development in German books, as apparently it was used there only in the form of leaf-catchwords.

The case is quite different in Italy. Although Wendelin of Speyer supplied his *Tacitus* with leaf-catchwords, Azzoguidi in 1472, and the printer of Cicero, *Rhetorica,* mentioned above in relation to signatures, put their catchwords only at the end of the gathering, and both agreed in placing them vertically. This form was the one usual in the seventies, but individual presses were never consistent in their use. Thus Johannes de Colonia and Johann Manthen in Venice, 1474, supplied a *Confessionale* of Antoninus de Florentia (Hain 1177), with leaf-catchwords, while a new edition of the same work in 1476 showed catchwords only for gatherings.

That they were really looked upon as a help toward the correct assembling of the sheets of a book is to be inferred, since many printers put them only on the first leaf of the sheet, so that the second half of each gathering was without them. Since however this method afforded no help in arranging the gatherings in proper sequence, the custom gradually became established of supplying the last leaf of each gathering also with a catchword. It is less clear on what ground Christophorus de Canibus in Pavia (Accoltis, *De acquirenda possessione,* 1494) placed catchwords at the end and in the middle of a gathering, an example which Laurentius and Franciscus de Cennis followed in reprinting the same work. It is still less comprehensible why Henricus de Colonia in Siena 1486 in an Accoltis, *De soluto matrimonio* (Hain 49), used catchwords only in the middle of the gathering, which procedure was imitated even in 1498 by Franciscus de Guaschis in Pavia in a B. de Benedictis, *Consilia* (Hain 2768). Christophorus de Canibus then proceeded to place catchwords at the foot of each page, a usage in which, however, Johann Schall in Mantua had once, 1479, preceded him.

Differences in the arrangement of catchwords are also to be observed. That they occasionally occur in a vertical position has already been noted. Stephanus Corallus printed on the lower margin of every leaf of his *Pliny* of 1476 (Hain 13091), the first two words of the following leaf, and in his Barbatia, *De Fidei commissis* of 1474, he placed the catchwords centrally below both columns. In general however we judge from peculiarities of their arrangement that there was no hard and fast rule as to their position.

THE BOOK · *Printing*

1. TYPEFOUNDING

WE have a very great deal of source material dealing with the technique of early printing, but it is neither so exhaustive nor so clear as to give us an exact and definite picture of the manner of printing the earliest books. Besides, these documentary sources do not of course go back to the beginnings of the art; hence it is inevitable that various students working with examples of early printing should have formed widely different opinions as to its method. Critical treatment of the matter belongs however to the history of the printing art; for the study of incunabula it concerns us only as far as it is significant for the right understanding of early books.

Gutenberg had already enlisted a number of co-workers in his earliest attempts at printing, and kept a carpenter and a goldsmith at work as well. From this we might conclude at the outset that a good many people were necessary in printing and that their duties need never have been quite identical. Printing however very soon learned to get along without outside aid so that such pieces of work as Gutenberg had turned over to a carpenter or a goldsmith were done by the printer himself. We learn definitely that the building of the printing-press, in itself a task for a carpenter, as well as type-cutting, which might seem a goldsmith's work, were carried out by the printer. Even so difficult a task as the making of a casting-tool belonged to the knowledge that the learned master of printing must have. In spite of this, even in the earliest days of book production (our information goes back nearly to the year 1470) a certain division of labor was followed in the management of the printing-shop. Scarcely a printing-house existed anywhere in which a single master alone looked after all the related processes without other help, for had he done so he would have been obliged to spend an extraordinarily long time on the production of each book, with the result that his work would not have been sufficiently profitable or that his customers would probably in time have lost their patience. Probably rather, just as was true of Gutenberg in his first attempts, it was always the case everywhere that several people were associated in the management of a printing-shop.

How large a number this was, naturally depended on the size of the undertaking. A printing-house which worked with four to six presses (and documents prove that several workshops of such a size existed, from which, so far as we know, scarcely more than one extensive piece of work issued) would naturally have employed a far larger staff than an establishment which possessed only one press. But even such a one employed at least four or five men. We learn for example of the shop of Johann Luschner of Montserrat that, although possessed of but one press, it employed besides the master himself, two

typesetters, two printers, a preparer of ink, and a press-
man, and that still more help was summoned when necessity arose, to procure new type. We learn moreover from
this and other sources, that all these duties were looked
after by the printers themselves, and that no longer, as in
Gutenberg's time, were other workmen called in to render
assistance of various kinds. That is true particularly of
type-cutting and type-casting, and the often expressed
opinion that even in the early period there were professional typefounders from whom printers could obtain
necessary stock, is in complete contradiction to what documentary sources disclose.[1]

It is not however impossible that, owing to the necessary division of labor in every workshop, one particular
journeyman should have been used chiefly as compositor,
printer or type-caster and should have gone from one establishment to another. The story of such an instance has
been handed down from the year 1476-7 when a certain
Crafto went in the capacity of type-caster to Rome after
having been similarly employed in Foligno and Perugia.
The matter of type-casting has a special significance for
the student of incunabula, since upon it depends whether
one must accept a decision as to the origin of an incunabulum based on the types used in it, as conclusive or not;
hence the necessity for establishing that type-casting in
the early period was the personal concern of each individ-

[1] Cf. my article: *Schriftguss und Schriftenhandel in der Frühdruckzeit,* in
Zentralbl. f. Bibl. 41, pp. 81ff.

ual printing-house, and that the printers with a few exceptions, as those who belonged only to the last years of the fifteenth century, were always actuated by an idea of investing the types they used with an individual character.

2. THE PRINTING-PRESS

While for our knowledge of typefounding we have relied entirely upon documentary sources, we have besides these still other valuable sources at our disposal for the investigation of the actual printing process. These are cuts of the printing-press or of the printing-shop which have come down to us from the fifteenth and sixteenth centuries.[1] The earliest representation of a printing-press is found in a book as early as 1499, as a printer's device of the "praelium Ascensianum," and a careful examination of that and of later cuts leads to the conviction that in these we have before us very carefully executed and exhaustive illustrations. Belief in this is considerably strengthened by the fact that as regards technical details, all agree perfectly and show that the kind of materials and method of use experienced almost no important change in the course of a century. Since however revolutionary changes in technique in the fifteenth and sixteenth centuries occurred much less often than in our day of rapid living, we may

---

[1] Enschedé, *Houten handpersen in de zestiende eeuw*. In *Tijdschrift voor boek-en bibliotheekswezen* IV (1906) pp. 195ff., 262ff. Madan, *Early representations of the printing-press*. In *Bibliographica* I (1895) pp. 223ff., 499ff.

infer from this similarity in the cuts that they also agreed with the contrivances of the previous half-century. That means moreover that they give us an approximately true picture of conditions in a printing-shop as it was equipped in the early period.

From it we can learn the following facts: The printing-press, made of wood, consisted of a table, the flat top of which was however not stationary but movable in a horizontal direction by means of a crank. This was the carriage, still to be seen in every modern press. On it was the place for the forme, *i.e.*, the setting of type ready for printing. At its end was the tympan, the function of which was to receive the sheet of paper, to fasten it in place with a number of pins, and then so to lay it on the type that its margins should be kept by a frisket from contact with the inked type. The paper, unsized in the incunabula period, was dampened before it was laid in, and therefore had to be hung up to dry after it had been printed. The carriage, on which was the tympan, was slid under the platen, which had a vertical handle. When this was turned, the frisket with the paper was pressed firmly down upon the type.

It is worthy of note that in the early cuts the platen was not the same size as the tympan but apparently covered only half its surface. This is decisive in judging whether work was placed in the forme by pages or by sheets. Printing by single pages was undoubtedly the original method. It is usually betrayed by the fact that the printer by this method seldom succeeded in bringing into absolutely the same position columns printed in turn on the same sheet, so that the letterpress of these pages was out of alignment both vertically and horizontally. Investigations in this direction have led to the discovery that printing by pages is not necessarily always a mark of the especially early printing of the work concerned. Rather, it has been proved that workshops with modest equipment in, comparatively speaking, later times, occasionally produced books printed by pages. This was probably because the carriage of the presses in question possessed only the single size platen so that a forme (two pages) could be printed by one impression, only if its size did not exceed that of the platen.

The dimensions of the carriage however made it possible for printing to be done by single pages even if the forme was double the size of the platen. It is therefore to be assumed that printing by single pages was customary much longer and used to a far greater extent than the early books reveal to us by discrepancies in the position of the type-page. It was, namely, possible to place two type-pages on the carriage at once, to run one page under the platen and then by a new pull of the lever, running the carriage further in, to print the second page. The paper was held in place firmly enough by pins so that pieces of work done in this way would register as well as if they had been printed by a single pull. Whether printing was done by pages or by formes is often traceable in some copies even to-day, for the powerful blow of the platen on the thoroughly dampened paper usually made a noticeable impression of the type on the verso, which lost this impress only by the counter-impression of its own type. It is thus usually easy to determine which page was a first and which a counter-impression. If it appears that all the rectos of one gathering show only one kind of printing, it must have been done by pages, for in printing by formes, recto and verso are done at the same time. These investigations have shown that until about 1470 the greater number of incunabula were printed by pages, as for example even the earliest quartos of Ulrich Zell in Cologne.

We must conclude from the cuts of printing-presses that this method of procedure was in vogue possibly to the end of the incunabula period, but that is not saying that printing was always done by single pages. With corresponding sizes of forme and paper, on the contrary, two pages could be printed simultaneously, or four pages, without moving the carriage.

3. POINTS

The points, *i.e.,* the sharp points of the pins intended to hold the paper in place during printing, naturally had to be set far enough away from the page of type; not to be disturbed by the blow of the press on the type. Their traces are now seldom perceptible in the case of books which have been much cut down. Points were at first used in greater numbers (six to eight) on all four margins of the type-page (the outer edges). In time their number was decreased when it naturally happened in practice that the locked-up forme could take over a part of their functions and not as many points were needed to hold the paper in the right place. Their most important task was that of keeping the register, *i.e.,* of taking care that when the verso of a sheet was printed (counter-impression), the paper was lying in exactly the same position on the type-surface as in the printing of the recto (first forme). The reversed sheet had to be laid down with its perforations exactly on the points. Holes however are to be seen only in the incunabula of the earlier period. The reason for this is probably that later the points were transferred from the outer edges to the middle of the sheet (the fold). Here however they cannot be discerned because of the fold of the sheet and the sewing of the binding.

A difficulty still not clear in regard to the points, is that apparently they were always pricked through from the recto to the verso. One would really expect the reverse, since if the pins were put not on the tympan but on the chase, the tympan would be useless and the lifting off of the printed leaves become very troublesome. The question is made still more difficult since traces of a substratum of matter on certain sheets of some incunabula'show that probably it sometimes happened that more than one sheet had been fastened on the points at the same time. The assumption that the sheet was first of all adjusted on the chase exactly with the type-page by the help of points, and then transferred to corresponding points on the tympan, would explain the matter, but the procedure seems so troublesome and its advantage so trifling that we cannot yet view this explanation as final. Further investigations may perhaps throw new light upon it.

4. COLUMNS

The arrangement of the type-page in long lines or in double columns was after all the result of an arbitrary decision on the printer's part. We may however assume that the form of his copy, at first in manuscript, later in print, materially influenced this, with the result that there was established for certain kinds of work a custom which no later printer would dare to break lest he thereby endanger the salability of his production.

Gutenberg printed the 42-line Bible two columns to a page, although with its large type a page in long lines would have been sufficiently readable. It is a peculiar fact that, perhaps influenced by this model, almost all texts printed with missal type were in two columns, even in so extreme a case as that of the printing of Guillaume Leroy, Lyons, in small folio (*Clamades*. Copinger 1653; *Quinze joies du mariage*. Copinger 5020), in which there was room for scarcely more than twenty letters to the column. Almost all Bibles and all liturgical books especially, were printed in double columns. The Psalter only is an exception to this rule, and of this even reprinted editions were almost all in long lines, although the same text as a component part of the breviary was always in double columns.

I make no attempt to decide whether the model of the Psalters of 1457 and 1459, or the custom handed down by manuscript tradition, was responsible for this.

The earliest Strassburg printers seem to have had a strong preference for printing in long lines, and the huge pages of their large folios are therefore less easy to read quickly. Sweynheym and Pannartz also, and, following their example, the majority of Italian early printers favored long lines, a practice natural enough since they published texts of Greek and Roman classics so extensively, for such works the single-column type-page being the almost unbroken rule.

On the other hand, from the first, in Germany as well as in Italy, legal texts with their commentaries were printed almost without exception in double columns, although in so doing, occasionally but few lines of text would be printed in a column.

In the later years of the incunabula period not only custom but the size of the type chosen for the book was the deciding element in the arrangement of the type-page. As printers more and more chose smaller types in order to issue less bulky volumes, they must also have come to the conclusion that their texts in quarto would not be easy to read if they set the pages in long lines. Italian printers realized this more quickly than the others, as they were by far the most enterprising in all printing procedure, while the French, the least progressive, issued quarto editions in long lines, with type so small that the reader's eye has great

difficulty not to stray from the line in returning to the beginning of a new one.

5. SIZES OF TYPE

The form of the type-page was in great measure influenced by the efforts of the early printers to economize on
paper. In Italy, where paper-making had grown to great
dimensions even before the invention of printing, resulting naturally in a reduction in price, the desirability of
saving paper was not felt as strongly as elsewhere. Among
German printers of the earlier incunabula period it was
a particularly expensive detail and their endeavors to be
sparing of it are often very obvious.

According to some authorities we should assume that
Gutenberg, who without doubt owned the types of the
36-line Bible in the beginning, first planned to print the
Bible with those types. It is well known that the first
leaves of the 36-line Bible form a single independent piece
of printing, while from the sixth leaf on, the remainder
proves to be a reprint of the 42-line Bible. Gutenberg however was influenced by the scarcity of paper to make a
new and smaller type for printing the Bible, and the presswork itself of the 42-line Bible affords incontrovertible
proof of the fact that the effort to be sparing of paper
played an important part. It did not begin with a page
of forty-two lines, the original type-page of the first sheets
having but forty. Experiments were then made on a few
leaves with forty-one lines, and only after eighteen pages

had been printed in this way in two different settings, did Gutenberg finally change to the setting in forty-two lines, in which parts of the original printing were then reset.

If one should investigate all the earliest books with like thoroughness, perhaps still other evidences of a similar procedure would appear. In any case there was an obvious effort on the part of the earliest printers to lessen the book's size by increasing the number of lines on the page and thereby to save paper. It is almost certain that of two editions of the same work that having the fewest lines on a page should be regarded as the earlier.

This endeavor to include more lines on a page was chiefly limited by the size of the chase, for the printers of incunabula in the earliest period designed their books with wide and beautiful margins for aesthetic reasons, and we have been able to discover almost definite rules for their proportions. Since the paper in the earliest books was not fully used, the margins enabled the printers in the beginning to enlarge the type-page without requiring a larger size of paper, although naturally this enlargement of the type-page was limited by the size of the press. This was true above all of the dimensions of the platen and the carriage which were not as easily expanded as those of the chase. The result was that, after a period of experimentation in which the wide margins peculiar to the earliest books had to be reduced, normal sizes again were produced, in which a better use of paper was attained only by a reduction in size in the kinds of type chosen.

Although the diminution of type-sizes in incunabula does not seem especially marked, yet, and this is more important, at the same time the custom was developing of cutting types in a great variety of forms. While the earliest printers at first usually printed only with a single size of type, as time went on it became more and more customary to use a great many types of various sizes in the same book. It was not long before certain printers were seized with the ambition to provide themselves with an extensive stock of the most varied kinds and sizes, a striking proof of this being the specimen type-sheet with which Erhard Ratdolt advertised his types after his removal to his native Augsburg. The printer's efforts now however took the form of putting these varied types into use as much as possible even in the same piece of printing. So it happens that occasionally, even in the incunabula period, we find from six to eight different kinds of type used in one piece of printing, and in a book of the sixteenth century (*Historia horarum canonicarum s. Hieronymi et s. Annae.* Augsburg 1512) Erhard Ratdolt could boast of having used not fewer than thirteen different types.

It is clear that the diminishing of type-sizes could not be advantageous as far as individuality of shape was concerned. Small types obviously offered less opportunity for the addition of ornamental touches. Moreover they made greater demands on the reader's eyesight, demands which it was undesirable to increase by the choice of shapes the peculiarities of which were perhaps unfamiliar. Hence, especially in the case of types of a smaller size, a certain uniformity gradually gained ground, with, in time, the result that even the larger, ordinary types, and finally even the display-types, became more alike, although the final achievement of this development did not come about until after the incunabula period. By the time of the Reformation, however, the printing of almost all countries was governed by a national style in which individual peculiarities gradually disappeared.

6. FORMES

So long as printing was done by pages, the order of the pages gave the printer no particular trouble, nor was there any increase in the use of double columns. But printing by formes, which were to be assembled from a number of pages of type, for the first time demanded greater attention, especially in the smaller formats. We know that this care was not always sufficiently exerted. In the Cicero, *De natura deorum* of 1471 (Hain 5334), Wendelin of Speyer mixed the last leaves of the first (preliminaries) and second gatherings; the error was however discovered before publication and the leaves were cut out and inserted on stubs in their proper places. Such errors are often found in but a part of the edition, proving that they were noticed during printing and corrected.[1] It is evident that when such mistakes occurred in the work of the early printers

[1] Further examples: *Repetitiones diversorum doctorum,* Wendelin of Speyer 1472 (Hain 9884. B.M.C. V p. 161); *Vita et transitus b. Hieronymi,*

they were for the most part less the result of clumsiness than of lack of pains. They did not always confine themselves to mixing the leaves of a gathering, and it happened repeatedly that the same page was put in the forme twice and the proper page omitted. In the *Herodotus* of Jacques Lerouge of 1474 the text of leaves 155b and 156a was printed a second time on leaves 161b and 170a, but this error involved no loss of any text. In the *Virgilius* of Baptista de Tortis of 1483 (Copinger 6049), however, leaves b4 and b5 were printed on both sides with the same text while the text of the versos was omitted. A similar procedure occurs repeatedly in the Mancinellus, *De floribus*, Venice, Tacuinus 1498 (Hain 10618), in which not only leaves E5b and E6a exchanged places but F2b and F3a were printed with the same text, so that in another place one page was omitted. That such mistakes could be made even in printing-shops acknowledged to be good, is seen in the *Mammotrectus* of Nicolaus Jenson of 1479 (Hain 10559), in which four pages of gathering 1 exchanged places. The same happened in gathering o of Diogenes, *Vitae philosophorum*, Venice, Locatellus 1490 (Hain 6202). All these mistakes prove naturally that the books in which they occur were printed by formes, and show for example that Jenson, in 1479, was able to pull four quarto pages at the same time.

Pasquale 1485 (Hain 8631. B.M.C. V p. 391); Ptolemaeus, *Geographia*, Turre 1490 (Hain 13541. B.M.C. IV p. 133); Petrarca, *Trionfi*, Piasiis 1492 (Hain 12773. B.M.C. V p. 270); Philelphus, *Epistolae*, Benalius 1493 (Hain 12942. B.M.C. V p. 375); et al.

The printers soon took up still other problems in regard to the type-page. In the manuscripts of Roman and Canon law the custom had become established of so writing the commentary that it surrounded the text on all three or on all four sides. In an effort to make the piece of printing agree as far as possible with the manuscript source, the printers imitated this method. As early as 1460 Fust and Schöffer in the Clemens V, *Constitutiones* (Hain 5410), understood how to achieve this arrangement in printing and this "modus modernus" as it is called in a contract of Andreas Belfortis, 1473, was soon as widespread among the German as among foreign printers.

But the development of the book was once more furnishing the printer with fresh problems regarding the type-page. Even in very early Mainz books large spaces were left in certain places at the beginning of a new section, which were obviously meant to be filled in as in manuscript, with painted miniatures.[1] Soon however the printers quite generally began to replace these paintings by printed pictorial cuts, although during the entire period of early printing the custom was maintained of coloring these by hand. It is a disputed question how far printing from wood-blocks merits a place among the first steps

---

[1] *E.g.,* Bonifacius VIII, *Sextus decretalium,* Schöffer 1465, 1470, 1473, 1476 (Hain 3586, 3587, 3590, 3593); Kesler s.a. (Hain 3585); Gratianus, *Decretum,* Eggestein 1471 (Hain 7883); *Biblia Ital.* Ambergau 1471 (Hain 3148); Berchorius, *Biblia moralisata,* C.W. 1474 (Hain 2795); Panormitanus, *Super II Decretalium,* Koelhoff 1477 (Hain 12323), Clemens V, *Constitutiones,* Koberger 1482 (Hain 5427); et al.

of printing. Close investigation has shown that most of the surviving xylography originated at a time when printing also had gone through its first stages of development. But since without doubt the beginning of printing from the wood-block antedated that of printing from type, the printer could pattern by the work of the manuscript maker in combining pictorial decoration and explanatory text in the same production. The first to undertake to ornament a printed book with illustrations was Albrecht Pfister in Bamberg, whose popular books of the years 1461 and 1462 were adorned with numerous cuts. Later the Zainers in Augsburg and Ulm and Bernhard Richel in Basel, among the earlier printers, distinguished their books by the use of woodcuts. Outside of Germany Ulrich Hahn was the first to follow this custom, in the *Meditationes vitae Christi* of Johannes Turrecremata, 1467, using in this a large number of illustrations copied, it is supposed, from the original paintings in the church of S. Maria sopra Minerva at Rome. The woodcut ornamentation of incunabula is one of their most interesting and attractive features and has been the subject of a number of publications in all countries. Recently A. Schramm has begun to reproduce all incunabula which have woodcuts, in a series of separate publications, each one dealing with a single printer. If this undertaking is completely carried out a marvellous collection of material will be assembled as a foundation for a scientific treatment of the subject.

The division of the text for printing gave the earliest printers far more trouble than did the arrangement of the type-page. The production of a book of large size demanded so long a time in the early period that it was not deemed expedient to print the entire work from beginning to end, one sheet after another, but instead to cut the text into a series of parts in which the setting and printing would be carried on at the same time. Although such a procedure very vitally increased the cost of production since it required a large stock of type, it was certainly used by Gutenberg in the printing of the 42-line Bible and was continued to the end of the incunabula period, at least by all the printing-shops which laid claim to a greater working capacity. That it appeared less conspicuously in the later period of the printing of incunabula than in the beginning, was partly due to the printers' increasing certainty in estimating the space necessary for a single section of copy, but in far greater degree to the fact that later printers no longer needed to calculate this by the manuscript, but were in a position to work from a printed text, which afforded them a very much easier and much more certain basis for the estimating of necessary space.

In the discussion of the make-up of the gathering it has already been pointed out that the occurrence of larger or smaller gatherings in a series of others alike in size justifies the inference that those irregularly made up were the last of a compositor's section. Only very considerable differences however could be thus adjusted, since the gathering could be changed in size only by the addition of a whole double sheet. Somewhat slighter variations could be arranged by inserting in the gathering, instead of a whole sheet, only a single leaf, of the unprinted half of which only a stub would then be left in the gathering. This procedure is responsible for the fact that it is not unusual in incunabula to find gatherings with an uneven number of leaves. Evidently, likewise, printers always tried to join sections smoothly by narrowing or widening the type-page. The great variety of abbreviations used in the earlier incunabula period made this easy and one often notes at the end of a section their occurrence in unusual numbers, or finds that in abrupt contrast to their noticeable use in other parts of the volume they have been entirely abandoned. The manner of employing this method gives us an opportunity to make interesting observations on the procedure of compositors in the various workshops. In Germany we almost always find abbreviations thus used to justify the type-page, only at the end of the gathering, from which we may infer that the setting followed the sequence of gatherings and leaves. In Italian books however we often notice that the type-page in the middle leaves of the gathering is noticeably adjusted.[1] Therefore we must conclude that in these establishments not only the

[1]*E.g.,* Aristoteles, *Ethica,* Sweynheym . . . 1473 (Hain 1746, B.M.C. IV p. 17); Maroldus, De *epiphania,* Gensberg s.a. (Hain 10779, B.M.C. IV p. 50); Antoninus, *Summa III,* Jenson 1477 (Hain 1243, B.M.C. V p. 177); Cicero, *Orationes,* Girardengus 1480 (Hain 5124, B.M.C. V p. 272); et al.

printing but even the setting was done in such a way that no longer was the type of the whole gathering set up before the printing began but that they first made ready only the outer leaves of the gathering (1 and 8, 2 and 7, etc.) as they were needed for the making up of the formes, so that if necessary they could print these, distribute the type and use it over again for the inner leaves set up last.

Yet if early printers had miscalculated the amount of copy beforehand, they did not hesitate to increase or decrease the number of lines on certain pages, although this number was usually uniform throughout the book. They occasionally resorted to the expedient of setting one page in another, smaller, type in order thus to get somewhat more text on this page.[1] If the compositor happened to omit a few words or even one or more lines he usually remedied it by adding or pasting these words or lines to the lower, less often to the top margin of the page. With single words this was usually done by stamping; even whole lines were thus supplied, but greater omissions in many cases were later printed in, so that the sheets in question had to go once more through the press. It also happened in some cases that such additions were written in, in the printing office, in the whole edition.[2]

More serious errors also occurred, however, for the correction of which the above mentioned means were not sufficient. In such cases the omitted text was printed on a

[1]*E.g.,* Guido, *Manipulus,* Fyner s.a. (Hain 8158).
[2]*E.g.,* Single words: *Lapidarius,* Brandis 1473 (Hain 1777), leaf 26b;

special slip of paper, a "carton," which was inserted in the proper place in the gathering, to which, if need be, the rubricator would refer. Such "cartons" are very varied in extent; sometimes they contain but a few lines[1], sometimes their text takes up a whole page or at least a whole column, and it occasionally happened that the blank parts of such a leaf were not removed, so that certain pages are only half printed.[2] The "carton" was naturally of use also (if the space for the text was ample), in providing a blank column to serve to expand the type-page. The earlier printers however often made their task much easier, for if the text was not long enough to fill the sheet they simply left a larger or smaller part of the page or even a whole page un-

Ursinus, *Modus epistolandi*, Hochfeder s.a. (Hain 16106), leaf 7a; Joh. Chrysostomus, *Super Johannem*, Lauer 1470 (Hain 5036), leaf 175a; *Breviarium Romanum*, Jenson 1478, (Hain 3896), leaf a_8a. Stamped lines: *Biblia latina*, Richel 1477 (Hain 3064), leaf 248a; Mela, *Cosmographia*, Venice 1477 (Hain 11015), leaf d_7; *Mammotrectus*, Jenson 1479 (Hain 10559), leaf y_8b. Printed: Bernardus, *Sermones*, Schöffer 1475 (Hain 2844) leaf 111b; Albertus M., *Sermones*, Greif s.a. (Hain 473) leaf C_6a; Harentals, *Collectarius*, Koelhoff 1487 (Hain 8366) leaves 31a, 40a; Cosmico, *Canzoni*, Celerius 1478 (Hain 5781) leaf c_1a; Pasted: Justinianus, *Codex*, Sensenschmidt 1475 (Hain 9599) leaves 144b, 226b, 239a; *Breviarium Ratisponense*, Ratdolt 1487 (Hain 3884) leaves 28b, 100a; Aristoteles, *De animalibus*, Aldus 1497 (Hain 1657) leaf KK_{10}b.

[1] Augustinus de Ancona, *De ecclesiastia potestate*. Schüssler 1473 (Hain 960) leaf 162b, at least 6 different settings known; *Doctrinale clericorum*, Lübeck 1490 (Hain 6318) leaf p_3; Guillermus, *Postilla*, Wiener 1476 (Hain 8254) leaf 163.

[2] *E.g.*, Rufus, *Historia Romana*, Sachsel . . . s.a. (Hain 14027, B.M.C. IV p. 55); Joh. de Imola, *In Clementinas*, Lerouge 1474 (Hain 9143, B.M.C. V p. 215); Bartholomaeus Pisanus, *Summa*, Girardengus 1481 (Hain 2329, B.M.C. V p. 273); *Valerius Maximus*, Wendelin of Speyer 1471 (Hain 15775, B.M.C. V p. 156); Albertus M., *De officio missae*, Zainer 1473 (Hain 449) leaf 92.

printed. But in order that the reader might not believe those blanks to be errors in setting they used in many instances in these empty spaces a printed notice, as "Hic nihil deficit," or "Hic non est defectus,"[1] or if they wished to be more circumstantial: "Hic non deficit quicquam et sequitur . . ." or simply "Sequitur," with the first words of the adjoining text.[2] Such blanks vary in the earliest books from a few lines to whole pages and more. But later they more and more disappear for the reasons above given.

7. TYPE

The most important characteristic of incunabula, and that which is of the greatest significance in their study, is the shape of the types. With their help we may almost always establish whether we are dealing with a book of the fifteenth century or not, and they enable us in almost

[1]*E.g.*, Thomas Aquinas, *Summa* II, 2, Mentelin s.a. (Hain 1454) leaf 181b: "hic nullus est defectus". *Summa* I, 1, Zell s.a. (Hain 1439) leaf 47a: "Hic nihil deficit." Rainerus de Pisis, *Pantheologia*, Zainer 1474 (Hain 13016) I, leaf 381b: "Nullus defectus." *Statuta prov. dioec. Coloniensis*, Koelhoff 1492 (Copinger 5614) leaf 75a: "Verte. hic nihil deficit." Landinus, *De vita contemplativa*, Nic. Laurentii s.a. (Hain 9852) leaf f8: "nihil deficit." Barth. Sybilla, *Speculum*, Silber 1493 (Hain 14718) leaf C₆: "Hic nihil deficit."

[2]*E.g.*, Joh. Chrysostomus, *Super Matthaeum*, Mentelin s.a. (Hain 5034) leaf 130a: "Hic nullus est defectus. Sequitur textus." Dion. de Burgo, *In Valerium Maximum*, Rusch s.a. (Hain 4103) leaf 270: "Sequitur Scipio quoque." Thomas Aquinas, *Catena*, Zainer [1474] (Hain 1328) leaf 211: "Sequitur. fructus: statim." *Rudimentum novitiorum*, L. Brandis 1475 (Hain 4996) leaf 176a: "Hic nihil deficit sed sequitur gloriosa iherusalem." Hugo de s. Caro, *Postilla*, Richel 1482 (Hain 8975) leaf 738: "Hic non deficit quicquam et sequitur Ideo patitur in festo." Hostiensis, *Summa*, Wild 1480 (Hain 8963) leaf M₆: "Sequitur nunc rubriaca de sponsa duorum."

[97]

every case to find out from which printing-house a particular example came. What an extraordinary significance
this must have for the study of early printing is easily
apparent, if one realizes that more than a third of all the
early printing known to us contains no information whatever as to where and when it originated.

In shaping his earliest types Gutenberg's purpose was
to make the books printed with them as much as possible
like manuscripts. Whether he was directly guided in this
by a thought of deception naturally cannot be proved. In
any case no one of the books which may with certainty
be traced back to the inventor himself (the question
whether the *Catholicon* of 1460 was Gutenberg's work
cannot even yet be considered definitely settled) bears any
mark which accounts for its origin. Fust and Schöffer,
in the 1457 Psalter, were the first to proclaim printing a
new art, distinct from manuscripts. In order to produce
printed work which should resemble manuscripts, Gutenberg loaded his typecase with at least a quarter more type
than was necessary. The scribes were in the habit of join-
ing letters in quite different ways, giving them different
shapes according to their position, *i.e.,* adapting them to
their neighbors. Gutenberg tried to imitate this practice
by making joined types as well as separate ones. In the
42-line Bible this practice was followed very scrupulously
although it occurred also in most of the other books
printed with the types of the 36- and 42-line Bibles. In
these, however, differences are often to be noted in the

observance of this practice, and judging by these differences it seems possible that not everything printed in the earliest Gutenberg types was produced by the inventor himself.

The types of the 36- and 42-line Bibles were known as missal, letters of considerable size, like the manuscript ones used especially in the books for the offices of the church. The remnants of the Gutenberg principle of joined letters long endured in these large types, even down to the earlier years of Italian printing, and traces of it occur almost to the end of the incunabula period. Meanwhile however printing had already gone on to imitate the smaller letters also, those in general use by scholars in their manuscripts. Probably Gutenberg himself shared in this attempt, earliest examples of which are the Indulgences of the years 1454 and 1455. These survive in a large number of different styles, which however fall according to their types into two groups, one of which is believed to have originated with Gutenberg, since in it the type of the 42-line Bible seems to have been used as display-type. Already in these types, which were as yet not really small, the use of joined letters was less practicable than in missal type. Instead, there occurred a greater number of ligatures, *i.e.,* letters combined on one body so that they could be handled as one letter. In a similar way Schöffer cut his earliest text-type (type for the printing of solid text). Ulrich Zell in Cologne went furthest however in producing ligatures to replace joined letters, for his typecase was filled with an abundance of separate forms surpassed by those of no other printer in the incunabula period, save those perhaps of Laonikos of Crete, who in 1486 in Venice assembled for his Greek books a thousand different forms of letters, accents, ligatures and abbrevations.

Manuscripts held their place as the authoritative models for the shapes of early types for different lengths of time in different countries. This has not been sufficiently taken into consideration in the study of the types of the early period. A closer comparison of printing types with their respective local manuscript letters will doubtless result in many interesting revelations; and through it we shall be able to account for the fact that the types peculiar to various countries have widely different forms and that we often find type-forms of very individual styles in the work of one printer.

Although this is true to a marked degree of the types of the earliest Dutch books, we would not be justified in assuming for that reason that they were a peculiar creation, independent of the Gutenberg invention. In this regard, the legend of the invention of printing by Laurens Coster in Haarlem which, as can be proved, did not arise before the second quarter of the sixteenth century, has caused serious confusion among scholars. In contrast to the unanimous evidences of the period of early printing itself, all of which name only Mainz, and Gutenberg as inventor of the new art, these belated discoveries can count on credibility the less, since they frequently contradict themselves quite plainly. It is a fruitless task to attack and try to settle this question technically. While various modern experiments have shown that books not inferior to incunabula in clearness can be produced by the use of wooden types or with those made by sand-casting, the fact that various procedures have led to the same result deprives the single experiment of the power to decide the matter. Over against the sand-casting process and the conclusions drawn from it in favor of the independence of the earliest Dutch printing, may be set the fact that, first, in all the sources at our disposal for the history of the earliest printing, no trace is anywhere to be found of any other process for the production of types than that with the casting-tool; that second, the production of types suitable for setting would by the sand-casting process have demanded so extensive a working over of each separate letter that the manufacture of a sufficient stock of type would have required a wholly prohibitive expenditure of time and labor; and third, that the old Dutch types which, it is claimed, originated from the sand-casting process, can scarcely have resulted from such a method, since their numerous points and ornaments could not possibly be as clearly impressed as we see them in early Dutch books, had the types been produced in quantity by sand-casting.

On the other hand, even the Dutch books, in which a small number of groups of closely related types persisted up to the end of the incunabula period, are a convincing instance of how largely the design of printing types was conditional upon the style of handwriting peculiar to the place in which they originated. A quite similar procedure can be proved to have existed in France, Spain and England, where we find everywhere, either types which are quite widespread, but alike in their local character, or individual types of such unusual kinds that we pronounce them, unconditionally, imitations of manuscript models.

There are many reasons why such a relationship between manuscript characters and type-forms is not recognizable to such a degree in Germany and Italy. In the first place we must take into consideration the fact that the production of printed work in Germany and Italy took on so much greater proportions in comparison with that of other countries that what seemed extraordinary and unusual in the latter, was in the former far less noticeable. Upon closer investigation it becomes at once apparent that some of the German types, widely imitated and very widely used, showed on their first appearance (just as did characteristic types of the Netherlands and France) a strongly marked individuality, of which we cease to be conscious only when we see them copied and produced in quantity.

Moreover the intimate relations which existed between Germany and Italy throughout the incunabula period formed a second step in this direction, for these furthered an interchange of ideas relating to printing. That period continually saw German printers flocking to Italy but always returning home again enriched by experience gained in the foreign country. And here we note that peculiar mentality of the Germans, an intellectual curiosity through which they realize to a degree beyond that of any other nation the value of achievements made by other countries, which they imitate in full consciousness. This is quite clearly expressed in a calendar of Günther Zainer in Augsburg which he printed in an italianized roman, where he boasts "ne italo cedere videamur," that it might not look as if the Germans were inferior to the Italians. To what extent national consciousness was lacking at that time among German printers is shown in this item more clearly than anywhere else. The humanistic roman was certainly of Italian origin as a manuscript letter, but as a type, roman by no means made its first appearance in Italy; rather, even before the first German printers had taken their craft beyond the Alps, a roman letter had already come into use in Strassburg in Adolf Rusch's scholarly folios, while its basic principles were to be recognized further, in the earliest roman types cut not only in Subiaco and Rome but even in the earliest printing-shop in Venice by Johann and Wendelin of Speyer. Nicolaus Jenson was the first to cut, in his universally admired roman, a type in which the German touch was finally displaced by the Italian feeling for form.

The German-Italian interchange, in some though very rare cases, resulted in the use of types of a markedly German character by German printers in Italy, but its most notable effect was the adoption in Germany of very large numbers of Italian type-forms. Not all attempts in this direction were crowned with success, and many a German printer was left solitary and without a following in his efforts to transfer an Italian type to Germany. On the other hand, there are certain type-forms, especially some from leading Venetian work-shops, so directly and so widely adopted in Germany that one loses the feeling of their Italian origin. The use of some of these types has been so general as to class them as international even in the period of early printing.

Although now local style and individual imitation have brought about a marked uniformity of shape in certain groups of types, yet during the incunabula period each printer strove by means of some peculiarity of shape to keep a certain individuality in his types. It is scarcely an exaggeration to say that up to the year 1500 there were no two founts of type the forms of which were absolutely alike. Such a statement demands but very few reservations, nor is it even contradicted by the fact that we often find the same type in the course of time in the hands of two or more different printers. In Italy, where the earlier German printers often worked at the expense of wealthy Maecenases, their types upon the dissolution of the printing-firm usually remained in the possession of the patron and occasionally were used again by him. Here however it was always a case of the same type. Also elsewhere a transfer of type from one hand to another by sale and renting is occasionally traceable. In these cases however it has been repeatedly proved that the second owner did not use the type entirely unchanged, but undertook either a recasting on a different body, or a supplementing of it by different, new shapes. An interesting example of this is the type of Lambert Palmart in Valencia, of which we know from documentary sources that it was in 1493 in the possession of Dr. Miguel Albert, but that in his hands it experienced such a number of minor changes that any confusion between specimens from Palmart's printing-shop and that of the type's later owner is not conceivable.

A similar case was that of Günther Zainer's type in Augsburg. He sold his earliest type in 1471 to Johann Schüssler and this was used again apparently without change. Zainer then cut for his own use a new, very heavy type with marked peculiarities, but in the course of time this too passed into other hands. The matrices were acquired by a printer (probably Caspar Hochfeder) who used it in Graz. His type however underwent an adjustment in height and thickness. Large quantities of type must earlier have been produced in Augsburg from the same matrices and sold after the sale of Zainer's workshop, for it was to be found in the possession of at least four other Augsburg printers. Yet the printing of these establishments could be clearly distinguished, for each printer made some characteristic variation in Zainer's letters, differentiating them from those in the hands of his fellow-users of the type.

In French books we find distinctly fewer of the peculiarities alleged to be common to all incunabula types. In Paris and Lyons in the last years of the fifteenth century books were issued by preference in small octavo, printed with small type of international character. This type, in itself of no very marked individuality, seems to have been used very consistently by a number of printers, especially in books which have no colophon, and of which perhaps a greater number must be assigned to a period later than that of incunabula than has heretofore been established. But this kind of type is also uniform in the case of various printers in dated incunabula of the years 1495-1500, so that here if anywhere one could consider it an instance of purchase of ready-cast type from a common source.

A common origin has been claimed likewise for a Netherlands group of types such as that associated in Antwerp with one Hendrik Lettersnider, whose name at the very outset would seem to support this theory. In this case however exhaustive investigation has established beyond dispute that although it is rather a characteristic instance of uniform national features, yet the types of this group show very small but clearly recognizable variations which make it inconceivable that they all originated in the same type-foundry.

Robert Proctor formed his system of exact type identification, an account of which has already been given in the historical section, upon the idea of the diversity of all early types. In his *Index of early printed books of the British Museum and of the Bodleian library* he took the measure of twenty lines as a basis for his system of distinguishing types. How he made a change in his method of

arriving at this measure has already been previously re-
counted.

In addition to the measure, which he unfortunately
did not think necessary to give consistently for each type,
Proctor attempted a closer characterization of types by
emphasizing single type forms which had certain peculiarities, but particularly by indicating the similarity existing between them and other related types. In so doing
he himself departed from the exact method, since similarity is something differently felt by each observer, and
cannot therefore be of universal value as a standard of
measure.

In the *Typenrepertorium der Wiegendrucke* I attempted
to overcome these inadequacies. The *Typenrepertorium*
retained the measure of twenty lines as Proctor had established it in his supplements, *i.e.,* from the base of the first to
that of the twenty-first line above. Its measures do not,
therefore, agree with those of Proctor's *Index.* For the
smaller kinds of letters these differences are on the whole
of but slight importance. The value of any type-scale is
naturally only an approximate one, for it must not be forgotten that the paper was dampened before it was put
in the press and that the printed sheets in drying not only
shrank differently according to the several kinds of paper
used, but that there could even be variations in the same
paper, according to the circumstances under which the
drying process took place. The differences resulting in this
way were not so important as really to impair the value

of the measure in the determination of types, but nevertheless they should not be disregarded.

Instead of comparing types with the form most nearly related to them, the *Typenrepertorium* introduces a second exact factor in the form of the capital M, which it considers of greater value than the measure, in the organizing of groups. It was this letter which in gothic type in the early printing period underwent the most numerous changes in form. The *Typenrepertorium* takes into consideration only one hundred and two different M-forms, for the reason that it seemed best to avoid drawing too fine distinctions, and therefore trivial variations occurring only in scattered types are not used in making up groups but are kept as special forms with the types to which they are most nearly related. Since, in consequence, there threatened to creep in again by some loophole (although to a very limited degree) the idea of similarity which, we have noted, shifts with the individual, an enlarged table of M-forms is given in Vol. 4, in which all special forms are presented in the places where they are incorporated according to their relationship with the several groups. In this way the number of M-forms is increased to two hundred and seven.

Within the groups also the arrangement of the M-forms was intended to follow one main system. It starts with the M of the old roman type (*i.e.,* that first cut at Rome by Sweynheym and Pannartz) and follows its changes and growth, either as it did actually develop or could theoretically have developed in the progress of type-cutting. It was unavoidable that in so doing, when one line of development was completed, it was necessary to go back again to the general starting-point, to trace the second line along which growth took place in another direction. The sequence of groups was by no means determined arbitrarily, however, but methodically, and the explanations with the tables try to throw light quickly upon the aspects represented.

Identification by M-form and measure frequently suffices for types of marked peculiarity. But there are groups so large that they offer considerable difficulty in determining the individual types in them with absolute certainty. As an aid therefore the *Typenrepertorium* gives for each type a brief indication of the characteristic forms which occur in the case of the other capital letters. In doing this a complete survey of all the letters of the alphabet was omitted intentionally, in order to avoid burdening the presentation unduly and thereby rendering it less clear; hence it would be ill-advised to draw conclusions from the lack of any information in the case of a form not particularly unusual. And only in instances where it seemed out of the question to characterize type sufficiently by the capitals have some lower-case forms (especially, following Proctor, the h-form) and such sorts as the hyphen, marks of rubrication, etc., been mentioned.

The *Typenrepertorium* takes Proctor's point of view even more closely in regard to roman than to gothic type.

The former offers far greater difficulties for a sufficient characterization. Even the distinction beloved by Proctor and taken over by the *Typenrepertorium,* of the Q joined to the u or separated from it, does not make nearly so easy a mark of identification as the gothic M-forms. On the one hand there are many hybrid forms in which, according to the printing, the Qu makes sometimes a joined, sometimes a disjoined impression; on the other, however, as is plainly seen, even in cases where the u extends over the tail of the Q, the connection did not result from the casting on a joint stem but from the fact that the Q merely was so shaped that the tail had room to slide under the type of the u.

It is true besides, that roman capitals only exceptionally exhibit peculiarities useful in determining types. One must accordingly make a much greater use of lower-case forms, which, especially in the abbreviations present in almost every Latin text for "et," "con," "rum" and "us," afford an excellent and tolerably reliable means of identification. The ascertaining of the origin of Italian texts is however an extraordinarily difficult matter. These were printed by preference in roman, but when they lack the letters most characteristic of that type, one's only recourse is to reproductions, and often even then a decision as to their source is not to be made with full certainty.

8. INITIALS

Initials[1] in use during a period of progress and development prove useful as a means of identification. In the *Typenrepertorium* therefore, a place (not with its tables, but in the survey of places of printing and printers) is assigned to initials as well as to the consideration of marks of rubrication, printers' devices, title borders and title woodcuts.

The custom of indicating the beginning of a new section by making its first letter an ornamental one or by emphasizing it by means of another color goes back to the practice of the scribes in the early Middle Ages. The most effective initials so executed are those of the *Evangelia,* which often fill a whole page by themselves alone. As the work of the scribes increased however, initials became plainer. While the twelfth and thirteenth centuries show still prevailing the large initials reminiscent of nordic styles, often carried out in four colors, in the fourteenth and fifteen centuries the plain rounded uncials (Lombards), which at most were painted alternately red or blue, became more and more customary.

These styles passed from manuscripts to printed books, and with few exceptions incunabula of the earlier period left at the beginning of a section a blank rectangular space several lines in height, where it was the rubricator's task to paint the proper initial in red or red and blue. Although

[1] Jennings, O. *Early woodcut initials.* London, 1908.

this task offered no especial difficulty in itself, the haste with which perfunctory rubricators did their work led now and then to a mistake as to the letter to be written. To prevent this and to lighten the rubricator's work, the custom was gradually adopted of printing in small type (usually in lower-case), in the space left to be filled in by hand, the initial which the rubricator was to add in colors. Printed "directors" appear now and then in Nicolaus Jenson's *Eusebius* of 1470 (Hain 6699), and in 1471 they are found in use by various Venetian and Roman printers. In Germany they appear almost simultaneously in 1472 in books by Koelhoff in Cologne and by Zainer and Bämler in Augsburg, but here they were not in general adopted as quickly as beyond the Alps.

The printing of initials was attempted very early, but at first found little imitation. The Psalters of Fust and Schöffer of the years 1457 and 1459 show extraordinarily artistic ornamental letters in many colors and in three different sizes, and the method of their production was long a puzzle to those concerned in the art of bookmaking. They belong to the most beautiful and the best achievements which the early decoration of books accomplished. They introduced no new stylistic features, it is true, but only reproduced in the most finished form what manuscript skill had accomplished in its choicest work of this kind—initials surrounded by foliage in different colors. The Mainz printers occasionally used some of these initials later in others of their books, and one or another of the old masters who naturally came from that school (*e.g.,* Johann Neumeister in Mainz, Jacob Wolff in Basel) made at least one attempt to copy them. Their imitation was not, however, general, and the attempt to transfer this style of initials from manuscripts to printed books was almost entirely abandoned. In the sixties, we note at most here and there a single Lombard capital printed in red or even stamped in. Not until 1472 were further attempts made to print initials at the same time as the text, no longer in red as in manuscript, but in black like the type. In a few calendar leaves by Günther Zainer in Augsburg we first meet with some initials printed in black, and immediately after, Zainer put an alphabet of initials into use for the first time. He included also, some exactly following a manuscript style, the so-called "Maiblumen" design, and so well did he know how to make them that they blended harmoniously with the text. In his German Bible of 1477, Zainer used a large number of pictorial initials, a procedure new in printing, and adopted by Bernhard Richel in Basel. Their example was soon followed more or less in widely separated places, and a few years later we find initials of all possible styles used in incunabula of very varied origin.

If one surveys the varied production in this field down to the year 1500 he will see at once that local characteristics impressed themselves even more clearly on ornamental initials than they had done on type-forms. "Maiblumen" initials are really found only in German work.

If one of them occurred in a book from the Hurus press in Saragossa, it was the result of the transfer of a German block to a foreign country. Just so the Lombard outlines were almost entirely limited to German printing. Italy, it is true, took the most characteristic style of her earliest initials, letters closely entwined with coarse foliage, from a German source, the press of Johannes Regiomontanus in Nuremberg. But after Erhard Ratdolt and his associates established these in 1476 in printing-shops in Italy they were scarcely used again in Germany. Venetian printing-houses developed the Italian style of initials in other directions. Characteristic of this style is the combination of human figures and animals, urns and trophies, putti and birds, more or less richly entwined with borders of leaves or flowers, in the beginning almost exclusively in white on a black ground, later also in black on white. The influence of Venetian printing in Germany, due in part to the return of German printers who had received their training in Italy, accounts for our finding this style here and there in German books.

Doubtless Lyons printers were also influenced by this style. At first in Lyons grotesque representations of fabulous animals and human figures were especially characteristic of the style of the initials of Guillaume LeRoy. Soon however they disappeared from the place of their origin, to take up later a prominent position in Paris printing, though in somewhat altered form. In Lyons a species of the Italian vine-branch initials came into use, their characteristic feature being a single flower standing out sharply from the general picture. Grotesques underwent a modification in Paris printing, for there they were almost without exception human heads interwoven in grotesque shapes in calligraphic initials adorned with flourishes. The brothers Hist in Speyer once tried to imitate these in German books; but in general this kind of initial was limited entirely to Paris. From it resulted the peculiar forms which the Paris, and following their example some other French printers gave the initial "L" when it appeared on title-pages. These cuts often are so large that one is doubtful whether they should be called initials or title woodcuts. They were based on a calligraphic initial, but were usually adorned with such a profusion of fanciful ornamentation mostly in a grotesque style, that the pictorial form overshadowed the meaning of the letter. When such an initial, in form individual, but in style in accord with the original, appears even once in a Spanish book (*Oliveros de Castilla*. Burgos, Fadrique de Basilea, 1499), it is only a proof of how much Spanish early printing owed to foreign influence from the most varied quarters.

Another characteristic initial-form originating with Paris printers is the letter decorated as with pearls, on a dotted ground (fond criblé). These initials were probably executed on metal rather than on wood, and attempts to imitate the dotted ground on the wood-block met with less happy results. Besides the Paris printers, Alfonso Fernandez de Cordoba in Valencia executed initials on metal with especial skill. His Latin and Hebrew letters done by this method were however quite independent of the Paris *criblé* initials and took a medium course between Italian and German styles.

Dutch initials also have national characteristics. They are often framed in fruits instead of flowers and foliage, and are marked by a certain heaviness.

It has already been mentioned that Spain took her stylistic features for initial-forms from various sources. Spanish printers were very much in the habit of ornamenting their productions with printed initials, and they added to foreign features a miscellany of original ones. Vegetable forms unlike those used elsewhere were for the most part the foundation of these. Paul Hurus composed a series of initials in plastic designs of stems and branches, and the flowers of Spanish initials are distinguished from the Italian and French by a characteristic cross-hatching in the single leaves.

The connoisseur of the miniatures of medieval manuscripts will at once recognize that the basic elements of manuscript decoration as it had developed locally, laid the foundation for most of these initial-forms. But only in very isolated cases has the imitation gone beyond an adaptation of the style, and although the anonymous printer of Salamanca once attempted to cut an initial in absolute imitation of the handwriting of the thirteenth century, this is the only instance of the kind of which I have any knowledge.

9. WOODCUTS

It is easy to understand why local characteristics are not as noticeable in the woodcuts[1] of incunabula. The woodcut had been adopted in bookmaking as early as 1461 by German printers, who took the new art to foreign countries; hence it is only natural that in the woodcuts of early printing even in those countries, the German touch is unmistakable. There is moreover in various places direct evidence that woodcuts used in incunabula in places outside of Germany were not cut in those places but that the blocks came directly from German printing-shops, and so it is not surprising that incunabula woodcuts only gradually took on a local character. Here again we see German adaptability, for in this field various influences were struggling for expression, and this conflict we may follow to a considerable extent. Sometimes it seems as if a German master had worked from a foreign design; sometimes as if he had tried to adapt a German original to the style of the country in which he was at work; sometimes as if a foreign artist had attempted with more or less success to remodel a German original after an artistic conception of his own. Out of all this one fact clearly emerges, that the claim occasionally made by art students, attributing a decisive significance to woodcuts in deter-

[1]Pollard, A. W. *Early illustrated books*. London, 1893. Schreiber, W. L. *Manuel de l'amateur de la gravure sur bois* . . . V. Leipzig, 1910-11. Kristeller, P. *Early Florentine woodcuts*. London, 1897. Essling, Prince d'. *Les livres à figures vénitiens* . . . Florence, 1907ff.

mining the origin of incunabula, is thoroughly unwarranted. Even where the woodcut has a marked local character, the possibility must always be reckoned with that the stock in question had passed into foreign hands.

It is naturally undeniable that in the last two decades of the early printing period especially, woodcuts can give incunabula research valuable hints. The artistic style of the cuts often makes it possible not only to distinguish an Italian incunabulum from a German, but also to establish whether the book was produced in Venice, in Florence or in Naples; hence it is desirable for the student of incunabula to seek to acquire also some knowledge of the history of art. But he must be very much on his guard against allowing himself to be guided by this alone in cases where more trustworthy resources might make possible a more secure foundation for his opinion.

That other reproductive methods than the woodcut existed in incunabula does not surprise us if we recall that Conrad Sweynheym in Rome 1473 deserted printing to engrave the maps for the edition of Ptolemy first published after his death, 1475, by Arnold Bucking. But engraving on copper was not the only method for map-making. Maps were often done on wood, especially when they occurred but seldom, and it occasionally happened that they were executed in the reverse way, *i.e.,* white on black. Engraving on copper seldom occurred in incunabula, except for map-making. The Florentine edition of Dante of Nicolaus Laurentii of the year 1481 (Hain

5946) is the only work in which an extensive use of engraving on copper was contemplated in book-illustration;
but it did not reach completion, for of the ninety-six illustrations planned[1], only nineteen were carried out, and this
number actually exist in but a few copies of the book;
most copies have even fewer of them.

The metal cut as well as the copperplate engraving
played a certain part in the incunabula period, and in
many cases it is not easy to distinguish whether an illustration is an especially fine and sharply done woodcut or
a metal cut. A peculiar process, which however was used
only a short time and to a limited extent, is that of metal
plates known as "planches interrasiles." The best known
example is in the illustrations for the *Meditationes* of
Johann de Turrecremata which Johann Neumeister twice
printed from the same plates, Mainz 1479 and Albi 1481.
The borders of the first edition of the *Ordenanzas reales*
in Huete 1484 and 1485 seem to have been executed by
a similar process. But there occur only very scattered
examples of this technique, a reversal of the "fond criblé."

Incunabula study assigns a special place to the title
woodcut. This term by no means comprehends all woodcuts which are found on title-pages of the incunabula
period, but indicates only such illustrations as were meant
from the very beginning not for the decoration of one
book alone but for a certain kind of books, or which at

---

[1] One canto each in the *Inferno* and *Paradiso,* and two in the *Purgatorio,*
printed without space for a cut.

least were used by printers in a number of books without regard to the subject represented. It is characteristic of them that they serve the purpose less of illustrations of one particular text than of evidences of origin. The most widespread examples of this kind of woodcut are the school-scenes ("Accipies" cuts)[1] which printers were in the habit of employing in the numerous editions of Latin school-texts. In this sense one might also reckon with this kind of woodcuts the Canon cuts of the missals, although they were usually not on the title but in the middle of the missal, preceding the Canon.[2] But most of these too were used as decorations in more than one book, and because of the similarity of type-forms employed in printing missals are often of valuable assistance to us in identifying them. There are, besides, woodcuts used by many printers which may perhaps originally have been designed as illustrations for a certain text, but later found a far more general use. Even though doubtless not exhaustively treated in the *Typenrepertorium* they are specially listed there beside the press in question, as identifying marks of the printer.

10. PRINTERS' DEVICES

With woodcuts we may consider also printers' devices.[3] Absolutely independent of the contents of the book in

[1] Cf. Schreiber-Heitz, *Die deutschen "Accipies" und "Magister cum discipulis" Holzschnitte.* Strassburg 1908.
[2] Cf. Heitz, *Christus am Kreuz. Kanonbilder der in Deutschland gedruckten Messbücher des XV. Jahrhunderts.* Strassburg 1910.
[3] P. Heitz has published a series of reproductions of printers' devices in

which they occur, their purpose is exclusively that of indicating the source of the book as that of a certain workshop. They go back doubtless to the house-signs which
the tradesman was accustomed even in the Middle Ages
to put as a sign of his ownership, not only on his house
but also on the boxes and bales which he sent out. Fust
and Schöffer had such a printers' device in the form of
two small shields with characteristic marks, first used in
the Latin Bible of 1462 (Hain 3050), and this design of
two shields was later adopted by a great many German
printers. The oldest Italian device, that which the Jenson company in Venice first used in 1481 in the Innocent
IV. *Apparatus decretalium* (Hain 9192), was probably
at first a trademark. It was however so often imitated
or adopted by Italian printers that it has often been regarded as the mark rather of a guild or corporation than
the device of a particular workshop. Printers' devices
always kept their character as identification marks of the
printer, but gradually gained in style as their designs
increased in variety and lightness of touch.

The printers' devices used by the French, and particularly the Paris printers, experienced a very marked development. In German early printing the device is almost
always to be found at the end of the book, either in connection with the colophon or used in its stead, but a few

which Bernoulli compiled the Basel, Kristeller the Italian, Haebler the
Spanish. Cf. also, E. Weil, *Die deutschen Druckerzeichen des XV. Jahrhunderts.* Munich, 1924. 4°.

printers made their devices of a larger size and used them as title woodcuts, or incorporated their initial letters or device or initials of the press, in the border. In Italy too the device was usually at the end of the book, but in Italian incunabula a publisher's device appeared even more often than a printer's, and these publishers' devices, like the Florentine lily of Lucantonio Giunta in Venice or the angel's figure of Johann de Legnano in Milan are found far more commonly on the title-page, although the printer often used his device also at the end. These occurrences, which were nevertheless infrequent in Italy, later became the rule in Parisian incunabula. The French printers' devices found in incunabula are for the most part large pictorial cuts in a rectangular frame occasionally made from borders joined together. Since the device usually left but little room for text on the page of small books which were so soon used especially in France, it was in most cases impossible to put it at the end of the book together with the colophon, and it made its way thence usually to the title-page. Here it fared better, since the title consisted of but few words. So it came about that almost all French devices, especially those of Paris printers, took on the character of title-woodcuts; but another change came about with the growing influence which the publisher acquired in dealing with Paris books.

Publishing very markedly antedates the invention of printing. The conditions of the classical period need not here be taken into account; but in the selling of manuscripts written by hired scribes, as in the scriptorium of Diebold Lauber of Hagenau, we see the appearance of the most basic features of a publishing-firm's business. The production of books, especially of those of considerable size, demanded the arranging of sums of money which were for the time very large, and because of the slow process of early printing, repayment, upon the sale of the completed books, would naturally also be delayed. Even Gutenberg did not have such capital at his disposal, and was therefore obliged to take Johann Fust as a partner. By far the greater number of the early printers were in the same position. All the German printers who found their way to foreign countries were almost continually in search of backers who would help them in making a profitable use of their knowledge of the new-found art.

The backer however by advancing the necessary funds by no means became the publisher; indeed the credit accorded was only in a few instances a personal matter. The many printing-contracts which we have, especially from Italian and Spanish sources, make the rule seem rather to have been that printer and backer entered into partnership, in which, on a scale regulated by agreement they undertook the risk of sale. The trade-partnership which quite universally governed the business life of the fifteenth century, served also as a model for printing-partnerships. When wealthy Maecenases, university professors, well-to-do scholars, bore the expense of setting up a press which should at their direction produce certain books, they did not in any sense become publishers. As long as the printer was dependent, with his percentage, on the proceeds of the sale, they were rather his partners, whose part was performed merely in another way than was that of the printer. A publishing relationship arose first at the moment when the printer was paid a certain sum for his work, while the profits were the backer's as was also the risk of gain or loss.

Although the printing-partnership was the form most usual in the period of early printing, real publishing was in existence very early. When the professors and law students of Bologna formed a company in 1473 to publish the *Repertorium juris* of Petrus Brixianus de Monte, concerning which we have information from an unusually large number of sources, and engaged for this purpose Andreas Portilia, Perdocius de Panzaresis and Stephanus Merlini, for a fixed remuneration, although they called the relationship "societas" they were actually publishers, for the printers had to deliver their productions to the organization financing them, and were entirely unconcerned as to whether the sale of the books was advantageous or not.

We are now of course in a position in only a very few cases to determine with certainty the connection between the printer and his backer. Even when the colophons of the early printed books tell us in so very many cases that the work in question was published by the printer at the expense ("impensis," "aere et impensis") of another, further evidence is necessary to establish the fact that a definite understanding as to publishing existed between printer and backer. Moreover, the scholars and patrons of the earliest printers lacked that characteristic so necessary for the publisher, an interest in the profits. To be sure the early contracts show that such a thing existed almost everywhere, even where scholarly interest in the utilization of the new art was indubitably the chief factor in establishing a printing-shop. We cannot therefore reckon as publishers in the true sense, Petrus Antonius Advena, Franciscus de Vincentio and Lazarus de la Penna, members of the Bologna "societas" of 1473. But there were publishers even before that time who, as backers, entered into partnership with printers and scholars solely from a commercial standpoint. Annibale Malpighi, whose printing-contract with Baldassare Azzoguidi in Bologna goes back to the year 1470, and who died in 1484 financially ruined as a result of his extensive publishing business, is likewise spoken of as a publisher with commercial interests, like Philippus de Lavagna in Milan, who, beginning in 1472, employed a number of different printers on a contract basis in order to have books printed which were to be sold entirely in his own interest. It was in Milan that a purely commercial publishing business originated and was developed to a point of marked prosperity. It has already been noted that we know of Milanese publishers' devices.

The earliest Paris press, established at the Sorbonne by Ulrich Gering, Michael Freiburger and Martin Crantz at the invitation of Guillaume Fichet and Johannes a Lapide, is likewise clearly a publishers' undertaking by which the printer worked on a definite, paid basis without at all sharing in the profits of the business. The same printers however from the year 1473 were their own masters, and it does not appear that among the earliest Paris presses there were any others which worked exclusively for hire. In the last decade of the early printing period there grew up however in Paris more than in any other place a purely commercial enterprise, which, while it employed printers to a very great extent as craftsmen, itself took absolutely no part in printing and was limited entirely to the activities of the book-trade and publishing business.

The publisher's ascendancy in Paris over the actual printer now found its outward expression in the make-up of the book. They strove to have the various printing-shops to which they turned over their commissions, work with types as much alike as possible; hence the disappearance of typographical individuality in the productions of French printers at the turn of the century. Their influence is further responsible for the fact that information as to the printing of the book became more and more inadequate, and was gradually suppressed entirely or supplanted by information as to the place where it was to be sold. We have the publishers likewise to thank, that printers' devices were replaced in most cases by publishers' marks or at least dislodged by them from the title-page and again placed at the end of the book. The publishers' devices of Jean Petit, Guillaume Eustace, Simon Vostre and the brothers Marnef now often appeared as the sole decoration and as the only marks of origin of the books, while the printers' names are in most instances scarcely to be discovered. It is true however that these conditions originated in the last years of the fifteenth century and that they were most prevalent only in the sixteenth. They add perceptibly to the difficulties attached to the study of Paris early printing, and to a certain extent, of that of France in general. A close study of these particular conditions in France at the turn of the century is imperative for an exhaustive knowledge of French early printing.

11. COLOR-PRINTING

In an effort to make his book as much as possible like manuscript, Gutenberg attempted in the first gatherings of the 42-line Bible to print the chapter headings in red. The experiment must however have failed to satisfy him (although the reason is not apparent in the book itself) for after the first gathering he gave it up. The mechanical difficulties, that the sheets must go twice through the press in as nearly the same position as possible, could scarcely have been the only reason, for in the Fust-Schöffer printing of the Psalter of 1457 they triumphed over this in a fashion never again equaled in the whole period of early printing. Some authorities believe that the magnificent Psalter initials may be traced back to Gutenberg himself; but the calligrapher Schöffer probably had more or less share in them. In the Psalters of 1457 and 1459 not only were all initials in the text printed in red but each psalm begins with a highly artistic initial, the letter of which stands out, being of a different color from the arabesques, which, in the case of the large capitals, extend down the margin for a distance of several lines. There is evidence that the letter itself consisted of small, thin, metal plates which were separately colored and then set into a wood-block, which had a still different color. These initials were differently colored at each printing, so that the same initial appears in various colors in the several copies of the book. Since the red of the large capitals is only rarely repeated in the central letter, the colors of the initials usually being still different, we are dealing in the Psalter with four-color printing, the cleverness of the technique of which is not lessened by the fact that the initials proper were very likely separately colored and inserted in the printing-block, and printed at the same time with it.

The reason why Schöffer himself in the earlier period made no further use of printing in red, even in his large books, was not in any case because its technique offered the printer vital difficulties, but because he naturally shunned the increase in labor which put him to the inconvenience of running twice through the press every sheet on which red occurred, *i.e.,* every such sheet demanded twice as much work as the single printing in black. Yet two-color printing in black and red very early became widespread and was practically the rule not only for liturgical books but also for editions of law books with commentaries.

When the early printers' stock of type became varied enough to emphasize the headline otherwise than by printing it in red, two-color printing receded somewhat into the background, and in its place came the stressing of the headline by a larger kind of type.

Gutenberg and his immediate successors (Pfister, Ruppel, Neumeister) used but one kind of type in their books. The Psalters of 1457 and 1459 had already, to be sure, been printed in two harmonizing missal types, so that it was anachronistically credited with using consciously a design for missal-printing, which was in reality not intentionally carried out until fifteen years later. But apart from this consideration, that this later method cast these types with the same bodies but with larger and smaller faces (the usual expression for this being linked fount), and that it appcared after so long an interval with no connection with the earlier processes, Fust and Schöffer also used but one kind of letter in their earliest book-printing. The effort to make an obvious separation between text and commentary in editions of books of Roman and Canon law brought about the first use, after 1460, of two different kinds of type in the same work of secular character. From that it was only a step in such books as required but one type for the text, to use the second to make the headline prominent, a practice more and more in vogue after the year 1473.

But this method could be fully developed only in gothic type, the forms of which were varied enough to offer harmonizing letters of different sizes, a procedure which was not practicable in the case of stiff roman types or at least would have resulted in a very ugly appearance. That is why in places where roman type was most customary, *i.e.,* in Italy, red was used somewhat longer for printing headlines and the portions formerly added by hand, until there too a substitute was found, namely, the headlines and other parts to be emphasized were set entirely in capitals. This custom was adopted also by those printers who used gothic, who however made less use of it because of the less readable qualities of gothic type.

Printing in red was confined then, generally speaking, to liturgical books. Breviaries and missals were not printed in great numbers until after the middle of the seventies, since in them the printer was too much burdened with the difficulties involved in making so many differentiations in the text, for it was necessary not only to separate the text of the lessons from that of the choral parts (attained by using linked fount), but, in both parts, to bring out so clearly the directions relating to the outer forms of worship, that these directions would be easily distinguishable from the text itself. Some printers tried at times to arrive at this result by using for the liturgical directions, a third type quite different from the other two (Breviarium Lubucense GfT 821). These were however only scattered experiments belonging to a comparatively later period. For liturgical books of the very early time and far beyond it, printing in red remained the usual method;

hence they offer the most favorable opportunity for studying its exact technique.

For printing in red in incunabula apparently at least two different processes are to be considered, the choice of which depended chiefly upon reasons of fitness.

It was probably a most uncommon procedure to insert in the forme after having inked them separately (as was the case with the elaborate Psalter initials) such words as were to be printed in red. The usual plan was to have the passages to be printed in red, set up separately, and space left for them in the proper forme by means of blocks. It was apparently customary to print the red first, and not until then, to run the sheet through for the black. Not a few proof-sheets survive which show only the red impression, but I have never seen a leaf on which the black printing appeared with blanks for the printing in color. The technical difficulty in this procedure was in putting the sheet already printed in color, on the forme in such a way that the color-printing should come exactly where it belonged. Points somewhat obviated this difficulty; but we must not forget that later, fewer points were used, and that their holes because of repeated placing and removal of the dampened paper would easily become enlarged. In any case the proofs that red and black printing did not register exactly are extraordinarily numerous; for as the paper shifted slightly to right or left, up or down, it would necessarily happen that one impression neither kept away from the other nor kept in line, so that one color overlapped the other considerably. The places which show this prove the use of this procedure for printing in red. This is however not the only method used during the early printing time. The plan of leaving some places in the forme blank while printing, could be accomplished by putting over the setting of type a so-called mask, a piece of stout paper, so cut that only enough was left to cover the places meant for the other color. This method had the one advantage, that both colors could be proved from the same forme. On the other hand, for the printing of a sheet in two colors it was necessary to prepare two masks, of which one was to protect the setting to be printed in red, from the black, the other, the black from the red. This procedure was also customary in incunabula, as is shown by the fact that in spite of every precaution, the printer could not always succeed in placing the mask quite exactly on the forme. Whenever it slipped a little, the letters nearest the other color were colored by it, hence the edges of letters bordering on the red sections show red, instead of their own black ink. In the period of early printing this seems to have been the less usual procedure and was adapted only to such books as those in which there was no great amount of printing in red. Signs of its use are however indisputably traceable.

As regards the type-page, we are considering only color-printing in black and red. Naturally, however, incunabula include whole pages and sheets, even some entire books, printed in red or rose instead of black, while as is well known, Erhard Ratdolt used gold to print the dedication in a number of copies of his *Euclid* of 1482 (Hain 6693). All these cases have been concerned with printing in one color, however, not in several.

All the instances noted thus far have concerned a type-page in color, and in that, in the early period, there were no further developments. On the other hand, color-printing afforded opportunities for presenting facts pictorially, an aspect which we must not neglect. Erhard Ratdolt and some other Venetian printers specialized in printing astronomical works with very clear illustrations of conditions in the heavens, such as eclipses, and for their representation the printers had recourse to color-printing. Lighted portions and shaded ones were differentiated by printing in black and yellow. Occasionally however it became necessary to print also in red, so that we often find printing in three colors on one page. Erhard Ratdolt also used three colors in printing his own arms or other heraldic designs.

Once only, to my knowledge, do four colors occur on one page in the early period. Ramon Lull worked out a system of logic in which he tried to illustrate by means of tables the relation of the concepts to one another. Such a table occurs also in the edition of his *Ars brevis* which Peter Brun and Nikolaus Spindeler printed 1481 in Barcelona (Hain 10321). In this he attempted to accomplish the illustration of the inter-relationship of three different groups of concepts by dividing the general field into sections, and printing in one color each section of the general field belonging to one group. The colors for the three groups are rose, violet and yellowish green. Below the tables stands the text, of course in the usual black type, so that this page has four colors including the type.

12. MUSIC

The early printers had a special problem in the printing of music.[1] They were confronted by it in the Psalter of 1457, but Fust and Schöffer evaded it at that time by printing only a few staves, on which the music could be added by hand. Many printers followed this practice even long after they had learned to overcome the difficulties of music-printing. Others made it still easier and printed no staves but left blank the space for the musical portion. That procedure was by no means confined to the printing of the earliest period, but occurred in the sixteenth century, even with printers who understood music-printing in other productions.[2] Hence one can scarcely speak of a development in music-printing in the early period, for that began only with the discoveries of Ottaviano dei Petrucci da Fossombrone. His privilege dated from the year 1498, but of his music-printing no fifteenth-century examples have even yet come to light.

[1] Riemann, H., *Notenschrift und Notendruck*. In *Festschrift zur 50 jähr. Jubelfeier der Firma C. G. Röder*. Leipzig 1896. Wendel, C., *Aus der Wiegenzeit des Notendrucks*. In *Zentralblatt für Bibliothekswesen*. 19. (1902) pp. 569ff. Molitor, R., *Deutsche Choral-Wiegendrucke*. Regensburg 1904.
[2] Space left to be filled in: *Missale Romanum*. Hahn 1475 (Hain 11364). *M. Benedictinum*. Sensenschmidt 1481 (Hain 11267). *M. Constantiense*.

Conrad Fyner in Esslingen printed in Johann Gerson's tract *Super magnificat* 1473 (Hain 7717), a series of five notes (probably by means of metal stamps) without staves, and in so doing evaded the problem of music-printing no less than did the printer who placed staves without notes in his books. The earliest real music-printing which we know, originated in the year 1476. It is in a *Missale Romanum* which Ulrich Hahn printed in Rome, in which the Praefationes were printed with their music, two columns to a page. The five-line staves were printed in red, the square notes in black, and thus the sheets had to go through the press twice. But few neums were employed, usually separated from their particular notes. The book is, technically, a very finished production, and the fifteenth century produced nothing else comparable to it.

The Hahn *Missale* was not much used as a model, however. Plannck printed notes of a very similar kind in a few liturgical books, but that is naturally explained by the fact that he succeeded Hahn in his printing-shop. And that the "nota quadrata" or "romana" was used by most of the printers in the countries speaking romance languages can be attributed as much to manuscript texts as to Ulrich Hahn's example. In any case there is absolutely no reason for the impression that the earliest German music-print-

Kollicker 1485 (Hain 11283). *M. Coloniense.* Quentell 1494 (Copinger 4116); Bungart 1498 (Hain 11282). *M. Parisiense.* Hertzog & Emericus 1487 (Hain 11340). Lines printed for music to be added by hand: *M. Moguntinum.* Schöffer s.a. (Hain 11332). *M. Treverense.* Wenssler s.a. (Copinger 4250). *Graduale Moguntinum.* Drach 1500 (Hain 14897). *M. Romanum.* Nic. v. Frankfurt 1487 (Hain 11389).

ing was influenced by Hahn. I pass over the fragment from the Tübingen Universitätsbibliothek, the place and date of which are not established. The oldest dated German missals with printed notes are those of Georg Reiser in Würzburg, which go back to the year 1481. Reiser's output of music-printing was very large, as evidenced not only by a series of missals but by a large *Antiphonar* in three volumes. In all these books, however, he used gothic notes on a four-line staff, so that in common with Hahn he had only the separate printing of notes and staff, the latter, as in Hahn's case, in red.

Just as the "nota quadrata" was usual in Italian, the gothic was the customary notation in German music-printing, but in the latter, individual printers varied very greatly with regard to the form and number of the neums in their stock of type. Gothic notation, however, was as much the rule for them as were the two kinds of Psalter initials, elongated slender Lombards and usually very heavy xylographic capitals. Until late in the sixteenth century no great difference is noticeable in the style of music-printing as compared with its earliest examples, for the fact that one printer printed his staff from the woodblock, another from cast plates, can scarcely be considered important.

This printing process however was not the only one used by the early printers. A French invention of the sixteenth century cast each note and its line together on one type and set up the notation all at one time from these types; but the custom of thus setting notes and lines together from these short separate pieces first gained currency later, and only, moreover, in Italy[1], while the earlier books were generally printed with staves which were made the requisite length for each line.

On the other hand, the music-printing of the fifteenth century was in great measure done on the wood-block. This procedure eliminated very simply all the difficulties connected with music-printing, since the wood-engraver could cut a music note far more easily than a pictorial design, a practice to which through the publishing of illustrated books, not to mention blockbooks and woodcut books, he was already long accustomed. This process did away with any question of the technique of music-printing, for the woodcut made it possible to cut notes and staff at the same time on the wood-block, from which music could be printed with as much ease as could any other design.

Although this method simplified the printer's task very materially, there is no proof that it preceded music-printing from type, in point of time. The Tübingen *Graduale* fragment (Molitor, Table II), which was produced from the wood-block, has an archaic appearance; but it might well be that it owes this more to the printer's awkwardness than to an especially early origin. Moreover, it is certain the earlier printers did not venture to attempt music-printing

[1] *E.g., Missale Romanum.* Venice, Emericus 1498 (Hain 11414). *Missale Ambrosianum.* Milan, Pachel, 1499.

of the size of the Graduals and Antiphonals. For the present we cannot in any case draw extensive conclusions from the Tübingen fragment.

On the other hand, another point emerges from a bibliographical examination of early music-printing; *i.e.,* that in books on the theory of music the notation was practically always done from the wood-block. But this does not mean that early books of this kind always had printed notes. That is by no means the case. The Guillermus de Podio, *De musica* (Hain 13151), which Peter Hagenbach and Leonhard Hutz published in Valencia 1495, show only staves (although before that, ever since 1491 in fact, music-books had appeared in Spain with full notation), and the Caza, *Tractato del canto figurato* (Hain 4819) which Leonhard Pachel printed in Milan, 1492, left all the places blank in which musical examples were to be inserted[1]. Nor is the reason for this lack the fact that the mensural notes of the *canto figurato* would in themselves have given the printer so much greater difficulty. The belief that the Gafurius, *Theoricum opus musicae doctrinae* which was published in 1480 by Francesco di Dino in Naples, had already contained specimens of figurate counterpoint printing[2] is erroneus; but apparently they are met with at least in the Ramus, *De musica* of Bologna 1482

[1] A *Missale Ambrosianum* by this printer, of 1478, does not exist; there is only that of 1499.

[2] This error arose from a confusing of the *Theoricum* with Gafurius, *Practica.* Milan, Signerre 1496 (Hain 7407) which contains mensural notes, while the *Theoricum opus* has only the familiar woodcut illustrations.

(Reichling 1343), and certainly in the Burtius, *Musices opusculum,* Bologna, Rugerius 1487 (Hain 4145), also in books which appeared a number of years before the *Caza* of Leonhard Pachel. But in works of musical theory, in which the musical portion was also printed, it was always done from the wood-block and never, to my knowledge, from type, although in some instances these musical examples were very extensive.

The procedure appears wholly comprehensible however only in the much larger liturgical books of the early printers in which the musical portion was done from the wood-block. The most noteworthy example of this is the *Obsequiale Augustanum* (Hain 11925) which after his removal to Augsburg 1487, Erhard Ratdolt printed for his native city. Only seven pages of this book, a small folio, are covered with music, but these were not only printed entirely from the wood-block but had two blocks to each page, for even in this woodcut printing the notes are black and the staves red. Far more extensive than this music-printing from the wood-block is that in the *Antiphonarium . . . ord. s. Hieronymi* by the Cuatro compañeros, Sevilla (Bibl. ibér. 18). In this imperial-folio volume of more than three hundred pages the entire musical portion was done from the wood-block.

13. TITLES

The earliest incunabula lack title-pages just as do manuscripts. Although one often finds in the description of incunabula "title-page lacking," such a statement is misleading, since in most cases these books never possessed one. Just as in manuscripts, the text of the work begins in the early books on the recto of the first leaf, and if it does not usually end on the verso of the last one, it does not follow that the printer consciously so planned it, but only that the text needed that number of leaves, yet was not long enough to fill both sides of the last leaf. So it happens that we very often find[1] at the end of incunabula a blank page or a blank leaf, or occasionally a number of blank leaves.

Very soon however the printer realized that the first leaf of a book was in a peculiarly exposed position. If the folded and assembled printed sheets of a volume could not be bound immediately (and that was probably far more often the case than one at first thinks), naturally the outer leaves would become badly worn and soiled. The printers at first tried to prevent this by leaving the first leaf or at least its recto, unprinted, beginning the text on the verso or the second leaf. Rusch did this in the Durandus. *Rationale* c.1464 (Hain 6461), as did Zell in Cologne from 1464 on, Zainer in Augsburg and Sweynheym in Rome from 1468, and Johann of Speyer in Venice from 1469, while from 1470 on, it became customary everywhere. But even so the beginning of the book remained without a title-page, the place in which (if at all) the

[1] The *Psalterium*, Koberger s.a. (Hain 13457) has five, and Theramo, *Belial*, Schüssler 1472 (Copinger 5791), *Buch der Kunst*, Bämler 1477 (Hain 4036), and *Psalterium*, Caesaris 1484 (Copinger 4919), have respectively three blank leaves at the end.

printer gave an account of the book's origin and contents being, like that of the manuscript, at the end of the work. At the end of the text the scribes had often, with a "Thanks be to God!" given some information concerning the book itself and their own share in its reproduction, and that custom was transferred to printing.

The conclusion or colophon[1] in incunabula took the most varied forms. Fust and Schöffer in the Psalter of 1457 were the first to give it a form which became the model for a great many German and foreign printers. They announced in it, in an original way, that the work was not written but was the result of a new process "adinventione artificiosa imprimendi et caracterizandi absque calami ulla exaratione," and boasted of being the first to spread it. As they recorded place and time of printing, their colophon contained all that later appeared on the title-page; in fact it gave more information than did some later title-pages. As I said, this form of the colophon was used by numerous later printers and was varied by them in different ways.

In spite of this, the establishment of a colophon partaking of the nature of a title-page was by no means the general rule at first; and in spite of Schöffer's example the majority of the earliest incunabula lacked one. Occasionally however the printers themselves must have considered the lack of any information regarding origin, as a shortcoming, for one can scarcely explain it otherwise on finding copies of the same book with and without

[1] A. W. Pollard. *Last words on the history of the title-page.* London 1891.

a colophon, *e.g.*, the Turrecremata, *In Psalmos.* Zaragoza 1482 (Hain 15706). Still more peculiar was the method of the Fratres horti viridis of Rostock who, in a copy of the Vincentius Bellovacensis, *De institutione ingenuorum* (Reichling 358) in the Stuttgart Landesbibliothek, printed over the two-line "Explicit" which contained no information as to the printing, another, in three shorter lines, in which 1473 was given as the year of publication.

Even after it became more and more customary to add a colophon to the book, its forms continued to vary. I am firmly persuaded that even the "GOD AL" of the Augustinus *De civitate dei* of Subiaco (Hain 2046) is to be understood as a kind of colophon, *i.e.*, that in it is concealed a printer's name (Godefridus Alemanus?), not so much because even the earliest book of the Subiaco press, the *Lactantius* of 1465 (Hain 9806), has a colophon, though an incomplete one, as because a few other early Rome printers, Sixtus Riessinger in particular, likewise placed only initials at the end of a work, and because Antonius Zarotus almost exactly imitated the "GOD AL" in his *Valerius Maximus* of 1475 (Hain 15777) when at right and left of the other information he printed as a colophon the initials "AN.ZA."

The first printers of Rome and Venice introduced the custom of giving the colophon a metrical form. These verses, which recur almost without change in their various books, were usually not their own work, but originated with the learned correctors who were entrusted with the supervision of the printing. This example found many followers, and metrical colophons were used, in some instances intermittently, in others regularly, by numerous early printers down to the close of the fifteenth century. They were occasionally of considerable length and so convincingly appeared to be an integral part of the work itself that they were reprinted as such, unchanged, in places where their information was no longer at all applicable, as witness the long poem at the end of the first edition of the *Confessionale* of Bartholomaeus de Chaimis of 1474 (Hain 2481), in which Christopher Valdarfer in Milan announces himself as its printer. These verses were reprinted by various men, retaining Valdarfer's name, thereby misleading bibliographers as to the source of the editions. Sometimes other printers have inserted their own names instead of Valdarfer's, not always to the advantage of metrical correctness.

A similar reprinting of the colophon, leading to like inconsistencies, is known also in prose, the Spanish printers almost always retaining in their editions of the often reprinted *Cronica abbreviada* of Diego Valera, the colophon of the original edition, Sevilla, 1482, which in an enthusiastic eulogy hailed the printing art, and the Germans as its propagators. But in this case, in retaining the text each printer put his own name below in place of the known originator, and thereby gained credit to which he was scarcely entitled.

From the above-mentioned instances we see how greatly

the form and contents of the colophon varied. While down to a comparatively late period some printers contented themselves with putting their names only in one single word under their work ("Castro" in Diaz de Montalvo, *Ordenanzas reales,* Huete 1484 and 1485, (Bibl. ibér. 214 and 216); "Centenera" in Mendoza, *Vita Christi per coplas.* Hain 11073), other printers furnished very extensive information not only concerning the production of the book and all who shared in it, publisher, editor, corrector, but often also concerning its author, composition and manuscript sources. We find for example a great deal of valuable information concerning the authors' life-history and academic work, in the colophons of volumes of legal discussions and commentaries by Italian jurists.

Colophons were not invariably placed at the actual end of the book. If the work was in several parts or if several articles were collected into one volume, each section often had its own; sometimes only a few of the collected pieces had a colophon, while the very last was without one. This was very likely to be the case when shorter additions stood at the end of a long work, *i.e.* when an editor's epilogue, a short comparison with another work, and the like, were so printed. In this case the colophon was almost always at the end of the main work, so that at first glance the book seems to lack one.

A similar proceeding often occurred when a very long table of contents, a "Tabula," preceded the work. In time it became customary to print such a long table in one separate gathering, and to place it before the text. As a matter of fact this preliminary matter was set up, as it is today, *later* than the text, at the end of the entire job, and so it often happens that the register and the colophon are found at the end of this preliminary matter.

It would have been only logical if an end-title in incunabula had developed from the colophon, but while it is true that a limited number of books of the incunabula period show such an end-title, it is not to be considered a consistent development, the books in which one occurs belonging without exception to a comparatively late period in which the beginning-title was also adopted. No printer in whose books we find it made an extensive use of it, however. It appears earliest in Dutch printing, and Arnoldus Caesaris was the first to use one in his Guillermus, *Rhetorica,* 1483, (Hain 8306). This was a xylographic end-title, but many of his Dutch colleagues copied it in type, sometimes with a woodcut, sometimes without one. Comparatively speaking, Dutch printers made the most use of the end-title. Of forty such which I myself have noted, fourteen belong to the Netherlands, five to the neighboring Cologne, eighteen to Italy and only two or three to Leipzig[1]. All these however were but isolated instances and had too little connection to warrant draw-

[1] Cf. for the Netherlands: Campbell. *Annales* nos. 32, 35, 96, 374, 751, 905, 908, 947, 1233, 1303, 1427, 1453, 1490; for Cologne: Voulliéme, *Buchdruck Kölns* nos. 41, 73, 137, 140, 1207; for Italy: *Hain* 1657, 2065, 3404, 5110, 5639, 6978, 8042, 10569, 12878, 12993, 13116, 13896, 14056, 14125, 14214, 14292, 15477, *Reichling* 748; for Leipzig: *Hain* 10826 and a book by Kachelofen the title of which I lack.

ing further conclusions from them. Perhaps in the majority of cases it was really only an attempt to make of some
use to the book a blank leaf left over at the end, as one
occasionally also finds a printer's device without any text,
on a leaf otherwise blank, at the end of the book.[1]

The headline was the earliest form of title in incunabula. It was much used even in the manuscripts of the
fifteenth century, and in the earliest books belonged to the
portions reserved for the rubricator. When printers were
able gradually to limit the work of the rubricators, since
they could print together with the text the parts the rubricators had been accustomed to add, the headline became
like a chapter heading, and furthermore, was, as it developed, likewise printed with the larger display-type (introduced as a substitute for printing in red). Even in
manuscripts the headline had, however, been a book-title
only exceptionally, but had usually indicated only the
chapter of the book which was found on the page in question, and in printed books also it retained the character
rather of an index than a title. Nor should it perhaps,
really be called a forerunner of the title-page, for it continued in use with the title-page after the latter was generally accepted.

Nor can we reckon as a forerunner of a title, a signature placed on the first blank leaf of a book, as in a book

[1] If such a leaf became detached from the rest of the book, such a peculiar
state of confusion might result as in the case of the *Celestina,* Burgos,
Fadrique de Basilea 1499, which was denied the quality of an incunabulum until I proved from the type that the book also belonged to Fadrique.

by Johann Schaur of Augsburg, his *Ecken-Ausfahrt* (Copinger 2138). Even though this leaf was thus designated as an integral part of the book, it nevertheless contained nothing which hinted at its purpose; nor was there much more of the nature of a title in an edition of the Vulgate which was printed 1487 in Venice by Georgius Arrivabene (Pellechet 2324), which bore in many copies the word "Biblia" on its first, otherwise entirely blank, page. Titles consisting of but few words continued the custom during the whole incunabula period. They were for the most part printed in large display-type, usually in red, extending occasionally the length of several lines; but they usually confined themselves to indicating the contents. Pieter van Os in Zwolle may have been the first to preface a book with a title-page, as in 1480 in a Psalter (Copinger 4953), but the date of this book is very doubtful, so we have perhaps to put in his place Johann Otmar in Reutlingen, with his Gruner, *Officii missae sacrique canonis expositio* of 1483 (Hain 6810); but in Germany as in the Netherlands, title-pages became more general only in the second half of the eighties. In Italy the custom spread still later, in spite of Ratdolt's start in the year 1476 (see p. 150). Only a small number of printers began toward the very end of the eighties to use title-pages, and their use was not general until the nineties. Generally speaking, the incunabula period was ignorant of the title-page such as those of the Reformation period, which gave the place and date as well as the title of the book and its author's name. Beginnings which were made in that direction in a few isolated instances will be mentioned further, but they merely serve to prove the rule.

In time certain printers bethought themselves that they could make this simple form of title somewhat more impressive, and hit upon the expedient of cutting it on wood rather than using movable types. In Germany the brothers Hist in Speyer and Peter Wagner in Nuremberg were very much in the habit of cutting a brief title on wood. Their xylographic titles were executed for the most part in letters which were not much larger than the missal or canon types with which other printers usually set their titles. In Italy and France this form met with very little approval but in Spain it became a fashion which found favor with almost all the more important printing-houses. These titles also consisted for the most part of but few words, which was all the more imperative as the Spanish printers were trying to cut the xylographic words of the title in letters as large as possible.

These woodcut titles were not different in character from title woodcuts, for the method of printing was the same whether words cut on wood or pictorial designs were to be printed; hence it is not surprising to find woodcut designs on the title-pages of incunabula at approximately the same time as xylographic titles. Woodcut decorations were usual from the first in books intended for popular use, and so it is only natural that a title woodcut also should be much used in books of a popular type. It is true it did not continue to be limited exclusively to these, but the number of scholarly Latin works which show such a decoration is very small in comparison with that of books in the vernacular, the editions of which are adorned with cuts on the title. The particular forms which title woodcuts took as printers' devices or as characterizing a certain kind of literature have been discussed in another place in connection with woodcuts.

So long as printing confined itself chiefly to reproducing literature handed down in manuscript and thus rendering it accessible to wider circles, book-titles remained limited to the more modest forms above characterized; but this changed when printing gradually put itself at the service of rapidly spreading humanistic scholarship. Italian humanists had begun very early to add accompanying verses to issues edited by themselves or their friends, or to their own productions. With the German classical scholars it became the fashion to place such verses on the title-page, and from them came the custom of weaving superlatively laudatory epithets concerning the author, into the brief information on the title-page, thereby enlarging without really enriching it or incorporating in it any new elements of great significance.

Yet in very early times a start had been made in the direction of giving fuller information on the title. In the year 1476 Erhard Ratdolt, Bernhard Maler and Peter Löslein published in Venice an Italian edition of the *Calendarium* of Regiomontanus, the first page of which cannot it is true be properly designated as a title-page, for it contains within a graceful three-part border only a panegyric in verse concerning the work. The last lines of this however name the author and below, "Venetijs 1476" and the full names of the three craftsmen. If one reckons the verses in question as a title, there is united on this page all the information as to place, date and printer which was later considered necessary for a book-title; and even at so early a date it is enclosed in an ornamental border such as we are accustomed to find in books of the Reformation period.

This example however remained quite without imitation. Ratdolt and his associates with their borders and initials, became models for their Italian colleagues, but neither they themselves nor any of their imitators ever again put upon the title-page information as to the production of the book. Nor did anything similar to this happen until the very end of the incunabula period in the year 1500. Even then there was apparently no connection between what the Venetian printers had attempted in 1476 and what Wolfgang Stöckel in Leipzig accomplished in 1500. He gave the edition of the *Exercitium super tractatus parvorum logicalium Petri Hispani* of Johannes Glogoviensis (GfT 527), which he published by order of Johann Haller in Cracow, a title in tolerably large and grotesque woodcut letters, but printed below with the types of the text the note: "ad impensas Johannis Haller . . . per baccalarium Wolfgangum Steckel monacensem concivem Lipsensem fauste impressum anno jubilei.

M.CCCCC. Cum privilegio . . ." He is to my knowledge
the only printer of incunabula in whose case information
concerning the book's typographical origin is to be found
on the title-page.

It is obvious that at the turn of the century the lack of a
title was felt to be a defect, for only so can we account
for the fact that here and there we find a title-page in
some volumes of an edition of a book which was first published with this preliminary leaf blank. The case of the
Biblia of Georgius Arrivabene of 1487 has already been
noted. A similar instance is that of Engelhusen, *Collectarius* (Hain 7784), which Moritz Brandis printed in
Magdeburg. A leaf was subsequently placed in front of
his first gathering, this leaf containing as a title only the
word "Collectarius." Also the edition of the works of
S. Bonaventura which was printed 1484 in Cologne partly
by Bartholomäus von Unckel, partly by Johann Koelhoff
(Hain 3463), was subsequently given a title-page by the
latter. One must not, nevertheless, confuse with these
titles added later by the printers, those title-pages taken
from later editions and used sometimes by very early
owners to complete early printed books lacking titles.
Only in cases where similar title-pages not originally
belonging to a book occur in several copies of it, may we
see in this completion the work of the first bookseller, as
for example the edition of the *Trionfi* of Petrarca which
Petrus de Piasiis published in 1490 (Hain 12771). Several copies of it are known which have been supplied

with the title-page of the edition of Bartholomaeus de Zanis (Hain 12776). This can scarcely be explained in any other way than by conjecturing that de Zanis procured a number of copies of the earlier edition and completed them for sale by adding a title-page. A Stuttgart copy of Guido de Baysio, *Rosarium Decreti* (Hain 2718) reveals a similar very interesting procedure which it is true overlaps into the sixteenth century. This book first appeared with a title-page which bore only the one word, "Archidiaconus," and had a blank verso. The colophon gave the information that the book was finished by Andreas Torresanus in Venice, 14. April 1495. A second edition however was issued with the same colophon, the title and heading of the text substantially enlarged, and with marginal notes on every page of text. Careful examination showed that the expansion of the heading, and the marginal notes, had been added subsequently, with different type from that of the copies of the original edition. Only a long preface which is printed on the verso of the title-page betrays that this only appeared in the year 1503. A similar instance is that of the edition from the same press, of the *Opera* of Campanus (Hain 4285). The original printing gave information as to its source but contained no date. When a copy in Freiburg came to light dated 1502, it was thought that the book would have to be stricken from the roll of incunabula. But it was proved that the edition with the date 1502 had only two leaves at the beginning, and the last gathering, ee, reset, while the rest of the large volume agreed perfectly with the undated edition which certainly belonged to the incunabula period.

Later, Torresanus again issued two editions with the same colophon, the editions not agreeing at all in other respects. But there is another far more noteworthy example of the fact that remainders of incunabula editions took on new life in the sixteenth century. In 1488 Erhard Ratdolt published in Augsburg an edition of the *Imitatio Christi* (Hain 9094) without a title-page. Copies of this book however exist, with a full title enclosed in a border of the Reformation period, known to have been owned by Valentin Schumann of Leipzig. Naturally one's first thought is that possibly a former owner had thus completed his copy. When proof is at hand however that not only was a copy in Fürstenberg made up in the same way but another in Petersburg, we must accept unquestioningly that here again is an instance of the utilization of a remainder edition.

The adaptation of the same book for several bibliographical purposes, as we have seen it in the *Baysio* in particular, was, moreover, a method of procedure quite common even in the earliest incunabula period. Even the formes of the Psalter of 1457 had to serve for the production of two different editions, and this procedure was repeated not only in later editions of the Schöffer Psalter but also in the *Missale speciale* which was printed with the small Psalter type. Here however the repeated use was intentional at the time and not an adaptation of a remainder edition. The circumstances are more dubious in another case. Tradition said that in the years 1485 and 1488 two missals were printed for the dioceses of Saragossa and Huesca, but that no copies of these were known to exist. In 1910 however they were discovered in the Cathedral church of Saragossa[1] and it then appeared that the *Missale* of Huesca consisted mainly of the same sheets as that of Saragossa and that it had been adapted for its new purpose merely by printing new preliminary and end leaves and a number of leaves printed only on one side. But since this was only three years after the printing of the first edition it is obvious that it was a remainder edition thus rendered usable.

Preliminary pages are to be distinguished from title-pages, the former being raised to the rank of decorative features by means of pictorial ornamentation. In such decorating the calligraphers of the Middle Ages had already displayed remarkable skill. While Flemish and French miniaturists were conspicuous for the fine execution of their pictorial designs, Italian craftsmen excelled in the artistic ornamental borders with which they adorned the preliminary or "show-pages" of manuscripts which they decorated for patrons of high rank, whose arms they quite regularly incorporated in the lower part of the border.

These manuscript styles which were handed down had a great influence on book decoration in the incunabula

---

[1] A. Lambert, *Notes sur divers incunables d'Aragon inédits ou peu connus.* In *Bulletin hispanique* XII p. 37.

period. Naturally an extraordinarily large number of incunabula were decorated at the order of wealthy booklovers, the miniaturists supplying just such artistically executed borders for the first pages as they had done for manuscripts. But very soon attempts were made to imitate this decoration by mechanical means, as has already been discussed in the section dealing with initials. Independently of printing in the literal sense of the word, however, an attempt was also made to produce such ornamentation mechanically. Some copies of the earliest Rome books show instead of hand-painted borders on the title-pages, more or less extensive borders which are not an integral part of the book and so were not necessarily executed at the same time as the printing, showing indubitably however an attempt to imitate by mechanical means the decoration usually found on the first page. The borders growing out of initial letters in the early German calendar leaves, which as has already been noted go back to 1471, were a transition from this decorative style to that developed from initials, and some German printers made very great use of it. Johann Zainer in Ulm, in particular, was the owner of a large number of initials, the arabesques of which were so executed that they bordered the top and left margins almost to the edge of the type-page, thus enclosing the corner of the text. Similar initials were used by other printers, but rather on broadsheets than in books. These border-like extensions of initials later however became separated from the letters and were used independently as borders for title-pages or first pages. The borders could of course just as well have been made independently of the initials in the beginning. If we usually find such corner and title borders in a still further developed form, it is because of the printer's attempt to come as closely as possible to the decorations of the manuscripts, and these took the form less often of single borders and corners than of a frame on all sides of the page to be decorated. So it happens that, at least in Italian books, corner-borders are but seldom found while complete frames occur very often.

It is scarcely possible to make a constructive distinction in these frames, between the borders only the lower and left sides of which take the full width while the top and gutter side are narrower, and those the two long sides of which have the same dimensions, while only the proportions of the top and lower borders vary, depending upon the composition[1]. The former may be traced back to the page-decoration; the latter are supposed to have developed from the pictorial border. There is no doubt that the borders of graduated widths conform more nearly to manuscript models than do the absolutely symmetrical designs. Nevertheless I do not believe that any fundamental difference in plan is revealed in this change in shape, but that the symmetrical border was only a continuation of the unsymmetrical, to which an artistic, freer style

---

[1] See R. B. McKerrow. *Border-pieces used by English printers before* 1641. In *The Library.* Ser. IV. vol. V. p. 8.

merely gave the impulse, for there are symmetrical borders which are quite direct imitations of manuscript page-borders, and unsymmetrical ones which fully agree in size with the plan of the frame-like forms. Also as regards period, the symmetrical frames designed from a purely draughtsman's-like point of view are manifestly an out-growth of the manuscript type, and do not have so indi-vidual a style that one could assume they had an indepen-dent origin. It is true however that under the influence of Renaissance art, the border of the first page deviated from manuscript models and developed independently, a pro-cedure paralleled in other arts as well.

This whole matter of the decorative border is concerned with the history of the title-page, only in that the develop-ment worked out on the opening page was transferred in the sixteenth century to the title, where it was developed further. In the incunabula period, however, it had not yet reached the title-page.

On the other hand, this kind of book decoration occa-sionally shifted from the first page to the inner leaves of the book. Even in a manuscript when several independent parts were bound together, the first page of each of these sections was sometimes decorated with a border. We find the same thing in printing also, where the parts of Dante's *Divina Comedia* or Petrarch's poetical works are decorated in the middle of the book with the same border which appears on the first page of the whole work. The page-border attained a still greater vogue in devotional books.

The art of the miniaturist in the fifteenth century in Flemish-Burgundian territory reached a very high point in the *Horae*. This approximately contemporary development of book decoration challenged the printer to keen competition, and while charming specimens of painting also were produced, the art of wood and metal cutting reached its zenith in this field in the last quarter of the fifteenth and the beginning of the sixteenth century.

In France the limits of artistic and mechanical book decoration were not sharply defined as in other countries. Antoine Vérard not only had some copies of his books more artistically illuminated as royal commissions but had certain copies printed besides, just to be decorated by hand. Books by Vérard in the Vienna Hofbibliothek contain leaves showing this. The special place held by French *Horae* in the field of early printing is due in fact to this competition with the miniaturists. The manuscript models obliged the printer also to enclose every page in a richly pictorial border for which his stock had often to supply a very large number of cuts of different sizes and kinds of material. Wood-engraving was not a good medium for producing small illustrations with the necessary clearness and durability, and so in these books engraving on metal was very much used, a procedure employed in no other country to a like extent in book decoration.

The printers of other countries had perforce to follow the model of the French *Horae*, the development of which was nowhere else however as extensive as in France.

Naples was most productive in this direction, but the prayer-books of Mathias Moravus and Christian Preller could in no case be compared for fineness of borders with their models, the French *Horae*. Moreover their borders were always engraved on wood only, as in Germany and in the other countries.

When such page-borders occur now and then in other than liturgical books, one may, generally speaking, assume that they are from stock which had previously been used in a prayer-book. And that these decorations for prayer-books, in which very profane cuts were often used also, were the sources for the title-borders of the books of the Reformation period, there can be no serious doubt, a fact which must justify my having given them a more detailed consideration in the present connection.

14. CORRECTORS

Correctors played an important part in the printing-houses of the incunabula period. The language of the present day attaches to the term "corrector" the idea of an official who chiefly watches over the technical correctness of the printing, but in the incunabula period his duties were far more inclusive and important. Proof-reading was, it is true, carried on in the fifteenth century almost as it is to-day, as is shown us on the proof-sheets of that period which have come down to us as binders' waste in book covers. Even the marks which the early correctors used to indicate errors in setting were almost exactly the same as those which are still used to-day for the same purpose; although it is obvious that as to the setting, a fifteenth-century book received a great many more corrections and that these went on for a longer period of time. If we make a page-by-page comparison of several copies of the same edition of one of the earlier incunabula we almost always find some places in which the text does not agree in the various examples. Often they are merely errors in printing which obviously escaped the corrector's attention before the printing began, and were only corrected after a number of sheets had been printed with these errors. In other cases however it is not at all a matter of a wrong letter merely, but the corrector gives another, corrected, or at any rate changed, text. A detailed investigation of the German Bible of Günther Zainer of 1477 showed an astonishingly large number of such scholarly corrections which were undertaken in the text even during printing.

It appears from this that the correctors of the incunabula period were not mere pressmen but must have been men of scholarly acquirements. In any case we should not think the intellectual level of the individual printers of the fifteenth century as humble as modern ideas might lead us to believe. The printers of incunabula were not in general merely craftsmen with an exclusively technical training. Printing in its early period was esteemed a profession, an art, and even if it was not exactly a scholarly attainment yet it was very near it. In the course of time printers were admitted on the score of their art to the circle of those belonging to the universities and shared their special privileges, which was then perhaps rather a recognition of their activity. But the majority of the early printers were themselves men of academic training, as is proved by the rolls of different universities, in which the men who are known to us as printers are to be met with not only as students but also as bachelors and as masters.[1] That in time the title of master was pretty generally given printers as a craftsman's title does not alter the fact that a great many masters of the early period acquired their titles quite legitimately as academic degrees.

Therefore we may literally believe the printing masters, when they boast in many a colophon of their great diligence and effort in making the text of the book as correct as possible, "quam emendatissime," and facts as to the labor of some of these masters are known to us in more exact detail. Praise was given Vitus Pücher in Rome, because he was indefatigable in the tracing of more correct manuscripts as sources for his editions, and from the correspondence between Johann Amerbach in Basel and Anton Koberger in Nuremberg one can see what efforts were made by such efficient printers as these to compile a more reliable text from a large number of manuscripts. But for the most part they did not undertake these cares alone, but made an effort to secure scholarly men for this purpose as correctors. While it devolved upon the corrector

[1] Cf. K. Steiff, *Beiträge zur ältesten Buchdruckergeschichte.* In *Zentralblatt für Bibliothekswesen.* III pp. 259ff.

to watch over the printing, his more important task was to adapt the manuscript sources. In general he would scarcely copy these himself. Larger printing-houses maintained their own copyists for this purpose and we learn more than once that the printer borrowed a manuscript valued for its correctness, to have it transcribed. Judging by such accounts, one may doubt whether there is a warrant for seeing in the early printed editions of classic authors the representatives of manuscripts which were destroyed by their utilization as printers' sources. Even after the introduction of the printing art the value of manuscripts remained very high. As is well known, printing was disdained by royal book-lovers, who scorned to accept printed books for their libraries and even into the sixteenth century employed skilled copyists to execute magnificent and richly decorated manuscripts of their favorite authors. Johannes Trithemius even inveighed against books printed on paper, because he did not believe them as long-lived as vellum manuscripts. The powerful impetus which the Italian revival of learning received in the second half of the fifteenth century was very advantageous to printing. A great many distinguished scholars placed themselves at the service of printing-houses as correctors, and the activity of the printing-presses thus aided the humanists very considerably in their endeavors for the spread of learning, an association in which both sides were very successful, for while the printers were powerfully strengthened by the sale of their books, the scholars, many of whom were in only very moderate circumstances, found work as correctors a very profitable practice. It is true we learn only occasionally as to the pay which they received for their services, and there were no well known men at all among those of whom we have any information. We do know however that the correctors of the *Repertorium juris* of Petrus de Monte in Bologna received one hundred and twenty ducats for each volume, besides a copy free, an amount which represents a very considerable sum for that time. We may however suppose that the fees of correctors did not permanently keep to such heights. Buonaccorso Pisano received in 1475 a contract from Philippus Lavagna for a whole year for two hundred and forty lire (sixty ducats); but according to the value of money at that time that still was a very satisfactory income.

It is to be assumed that circumstances were not far different in Germany also, although we have not at our disposal as much authentic information in that direction. The revival of learning did not cross the Alps until somewhat later and in Germany took a rather different form from that on classical Italian soil. Still, we know that even up to the beginning of the sixteenth century Erasmus of Rotterdam occupied the position of corrector in Basel for Johann Froben for a number of years. Generally speaking, however, these positions gradually lost their original significance. A majority of the printers no longer took pains to edit the texts in conformity with manuscript sources, but the printed texts followed earlier editions more or less slavishly. The German scholars, as Celtes, Lange, Busch, Brant, occupied themselves, however, less with editing classical texts than with copying them in their own productions. They were also closely associated with the printers of their books and watched over their work. So the author-corrector took the place of the corrector of another's texts, and this arrangement gradually took on the form which it has kept to the present time.

It is obvious that in this way, in time, the printer's care waned as to the faultlessness of his texts. Hence printers were often obliged to correct, subsequently at least, the grossest errors which had crept into the printing, thus giving rise to the errata list. Although Berthold Ruppel in Basel about 1468 put one in his Gregorius. *Moralia* (Hain 7926) it is an isolated occurrence, but toward the end of the eighties and in the nineties it became increasingly customary for the printer to accompany his books with a list of errors, wherein at least the most glaring oversights were rectified. The printer himself was as a rule fully conscious of the defects and therefore usually closed with the words "reliqua ipse lector corriget."

15. DATING

It has already been stated that a very considerable number of the incunabula which have survived have no colophons, and incunabula study must assign to these books a place and date, based upon a comparison of their type, initials, etc. with those of known origin.

The dated books, however, are by no means free from difficulties. Even when the place and date of an incunabulum seem to be known, the student is not entirely spared the trouble of certain proof. There are in particular two sources of error which he must keep in mind, the reprint and printers' errors.

The later incunabula period made a very convenient arrangement as far as book production was concerned, by simply reprinting from previous editions works for which a demand either existed or was to be expected. Naturally this was done by different printers with varying degrees of care, and so it happened that even statements which referred to the time and place of issue of a book, were taken over in reprinting, in the new edition, although they were no longer applicable to it. This danger was naturally especially great where such information was concealed in obscure places. A small number of books which appeared in numerous undated editions contain, though obscurely placed, a note from which their date is to be inferred. The best known example of this is the *Postilla* of Guillermus, the chapter "Commune sanctorum," where from the Parable of the ten wise and ten foolish virgins, the time of waiting, the "mora," is explained thus, that "mora" meant the interval elapsing since the Saviour's act of redemption. At this point, where really the number of years should stand which had passed from the death of Christ to the writing of the work, almost all printers have simply set down the year of the date of printing, so that accordingly one can determine at least approximately, the time of even the undated editions of the *Postilla*. Quite similar places occur in the tract *Bewährung dass die Juden irren* (Hain 3023ff.), in the *Textus sequentiarum* and probably in other works also. But naturally through carelessness in reprinting, the usual correction of the date was in many instances not made at all, the date of the original standing unchanged.

In like manner a number of editions of the *Confessionale* of Bartholomaeus de Chaimis were erroneously attributed to Christopher Valdarfer in Milan. A long poem was included at the end of the first edition of the work and in it Valdarfer was named as printer. This poem was however repeatedly considered by later printers a part of the work and printed with it unchanged, so that now the misleading statement as to its origin remains in reprints by other hands.

A more careless reprint also sometimes occurs in a case where a circumstantial colophon makes an exact statement as to the origin of the book. Of the Andreas de Isernia, *Super feudis* (Hain 16249) there are two quite distinct editions, one in roman, the other in gothic type, both stated to have been printed 5 February 1477 at Naples by Sixtus Riessinger. This is however only another instance of a thoughtless reprint. The edition in gothic was published by Ulrich Scinzenzeler in Milan (whose practice as to reprints we have elsewhere considered), who, merely through carelessness, took over with the text, information as to the printer which did not belong to the reprint. While such instances occur but seldom, the student of incunabula must reckon with the possibility, and be on his guard.

Far more common is the case in which a misprint has crept into the date incorporated in the information given by the printer. These are comparatively easy to recognize when they relate to the century. It is naturally easy to see that the number in the hundreds has fallen out and been wrongly supplied when a Bernardus, *Flores,* is claimed to have been printed by Philippe Pigouchet in Paris in the year 1099. On the other hand considerable difficulty is occasioned in determining where an X was omitted. At the turn of the century it is especially fateful. Earlier bibliographers have wished to save for the incunabula period almost all those books which were wrongly dated as published in the first years of the fifteenth century, and they have accordingly set up the hypothesis that an XC had fallen out in front of the final figure so that the book belonged in the nineties of the fifteenth century. This possibility can naturally not be contested without more investigation. It is far more probable however that such books belonged to the first years of the sixteenth century, and that in the date only a C is to be restored.

But other printers' errors occur also in the date, and several of these have attained a certain renown. For many years there has been bitter strife among bibliographers as to the *Decor puellarum* of Nicolaus Jenson which bears the date MCDLXI. Venetian writers took enormous pains to establish by it the priority of their native city over Subiaco as regards the introduction of the art of printing into Italy. The strife ended only with the proof that the *Decor puellarum* belonged to a group of books all of which appeared in the year 1471 in the same size and quite similar in style.

A like case occurred in the Bartholomaeus Mates, *De condendis orationibus* with the colophon "Barcynone nonis octobriis . . . M.cccc.lxviii." Here also the supporters of the printed year raised the claim of the discovery of the first example of printing in Spain, and this claim has not been completely abandoned down to very recent times, although the book, with its signatures, clearly shows that it is an example of a later period. Through documentary evidence that the printer of the *Mates,* Johann Gherlinc, was staying in Barcelona about 1489, the question is decided that the date is a printer's error for 1488.

Such errors in date occur in great numbers, even though usually their significance in the history of printing cannot compare with that of the two foregoing examples. But even when there is no possible doubt as to the correctness of the date in the colophon, problems come up for investigation, the solving of which is often in the highest degree difficult and complicated. This is because the various countries and cities were not in complete agreement as to reckoning and indicating the beginning of a year. When colophons reckon the year "a nativitate" or "ab incarnatione domini," there is clearly no way of establishing with absolute certainty whether the printer reckoned in reality by the Nativity-style or by the Incarnation, which differ from each other not, as one might assume, by nine but only by three months. In many cases however a style was not specifically mentioned and it remains for the investigator to determine which style reckoning was used in the city where the book originated.

But seldom was a printer as painfully exact as Thielman Kerver in Paris. In one of the numerous books which he published at the turn of the century he furnished a double date, according to the native reckoning and "more gallico." It is almost always left to the reader's keenness to determine which year is meant by the date furnished according to the local way of dating. Incunabula study therefore even down to recent times has endorsed the policy of assigning to every book the date which it bears, for the difficulties of bibliographical research are enormously increased if we designate a piece of early printing by a date which, while it agrees according to the modern method of reckoning with the actual date of its issue, does not agree with the date which the printer himself gave it.

In spite of all the difficulties connected with it, the problem would still have been capable of solution if the printers themselves had been consistent in using local styles. But this was not the case, and in many places it was, it is true, clearly an impossibility. The papal court itself in the heart of the Christian world, in Rome, did not use a uniform designation for the year in reckoning but used one style in its Bulls, another in its daily records. Here

the custom of counting by papal reigns offered opportunity
for a certain stability, but it afforded also a loophole for
uncertainty, since although in general, reckoning was from
the day of enthronement, errors as to that occurred even in
the publications of the papal chanceries. Most of the printed
books had nothing to do with these official styles, however.
The German printers at home were accustomed to Nativity-style, but it is always uncertain whether they were in
the habit of reckoning this strictly from December twenty-
fifth or, as gradually developed, from January first. They
obviously often remained loyal to it even when they were
working in a place in which another method of reckoning
obtained, although they often adapted themselves to the
local custom; hence it happens that the date agrees sometimes with one, sometimes with the other style, even in
the case of the same printer.

We do not yet know of any definite rule which governed
printers in using the January kalends as a date, *i.e.,* whether
they connected with them the year which began with the
kalends or whether they chose as the date the year to which
the day in December belonged, on which the book was
finished. The latter seems to have been more usual, but
the matter is very uncertain.

For all these reasons it is unconditionally recommended
in citing quotations from incunabula and in comprehensive lists, to let each book appear under the date which it
itself gives. In any case it is not feasible even where the
real date is easily apparent, to quote a book only by its

corrected date without regard to the form given in the book itself. There is no question that numerous incunabula have thus been assigned a date which does not agree with their actual year of issue. General lists of incunabula and catalogues of single collections are primarily intended to meet all practical requirements; one cannot expect in using them to have at his disposal that degree of specialized knowledge which would be necessary to determine in each instance the true date of an incunabulum. In such works it is imperative that the incunabula be entered with the date which they themselves contain. On the other hand it is the self-evident duty of an exact investigation of an incunabulum to establish wherever possible its true date. For the history of individual printing-houses, for the relationship of different editions of the same work, the true date of certain books can be of extraordinary value, but in any case whoever does close work with fifteenth-century books must make up his mind that the date, so far as it appears in the book, cannot always be accepted as valid without further investigation.

16. EDITIONS

Concerning the size of fifteenth-century editions, a very considerable amount of information is at our disposal, coming in the majority of cases from Italian sources.

All kinds of calculations have been made with the purpose of finding out how large the editions of the earliest books by Gutenberg, the 42-line and 36-line Bibles, may have been, but it does not seem to me that the results are convincing. We know too little as to the technique of Gutenberg's book-making to be able to draw conclusions based upon it.

The most modest editions of which we have definite information amounted to one hundred copies. Although in the year 1480 Schöffer produced only forty-five copies of a book of documents it was not an edition published for sale, but only as many copies were prepared as the patrons thought necessary for their particular purposes. In Venice in 1469 Johann of Speyer printed one hundred copies of his first edition of the *Epistolae ad familiares* of Cicero, and a few books from the press at the monastery of Sant' Jacopo di Ripoli in Florence reached the same number—a *Confessionale Antonini* of 1477, as well as a *Statius* of 1480. Such small editions however were published only where the beginnings of printing were modest, or the contractors lacked business experience, or where commercial interest did not at all enter in, as for example when the chapter of the Cathedral of Santiago de Compostella in the year 1483 commissioned the printers Alvaro de Castro and Juan de Bobadilla to print only one hundred and twenty copies of a *Breviarium Compostellanum*. The editions must also have been very small which were issued in Rome from the establishment of Johannes Philippus de Lignamine. Following the example of Sweynheym and Pannartz he rendered in 1472 an inclusive account of work up to that time, in which he gave the titles of his publications and the whole number of copies published. The average size of his editions was only about one hundred and fifty copies, which is all the more remarkable since he could scarcely have been ignorant that his rivals Sweynheym and Pannartz were in the habit of printing considerably larger editions. They began with a 'Donatus' of which three hundred copies were published, and the normal size of their editions was, up to the report of 1472, two hundred and seventy-five copies. No doubt they further reduced their editions, for their later books are much rarer than those of their first years. Later the Rome printers also seldom printed fewer than three hundred copies. If we hear of smaller editions, they were usually first attempts. Johann Neumeister in Foligno made only two hundred copies of his *Cicero,* and the edition of the *Institutiones* of Justinianus published by Andreas Belfortis, 1473, was issued in the same small size. In the seventies the editions which were fewer than three hundred copies were exceptional, and some books even before 1480 were put on the market in considerably larger numbers. The Bible which Leonhard Wild printed for Nicolaus von Frankfurt, 1478, was to be published in nine hundred and thirty copies, while Wendelin of Speyer had ventured even in 1471 to bring out a thousand copies of a Panormitanus, *Super I. II. Decretalium.*

Such large editions were however very rare and their disposal may at that time have raised difficulties. It is known that Wendelin was soon financially embarrassed.

Generally speaking, in the middle of the incunabula period editions usually consisted of four to five hundred copies. Peter Schöffer issued four hundred copies of the *Missale Vratislaviense,* 1483, one of the few details concerning German conditions which have come down to us.

A peculiarity of the incunabula period was that the size of the edition was increased even during printing, a new impression usually having to be made of the sheets already printed. The edition was often so large that it seemed almost inconceivable that the printer could make it pay. Obviously however it is to be assumed that the printer understood how to interest the public in the forthcoming new issue even during the printing, and to collect orders for it, although it is doubtful whether that often happened. A single instance of a genuine subscription has however survived. In Saragossa the German printers Heinrich Botel and Paul Hurus in 1476 published an announcement in which they asked that orders and deposits for an edition of statute laws be left with a notary. They volunteered if sufficient interest were shown, to deliver the books within six months, and this edition of the *Fori Aragonum* (Bibl. ibér. 278) actually (but with no typographical data), did appear. It is obvious that the demand was steadily increasing, although, it is true, in the first half of the seventies the book-market at least in Italy became glutted with classics, which considerably hindered their sale. After this crisis was past, mainly because printing took up other hitherto somewhat neglected branches of literature, the demand for books grew uninterruptedly. In consequence of this it became more and more customary for the printers to dare to offer for sale surprisingly large editions. Hans Rix in Valencia ventured even, by 1490, to print an edition of more than seven hundred copies of a literary work, the romance of *Tirant lo Blanch,* and this was not a popular book of handy size adorned with cuts, but a heavy volume of seven hundred and ninety-two pages. Of so serious a publication as the works of Plato, Lorenzo di Alopa in Florence in the middle of the nineties ventured to print an edition of 1025 copies. Even this is not the highest point reached in the incunabula period. By 1491 Matteo Capcasa in Venice printed 1500 copies of a *Breviarium;* and in 1489, Mathias Moravus in Naples published 2000 copies of the *Sermones de laudibus sanctorum* of Robertus Caracciolus. He had, it is true, undertaken this risk only by relying on the safeguard of a privilege which forbade the reprinting and importing of another edition of the work as long as his edition was on sale. Still more astute was Baptista de Tortis in Venice. The numerous editions which this publisher brought out in quick succession, of the texts of Roman and Canon law, gave rise to the conjecture that his editions might not have been very large. That however was not by any means the case. The edition of the *Codex* of the year 1490 was issued in not fewer than 1300 copies, and he issued the *Decretales* of Pope Gregory IX not only once, but, within three years (1491-4), in two editions of 2300 copies each.

17. SALE

In order to dispose of such numbers of books, naturally an extensive commercial organization was necessary. Unfortunately there again we derive almost all our information from Italian sources alone.

The book-trade antedated the invention of printing. The scribes of the Brothers of the Common Life and private contractors like Diebold Lauber, for example, must naturally have sought a market. At that time Paris was probably the greatest centre of the book-trade, and the University there even in the later Middle Ages had its four "libraires jurés," who indeed would scarcely have made a living aside from their business with the Sorbonne. Paris was a market also for printed books in the earliest times, as seen by the journeys thither of Fust and Schöffer, for in the sixties they visited Paris repeatedly with their productions and maintained there a permanent agent in the person of Hermann Statboem, who at the time of his death in 1474 had Schöffer's books in his possession valued at 2425 crowns. We know also that Schöffer maintained a travelling agent to sell his books.

We may infer from their complaint in 1472 that they were in danger of being ruined because so many of their books were left unsold in the warehouse, that Sweynheym and Pannartz were not especially talented as dealers. But even they seem often to have sent travelling agents as far as Germany, for Hermann Schedel in 1471 bought a *Caesar* in Nuremberg from one Bernardus de Merdingen who was known as an employé of the Rome printing-house.

It is very interesting to trace the means by which even the modest workshops of smaller Italian printing-towns endeavored to sell their productions. Johannes Reinhardi of Eningen probably printed barely one book of large size in Trevi in the year 1471, but from the final settlements with his Italian partners it appears that copies of it were on sale also in Perugia and in Rome in the very year of its issue. Similar information came from the association of printers in Perugia. This was active throughout a number of years (1471-6), and published a very large number of editions. But even in the year 1474 they had warehouses for the sale of their books in Rome, Naples, Siena, Pisa, Bologna, Ferrara and Padua. Even so ephemeral a press as that of Antonius Mathiae and Lambertus of Delft in Genoa, according to an agreement of July 8, 1471, sold their books not only in the other towns of Lombardy but even in the kingdom of Naples. Francesco del Tuppo, on the contrary, after having taken over the printing-shop of Sixtus Riessinger in Naples, about 1480 sent a number of agents around the country to look after his business in that way.

The book-trade of Venice was naturally very highly organized. Even some years before his death Nicolaus Jenson seems to have somewhat withdrawn from actual printing and to have laid his greatest emphasis upon book-selling. To this end he formed a trade partnership with a few German dealers under the name "Firma Nicolaus Jenson sociique," and thus maintained establishments or agents in a number of cities of upper Italy, Milan, Pavia, Perugia. Under Jenson's name was issued a large and splendidly printed advertisement of which unfortunately only fragments have been preserved.

Such advertisements of books[1] were however by no means first issued by Italian printers nor were most of them from this source. It was rather the German printers Mentelin in Strassburg, Günther Zainer in Augsburg, and the like, who made use of this means of advertising, either as preliminary announcements of their publications or as praises of their firm. From these very book-advertisements we learn also that many of these printers maintained travelling agents for the sale of their productions, for they often close with an invitation to the buyers to look up the representative of the "Firma" at a certain inn. The name of the locality however was added by hand, an indication that the agent took these blanks with him on his travels.

Perhaps Nicolaus Jenson's attention was first called to advertising methods by his German partners. Immediately after Jenson's death the company was greatly enlarged by a union with the establishment of Johannes de Colonia and Johann Manthen. This partnership entered into, May 29, 1480, for five years, must have made a very powerful

[1] K. Burger, *Buchhändleranzeigen des XV. Jahrhunderts.* Leipzig 1907, and E. Voulliéme, *Nachträge zu den Buchhändleranzeigen . . .* In *Wiegendrucke und Handschriften, Festgabe K. Haebler.* Leipzig 1919. pp. 18ff.

book-trade organization. It appears that Johannes de Colonia in 1480 put 4776 ducats and 477 lire into the new company and that by 1482, 3391 lire had been credited to him, and by 1483, 3411 lire.

Although in the nineties printing had spread so widely over all the European countries, the foreign business of printers and booksellers continued to the end of the incunabula period. Its methods changed only in that besides the travelling agents of individual printing-houses, professional booksellers gradually became interested in the exchange of printed books between towns or between countries. Anton Koberger from the year 1506 on, maintained his own warehouses not only in Paris and Lyons but even in Toulouse; Johann Parix, printer of Toulouse, in 1491 sent one of his own agents to Spain, to settle accounts with his managers there, and in 1489 Hans Rix of Chur sold in Valencia the books of various Venetian firms, installed agents for them and for himself in the Spanish provinces, and himself engaged in the printing and publishing business.

With the rise of publishing on the one hand and that of the retail book-trade on the other, most printers dropped to the status of mere workmen, and when that happened, the great period of printing was over. This development began even in the incunabula period, at the great centres of commercial intercourse. Only where the living pulse of development was felt less perceptibly, did the printer maintain his place after the year 1500 as an artistic craftsman. When he relinquishes this, his work ceases to be of special interest in the study of incunabula.

18. PRICES

We are not entirely ignorant of book-prices in the incunabula period, although the numerous quotations of prices to be found in some copies of incunabula should be used with great caution. Only in isolated cases is the time also given to which the price in question was applicable, and even then it can never be determined exactly whether the book was sold directly by the bookseller or only at second-hand by an earlier owner. This is in fact the only plausible explanation of a great many quotations as to prices the amounts of which are out of all proportion to the size of the book concerned. A further difficulty in profiting by these details is occasioned by the monetary conditions of the time. In the fifteenth and sixteenth centuries each territory had its own currency, the rate of exchange of which was, into the bargain, subject to many variations, and these trying conditions have not yet been exhaustively explained. Only after long and laborious research therefore can one hope to arrive at a definite conclusion in each case. Fortunately however we have access to a certain number of other sources of information, and with their help we may be able to form a true conception of the book-prices of the incunabula period and the course of their development.

In the first years after the rise of printing, it is clear that prices were still wholly arbitrary and in no way uniform. That is reflected in the tradition that Johann Fust was persecuted in Paris as a swindler because he allowed his books to be sold to different customers at quite different prices. Nor can we make much use of the information that a copy of the 42-line Bible bound and rubricated was valued at one hundred ducats or that a vellum copy of the *Catholicon* of 1460 was sold at 2460 groschen and an Augustinus, *De civitate dei* of Subiaco, 1467, at eight and one-twelfth ducats (about twenty-seven Rhenish florins). The latter statement agrees fairly well with the fact that a copy of the Strassburg edition of the same work which was published by Adolf Rusch was valued at thirty-six marks (about twenty-four florins). That for a long time in Germany there were no fixed book-prices, we may infer because the early book-advertisements were drawn up entirely without information as to prices, but instead almost always promised that the customer would find a generous dealer.

The earliest trustworthy source for prices of incunabula is that of Hartmann Schedel's price-list of the earliest books printed by Sweynheym and Pannartz in Rome[1]. This is concerned with actual booksellers' prices and the information is extensive enough to enable us to draw from it fairly trustworthy conclusions. To be sure, inspection of it shows that the prices were arrived at in no very methodical way. The seventeen works printed in the years 1469

[1] K. Burger. *Buchhändleranzeigen des XV. Jahrhunderts.* p. 6.

and 1470, the prices of which are given, fall into three groups, large folios of about 265 x 170 mm. type page, small folios of about 220 x 130 mm. and quartos of about 185 x 115 mm. As a matter of fact the quartos even then usually cost nearly half as much as a folio of similar bulk, a price which later became the rule. On the other hand a definite basis for calculating as between large and small folios apparently had not yet been established. If one should use in Sweynheym's books the method later in general use among Italian booksellers, reckoning how many quinternions were sold for a ducat,[1] it would be found that the number in the large folios shifts from 4.7 to 6.7, averaging about six quinternions for the whole production. In the small folio the corresponding figures are six to eight quinternions and the average for the whole amounts to seven.

The next available statement as to book-prices occurs in the year 1480[2]. In the period from February 27 to May 5 Antonius Moretus delivered to Domenico di Gilio in Padua, in three lots, about one thousand volumes of approximately two hundred and fifty different works, the prices of which were quoted. Naturally we cannot be absolutely certain which editions were meant as regards all of these books. But if one assembles by size the books alike in price, so surprising a uniformity results that we

[1] I have omitted one estimate as to quarto and one as to small folio which differ surprisingly.
[2] Fulin, in *Archivio veneto* XXIII pp. 395ff.

may well trust the scale obtained by this means. It appears in this way that of the books of large size nearly twelve, and of the usual folio size about twenty, quinternions were to be sold for a ducat. The prices have dropped more than half, as against 1470.

There appeared at approximately the same time the long list of gifts of the Basel printers to the Charterhouse at Basel which as is well known continued into the sixteenth century.[1] It also must be used with caution, as it gives different prices for the same book. In spite of this however one can gain certain facts from it. First of all, two facts stand out, that Basel book-prices were lower in general than the Italian, and further, that price reduction steadily continued in the last twenty years of the fifteenth century. If we take into consideration that the ratio of the Rhenish florin to the Roman ducat was approximately three to two, it is very surprising to see that Amerbach's large books were valued in 1484 at about seventeen, but in 1489 at twenty-seven quinternions and more, for a florin. Likewise the books in the usual folio size dropped from twenty-two quinternions in 1481 to thirty-three in 1493. In the case of Nicolaus Kesler price changes were less continuous but the expansion was still greater, for while in 1486 twenty-three quinternions were sold for a florin, the number in 1493 advanced to nearly fifty. Froben's books showed a like result, even though in narrower limits, *i.e.,*

---

[1] Zedler in *Beiträge z. Bibliotheks- und Buchwesen, P. Schwenke gewidmet.* pp. 276ff.

from thirty quinternions in 1496, the equivalent of a florin a few years later mounted to thirty-seven and one-half. In the Basel book-trade at this time the policy seems to have been carried out, that, roughly speaking, of quartos, double the number of quinternions was sold, while the price of an octavo was apparently somewhat more than half that of the quarto.

As is well known, the Basel printers gave their Charter-house not only their own productions but also books of foreign origin, some contemporary with theirs, some earlier, and it is very surprising to find that these early editions were evidently still valued according to the standard prevailing at the time of their issue. The price of the *Epistolae* of Hieronymus in Schöffer's edition of 1470 was set at seventeen florins, a rate of two and one-half quinternions to a florin. That agrees very well with the fact that Mentelin's printing of the *Summa* of Astesanus was valued at thirteen ducats, about three and one-half quinternions to a ducat. The tendency toward a decline in prices was plainly apparent when Schöffer's Turrecremata, *Super Psalterium,* 1474, his Bernardus, *Sermones,* 1475, and Wenssler's Durantus, *Rationale (ca.* 1478) were sold at the rate respectively of eight and one-half, eleven and three-quarters and thirteen quinternions to a florin. An important insight is afforded us by the information that Koberger's German Bible of 1483, so richly adorned with woodcuts, was sold at nine and three-quarters quinternions to a florin, while the *Repertorium* of Bertachinus which appeared in the same year,

and the *Biblia cum postillis* of 1485 were valued respectively at fifteen and one-third and twenty-seven and one-
third quinternions to a florin. The shift in prices of the
Cologne and Strassburg books was similar, in 1483 about
twenty-one, in 1489 however about forty quinternions being reckoned at a florin.

A still larger group of book-prices, *i.e.,* those of missals[1] is at our disposal from German sources. These prices
were not however reckoned in as commercial a way as
those mentioned above, and did not undergo a reduction
corresponding to that of the general shift in prices. On the
contrary, the great uniformity in missal prices in different
places indicates that a standard had been established in
this field. When the *Missale Herbipolense* was reduced
between the years 1491 and 1499 only from four to three
and one-half ducats, it was, certainly, considerably below
the market value; but when the price of thirteen German
missals fluctuated only between three and one-half and six
ducats, an average of about 4.4 ducats, it was indubitably
characteristic of this particular branch of book production.

We have access besides to some foreign information
concerning missal prices. We may assume that the prices
agreed upon with the printer for missals and breviaries
served as a standard also for the valuation of these liturgical books in the dioceses. According to that, the missals in
Italy and Spain were considerably cheaper than in Germany. Even in 1475 the *Missale Ambrosianum* of An-

[1] Molitor, *Deutsche Choral-Wiegendrucke,* p. 38.

tonius Zarotus was reckoned only at one ducat, and a missal (now lost) printed in 1500 in Messina cost the same amount, as did in 1499 the *Missale Benedictinum* from the press at Montserrat. Still lower was the *Missale Vicense* of 1496, copies of which did not even cost a Barcelona pound, the value of the pound being about nine-tenths of a ducat.

In the year 1488 a settlement took place between the Milan bookseller Ambrosius de Chaimis and the widow of Peter Uglheimer, who was agent[1] in Milan for the Jenson-Colonia company. Here for the first time no individual prices were made, but it was agreed that twenty-eight quinternions of each large book, fifty-six quinternions of each small folio, should be reckoned at a ducat. Quite the same plan was established in the partnership-agreement between Petrus Antonius de Castelliono and Philippus de Lavagna in the year 1490, except that in that case thirty and sixty quinternions respectively were sold for a ducat, and that it was expressly stipulated that the smaller sizes, quarto and octavo, were to be valued in accordance with their relation to the folio sheet. The contract contained, besides, the noteworthy memorandum that these prices applied only to local sales and that when the books were sold afterward abroad, they would be advanced about ten soldi per ducat (about ten per cent). Prices remained thus approximately to the end of the incunabula period, for even in 1507 the firm of Moretus-Giunta-Arrivabene-

[1] Motta. In *Archivio storico lombardo.* 1898. pp. 66ff.

Tortis expected to put its productions on sale at the price of thirty to thirty-five quinternions per ducat.

That there were special prices for special books the Koberger illustrated Bible has already shown, and this is confirmed by the advertisement of Aldus Manutius, which was provided with information as to prices.[1] This shows that the books printed by Aldus in roman type were priced approximately like those of other establishments. The number of quinternions to be sold per ducat advanced from twenty in 1495 to twenty-four in 1497. On the other hand the books printed entirely in Greek type were somewhat higher in price and also maintained those prices longer. The different volumes of the edition of *Aristotle,* although they appeared from 1495 to 1498, were clearly priced alike at about fifteen to sixteen quinternions per ducat, and the *Aristophanes* of 1498 in somewhat larger size even sold at only fourteen.

To the books specially gotten up belong, above all, the copies on vellum. Peter Schöffer, in particular, published many vellum copies but unfortunately we have no statement as to the prices of his productions. Missals are for the most part the books which enable us to judge as to the relative prices of paper and vellum copies. In general the price of a vellum copy seems to have been approximately three times that of one on paper. But there were greater differences, as an actual fact. For example, the *Missale*

---

[1] According to Omont's facsimile reproduced by E. Voulliéme. In *Wiegendrucke und Handschriften, Festgabe K. Haebler.* p. 32.

Benedictinum of Montserrat on vellum cost ten times as much as the paper, and the same ratio seems to have prevailed for the *Breviarium* of the same provenance. Probably however special reasons determined that, for in the case of the *Processionarium* from the same press the ratio between paper and vellum copies was only as one to four, not very different from the German prices.

All these prices concern the independent book-trade, in which limits were apparently nowhere drawn by special regulations as to price-making. Only in Spain in the sixteenth century did the book-trade submit to special price-fixing. Precursors of this occurred even in the incunabula period, when for such books as were protected by a privilege the government dictated the price at the same time. Prices in Spain show that the level there agreed approximately with that of the Italian book-trade, for the *Leyes por la brevedad de los pleitos* (ten leaves) was priced at twelve maravedis (about one-thirtieth of a ducat) and the Gutierrez, *Cura de la piedra* (eighty-eight leaves) at seventy-five maravedis (about one-fifth of a ducat).

Missals reveal some information as to the price of book-binding. In the case of the Würzburger Missals of 1497 and 1499, to the price of three and one-half florins, half a florin was added for the binding, while the *Missale speciale* of the same diocese cost one florin, and bound, one and one-quarter florins. Some manuscript notes found in incunabula show that the binding of a book of from one hundred and fifty to two hundred leaves fluctuated between four and six groschen, and a volume of three hundred and fifty pages once cost eight groschen. On the other hand estimates which we can gather from Italian sources concerning bound books appear very high. Of course they apply only to a few works of small size, so that one perhaps should not generalize without further knowledge. Moretus in 1480 as well as Siliprandus in 1483 published 'Donatuses' bound and unbound. In the case of Moretus in a book-price of ten soldi the difference amounted to two lire *i.e.,* almost fivefold, and in the case of Siliprandus approximately threefold. Moreover as regards the *Macer* of Siliprandus (assigned to Ovid), the price of binding was almost two lire, while it amounted also in the case of the *Officia b. Mariae* of Moretus to more than one lira (one-fourth of a ducat).

It is apparent from other sources as well that the cost of binding in Italy must have been very considerable. But bookbinding bills, as that of Schariglia in Naples, in which the books were not specified, do not help us much. More useful is the entry in a Stuttgart copy of the *Biblia Latina* of Franz Renner of 1475, which cost seven ducats, 'ligata et miniata'. Since the book contained forty-five and one-half quinternions, one might allot about three and one-half ducats for the book itself. The rubricating is very abundant; the large initials are on a gold ground while the small ones are on a background of vines in color, and the first page is enclosed in a richly colored border. But if one should reckon even two and one-half ducats for that, one full ducat would still be left for the binding. The costs of rubricating are difficult to compare because it differed very greatly as to the richness with which it was done. We have copious information from manuscript times in the bills of Taddeo Crivelli concerning his business with Nicolaus Laurentii, for whom in the years 1451-56 he acted as rubricator.[1] According to that, he received for a thousand painted initials (parafi) three soldi; for a hundred ordinary initials, from three to four soldi; for one initial on gold from three to six soldi and for a miniature (Crucifixus) two lire. In the beginning, prices for rubricating printed works were probably similar.

In the case of Moretus, 1480, a rubricated 'Donatus' was valued at sixteen soldi as against a plain one at ten. In the contract of Lavagna and Castelliono in 1490 the *Decretum Gratiani,* not rubricated, was reckoned at one and one-half ducats, and rubricated (miniato), at two. Another case approximating this is the *Vocabularius* of Reuchlin, Basel 1482, which was valued at one and one-quarter florins, but bound and rubricated, at two and four-fifths.

That the appreciation of early editions apparent in the gift-list of the Basel Charterhouse was not universal, is shown by a book-tax of the year 1511 in Venice. A shareholder of the Jenson-Colonia company, Hieronyma, the widow of Aloisius of Dinslaken, had her dowry appraised at that time. In it was a stock of one hundred bales of books,

[1] Fischer, *Die Entdeckungen der Normannen in Amerika.* In *Stimmen aus Maria-Laach.* Suppl. no. 81. Freiburg, 1902. pp. 121ff.

comprising 4173 volumes, most of them of large size. In spite of this the entire stock was valued at only six hundred and sixty ducats, a minimum amount in contrast to the single prices which have come down to us for some of the books contained in this stock. In this rapid depreciation is to be seen the result of mass-production in Venice.

19. PRIVILEGES

We must be careful not to confuse conditions of the incunabula period as to reprinting, with the matter of copyright in our own day. Doubtless a very large number of incunabula were simply reprints of extant editions, but the fifteenth century saw nothing at all objectionable in this practice. In manuscript times it was an unqualified gain to reproduce a much-desired text by copying it, and it was deemed equally commendable in the incunabula period for printers again and again to put on sale in new and, when possible, cheaper editions, books for which there was a particularly heavy demand. Naturally however a printer could sell more cheaply who simply reprinted a well corrected edition of his predecessors' than the one who had a manuscript text edited by scholarly correctors. For all that, even down to the sixteenth century there were high-minded printers who did not shun the trouble and expense incurred in producing with the help of scholarly collaborators, a text as free as possible from errors; but they did not consider it wrong for a printer to use as the source of a new edition a printed text instead of a manuscript.

PRIVILEGES

The period of early printing on the whole did not recognize an author's right, and even when a privilege had once been given an author for the printing of his books, it had to do, not with his intellectual production, but the material expense incurred in printing his work.

The earliest privilege known to us is of an entirely different kind. The protection given by the Signoria of Venice to Johann of Speyer in 1469 was rather a trade monopoly than a printing privilege, not insuring his books against reprinting, but granting him the exclusive right to print books in the territory of the Republic of Venice for the next five years. The privilege lost much of its significance as Johann of Speyer died during the first year of his monopoly. A privilege of this kind was limited very strictly to the person invested with it and lapsed immediately upon his death, and so far as we know, Johann's successor, his brother Wendelin, did not try to have it transferred to himself. Moreover, the Republic of Venice never twice granted such a privilege to any one person, although this was done elsewhere, though to a very moderate extent. Adam von Rottweil upon transferring his business from Venice to Aquila in the Abruzzi in 1481 wanted the exclusive right to print and sell books during his stay there. The town councilors could not grant both requests but awarded him the printing monopoly. Efforts to obtain a similar monopoly were known elsewhere as well. At about the same time that printing was introduced into Venice, like negotiations were pending in Milan, and there too the applicant sought a monopoly. He learned however, at once, that other rivals were willing to renounce such a privilege. It was probably because of the great number of bids from printers that a printing monopoly was evidently never granted twice to any individual. Neither the printing monopoly nor the special privilege should be confused with the privilege shared by individual printers. Landgrave Friedrich I gave Heinrich Eggestein a special privilege in 1466 with a promise of protection, but this security did not cover the reprint. Sometimes this special protection was awarded to printers but more often they applied for it. Antonius Planella in negotiating with the envoy of the Duke of Milan in 1470 for the right to establish a press in that city took as a precedent the fact that he had been assured free lodging and other advantages in Venice and demanded the like as a condition of his removal to Milan. Clemente Donati also asked the Duke of Ferrara in 1470 for the assurance of free maintenance for three years, upon his coming to Ferrara to set up a press. When Ferdinand and Isabella called Meinard Ungut and Stanislas Polonus to Seville they assured them exemption from taxes as well as other privileges. Although this may often have happened in the early days of printing, it was quite a different matter from the protection of their work by a privilege.

A printer's privilege had very limited power however in the fifteenth century. One issued by the Signoria of Venice was good only for its own territory, losing its efficacy on the mainland, and therefore becoming of no effect in Milan or Pavia. Territorial division in Germany as well as in Italy lowered the value of such a privilege, which could guarantee really effective protection only in countries in which centralization was already far advanced, as in France or Spain.

Many years passed before a legal protection was sought for single pieces of printing, and the earliest privileges of this kind appear to have been granted in Milan rather than in Venice. Under the date 6 July 1481 the bookseller Andreas de Bosiis was given a privilege for the printing of the *Sforziade* of Johannes Simoneta, who had it printed (without a colophon however) by Antonius Zarotus. Some years later a similar transaction as to a privilege took place in Milan and we learn that in that case at least the protection was really effective. On March 15, 1483, the Duke had granted Petrus Justinus of Tolentino a privilege for five years for the printing of the *Convivium* of Franciscus Philelphus. The book made something of a sensation at the time; hence Antonius Zarotus also wished to bring out an edition even before the authorized one was on sale. The Duke however under the date of November 10 strictly prohibited this, referring to his previous grant. Milan privileges were conferred however in many other instances, the printers often accompanying the books so protected, by a reference to the document concerned.

In both cases mentioned above, the publishers, not the printers, were the holders of the privilege. In the one granted August 22, 1489, by King Ferdinand of Naples for the printing of the *Orationes de sanctis* of Robertus Caracciolus, two names appeared, first that of Joanmarco Cinico of Parma, and beside it that of the printer Mathias Moravus as owner of the patent. This was not limited to a certain number of years, but was to end only when the unusually large edition of 2000 copies was sold. A second Neapolitan privilege, December 5, 1492, was granted to Aiolfus de Cantono, a printer of Milan, for the *Formularium instrumentorum,* and it is natural to deduce from these grants that the protection already enjoyed by these printers at home had some weight in the matter. The second privilege was however granted for three years only. At about this time printing privileges were granted in Venice as well, but on a much larger scale than elsewhere. The first candidate was Bernardinus Benalius who successfully applied for one, August 17, 1492, for a projected edition of Bernardus Justinianus, *De urbis Venetiarum origine.* Two days later, August 19, 1492, Hieronymus de Durantibus was granted one for two separate books, both of which however were commentaries on Aristotle's works. In this case the privilege, which naturally included the whole domain of the Republic of Venice, protected a printer outside the city, his workshop being in Padua. Inclusive privileges in Venice soon became more common. As early as February 15, 1493 (4), Lazarus de Soardis was granted a general privilege for five books on different subjects. One of the most extensive privileges in which the several books to be protected were itemized was that of

Bonetus Locatellus, April 19, 1497, which covered four-teen different works. Another was that of Alexander Cal-cedonius, May 4, 1499, which listed eighteen. From the latter's privilege we see that also in Venice it was not always printers alone who sought a grant, but publishers and other interested persons as well. There were a large number of applicants who either designated themselves as book-sellers or are known to us as such. Occasionally it was not merely business men who applied. For example, on Feb-ruary 20, 1496 (7), Andrea Manio, "professore di gram-matica," applied for protection for a number of grammati-cal works and school-editions of the classics, thus combin-ing the scholastic and the commercial.

In a large number of cases the authors themselves applied for privileges but they did so apparently to protect their commercial rather than their intellectual rights. Usually, in any case, their work was not original but consisted of new editions of scholarly texts chiefly in classics and medi-cine, which they had supplied with commentaries. An-tonius Sabellicus received a true author's privilege August 27, 1497, for his *Historia omnium gentium,* as did Bartolo-meo Pasi June 3, 1500, for his well-known *Tarifa* and an *Abaco*. One akin to these was that granted a widow, Antonia Kolb, October 30, 1500, for the great *Veduta di Venezia* which with its woodcuts had been the result of several years' labor on her husband's part.

Sometimes a privilege was not granted for a book but for some peculiarity of the method of its production. Some printers applied especially for a privilege covering the change in size, when they planned to issue in smaller format works in great demand which had previously appeared only in thick folio. February 16, 1494 (5), Girolamo Biondo and Company applied for a privilege for printing the *Evangelia et epistolae* in quarto and octavo as well as in the usual folio. Antonio di Zanobi asked for more than this, on May 20, 1497, and January 11, 1497 (8), *i.e.,* for a privilege for breviaries in quarto and octavo, for missals in folio, quarto and octavo, for *Officia* in five sizes, folio—32mo. and for some with illustrations (historiata). In the same category belongs the claim made by Aldus Manutius, February 25, 1495 (6), for a special privilege for his Greek books; and Nicolaus Blastos made a similar one, September 21, 1498, for books in Greek, as did Democritus Terracina, July 15, 1498, for those in Arabic and other Oriental languages; and not unlike these privileges was that received by Ottaviano dei Petrucci, May 25, 1498, for his music-printing, of which mention has already been made. Andrea Corbo proposed a similar one, January 21, 1498 (9), for his *Litterae chorales.*

In Venice it is possible to observe the question of privilege quite extensively from documentary sources, but this does not finally dispose of the matter, for many Venetian incunabula contain the note "cum privilegio," which we fail to find recorded in any documents. Our information as to the granting of privileges in other countries is still more meagre. On account of the variety of legal authority in the

German Empire it was not until the end of the incunabula period that legal protection began to flourish there. The imperial power was not far-reaching enough to grant actual protection, and the territorial lords individually were of scarcely any assistance to printers and booksellers, since a certain freedom in the book-business had developed very early throughout the whole empire. By the sixteenth century it was more usual to hear of German books which were protected by a privilege. Conrad Celtes was one of the first (after 1501) to show the possession of one, and it is interesting to find that one of his privileges was for the *Frankfurter Messe.*

Charles IV issued a privilege in France, and in Spain a number were granted by Ferdinand and Isabella. These were obviously formed on Italian models. It is a characteristic of Spanish privileges that at least in the case of official printing, the sale-price was prescribed at the same time.

20. REPRINTS

A closer examination of the privileges of the incunabula period as a whole convinces us that they were directed less against the reprint in itself than against foreign competition; *i.e.,* that they were to act as a protective duty and could in fact scarcely serve any further purpose since they were so locally restricted. It was impossible either to debar or even to combat the reprint; its power in the field of book production was of the greatest, and every author whose fame penetrated beyond his narrow political boundaries had to be prepared to see his writings, even though protected by a local privilege, reprinted in every other place where a printer or an interested publisher hoped to do a good business with a new edition. In all this, the reprint was not an infraction of the law, but rather an acknowledgment of the value of the literary performance.

It became unlawful only when it presented facts in a false light. The earlier school of incunabula study has not been able to recognize that this was already the case to a certain extent. Only the more exact comparison of types made this possible, and it is by no means certain that we have yet come to the end of all the really demonstrable cases.

In discussing the colophon we have already seen that a reprint was occasionally so carelessly done that the colophon was transferred to the new edition without changing the data as to printing. This was done with no intent to deceive, although a book with a false colophon was the result. We know, for example, two different books of Andreas de Isernia with the colophon "Naples, Sixtus Riessinger, 5 February 1477," and two different editions of Angelus de Aretio, *De maleficiis,* of May 22, 1494, both of which claim to have been issued by Baptista de Tortis in Venice. Naturally in both cases only one of the two editions could be genuine, and an investigation of the type settled with no effort, which that was, and from whom came the reprint with the false printing data.

Both original edition and reprint have not always sur-

vived, however. We know a few incunabula only in one
edition, with a colophon; but an exact investigation of the
type has shown very easily that the information as to
origin could not be correct. There is a volume called Catullus, Tibullus, Propertius, *Carmina,* printed 1475 "opera
et impensa Johannis de Colonia & Johannis Manthen,"
which could however as little have been issued by those
printers as could the book Philelphus, *Consolatio,* printed
with similar type on the same paper, with the information
in the colophon "impressum Romae." The sources of both
these reprints are not yet known.

In all such cases it has always been very difficult to
decide whether the one responsible for the reprint was
intentionally deceitful; naturally, however such a conviction is scarcely avoidable when one finds several such im-
prints clearly from the same source, which happened in
some of the reprints just mentioned. The *Andreas de
Isernia,* like the *Catullus* of 1475 and the "Rome" *Philelphus* were Milan products, judging by their types. It has
been decided that the *Isernia* was printed by Ulrich Scinzenzeler in Milan, and while the type of the other two
books has not as yet been found with any other printer,
its Milanese character is certain. The press of Leonhard
Pachel and Ulrich Scinzenzeler was well known for its
many reprints, some very carelessly done, one in fact pos-
sessing a double colophon, that of the original as well as
that of the reprint. In spite of that it is impossible to
absolve them wholly of intent to deceive, judging from

the edition of Roman law sources with commentaries by Baldus de Ubaldis, issued from their workshop in the years 1494 and 1495. This was collected from a large number of volumes which had appeared separately, some of which suppressed the place of printing, some alluded to it as Venice, some named Venice plainly.[1] The printers had already published the same texts with the correct imprint, just as they did later; hence it was not a mere oversight, nor can we assume that the information could be authentic, for in all their other publications there is no trace of an establishment of Milan printers in Venice. We must think of it then as the performance of self-interested men desirous of passing off their careless reprints as the work of Venetian presses famed for the care bestowed upon their books. We are not unjust in saying this, for other books were issued from their workshop with false colophons.

21. CENSORSHIP

During the Middle Ages the Roman church had already claimed the mission of taking care lest any heretical or other false doctrine should be adopted in Christendom in either speech or writing, a task which, comparatively easy in the days of slow circulation of literature in manuscript, became extremely difficult when the printing-press introduced production in quantity. Whether because of the

---

[1] Haebler, *Der Nachdruck im 15. Jahrhundert.* In *Collectanea variae doctrinae L. Olschki oblata.* pp. 113ff.

realization of the uselessness of struggle, or from other considerations, the church obviously let things go on for some time as they were. Such tolerance on its part was made easy, for the earliest printing was of course limited to reproducing books long circulated in manuscript upon which the church's judgment had already been passed.

The earliest trace of censorship is perhaps in Franciscus Niger, *Contra judeos,* which Conrad Fyner printed in Esslingen 1475. The passage in which this occurs however says only that the author submits his work to the judgment of his spiritual overlords. Such turns of expression are often found in widely different authors; but submission to the church's judgment was of course different from approbation by the censor, of the printing or at least of the circulation of works.

The earliest accounts from which we learn of actual interference by censorship are some years later. They originated in Cologne, which was not only one of the earliest printing centres of Germany but was also a stronghold of orthodox ecclesiasticism. There, several times in the middle of the seventies, a *Disputatio s. dialogus inter clericum et militem super potestate ecclesiastica* was printed in which superiority was denied the spiritual and adjudged rather to the temporal power. This publication, it seems, caused the ecclesiastical authority September 21, 1478, to call for and receive the help of the town council of Cologne in taking measures against the printer of the *Dialogus.* Nothing is known of the further course of the matter. The circulation of the *Dialogus* was however actually stopped; then after a number of years it was again published.

In principle, the church held fast to its right of examination of printing, nor was this true in Germany alone. For example, an edition of the tract of Johannes Carthusiensis, *Nosce te ipsum,* published by Nicolaus Jenson in Venice about the middle of the year 1480, bore at the beginning a number of approbations by the clerics commissioned to examine it. This is an isolated instance, perhaps because the force of the approbation was not strictly maintained, perhaps because the printers did not find it necessary to make public acknowledgment of the resulting sanction.

Matters did not become pressing until the year 1487, when Pope Innocent VIII was moved to address a Bull to the whole of Christendom, in which on behalf of the church he laid vigorous claim to the right to examine all newly issued books, even indeed before their publication, enjoined upon the ecclesiastical authorities a strict fulfilment of this edict, and demanded the aid of the temporal power in executing sentence, this demand made more attractive by a promise of half the fine imposed.

In consequence of this Bull approbations were placed in a few Spanish books in the years immediately following its issue. For example, a general one occurs in the Fernando Diez, *De la sacr. conceptio,* Valencia, Palmart 1487 (Hain 6163), and a more specific one naming the clerical examiners in the *Janua artis Raimundi Lulli* of

Petrus De Gui, Barcelona, Posa 1488 (Hain 10323). As a consequence of the Bull, censorship became active again in Italy. In 1491 the ambassador Franco, Bishop of Treviso, referring to the papal Bull, forbade the printing of books in which ecclesiastical matters were touched upon without having been previously approved, and censored the well-known theses of Johann Picus de Mirandola and Rosellis' *Monarchia.*

But the claims of the church did not lack opposition. The Imperial Diet at Worms in 1495 issued a demand that censorship should be transferred from the hands of the ecclesiastical authorities to those of the Emperor and the princes of the empire. This proposal cannot have been acted upon for at least in the year 1499 it was the ecclesiastical authorities in Cologne who, remembering the Bull of Innocent VIII, based their right of examination upon it. In some other places the universities claimed the right of censorship. In 1495 the *Expositio missae* of Francisco de los Santos appeared in Salamanca and was examined by order of the rector and approved by two professors; and in 1498 the University of Paris censored the *Enterrement de Charles VIII* (which a Paris printer had issued without previous examination) for an inexact statement as to the procession, by which it felt insulted. I have alluded to a demand for the right of censorship on the part of the German princes. This seems to have been attained in Spain, for the *Regula s. Augustini declarata per Hugonem de Sancto Victore* which Hans Giesser printed in Salamanca about the year 1500 was approved "ex commissione regum hispanorum." An ecclesiastical right of approbation must also have existed, for at the same time were entered other marks of censorship traceable to the ecclesiastical authorities in Saragossa and Seville.

22. LOST INCUNABULA

Approximately 30,000 incunabula have been preserved to the present time in at least one copy each, but it would be a great mistake to assume this to have been the entire output of the fifteenth century. It is unlikely that large folios on scholarly subjects would have been destroyed to the last copy, yet on realizing how many incunabula we know in but a single example we must admit the plausibility of the assumption that in many cases even one such single copy may have been destroyed in the course of the centuries. It is a fact that documentary sources reveal editions of a large number of incunabula of which not a trace now remains.

A kinder fate seems to have watched over the productions of Sweynheym and Pannartz. With the exception of the 'Donatus,' of which three hundred copies were printed, not a single one seems to have been entirely destroyed of the twenty-eight books specified in their petition of 1472. Their rival Johannes Philippus de Lignamine was less fortunate. Two of the books printed at his workshop up to the year 1472 are still missing although in that period he issued only nine in all; and, moreover, the two missing ones were scholarly books of large size, a *Lactantius* and a *Horace*. If by means of surviving incunabula we try to check the fifteenth-century booklists which survive in the form of book-advertisements, price-lists and inventories, many titles remain unchecked, and of these many are undoubtedly entirely unknown works.

This does not always mean the irretrievable destruction of such books. Every year, catalogues of collections and those of secondhand book-dealers surprise us by revealing the titles of unknown or lost incunabula; nor has even the general inventory of incunabula in Germany put an end to this, for since it was made up almost exclusively of collections of books in public ownership it comprised in most cases only such works as were listed as incunabula in these collections. It is almost unavoidable that even in the best arranged of these, an incunabulum may come to light which either has been hidden in an unexpected place, among manuscripts or documents, or has not been recognized as an incunabulum in cataloguing.

An equally systematic listing of early printing has not on the whole, however, been made in any other country. Attempts to appraise the contents of certain Italian libraries, the Nazionale in Florence, the Borbonica in Parma, the Communale in Siena, have led in every instance to the discovery of a number of books which had escaped even Reichling's notice. It must be remembered moreover that these investigations covered only such works as were already listed as incunabula in the catalogues of their libraries. Each volume of the *Catalogue général des incunables des bibliothèques publiques de France* revealed a great many incunabula of the existence of which no one up to that time had had any suspicion. Nevertheless Anatole Claudin's *Histoire de l'imprimerie en France* long ago proved how incomplete the *Catalogue général* must always be since it was limited to public institutions only, while the extremely rare pieces of early printing are often found in private collections.

Still greater surprises occur continually in the Spanish book market. Every year the most remarkable discoveries are made, proving that in the Spanish peninsula there are even yet many incunabula which neither special attempts at listing nor the research of zealous booklovers have succeeded in discovering.

Such research has been very difficult since not every librarian, especially in foreign countries, has had at his disposal the necessary expert knowledge by which to recognize incunabula among the books in his keeping. Early books have also been very often in a condition which has not only made the establishing of their identity a weary task but has caused them at first glance to look like waste paper. How many valuable early books may have been lost, since, first put aside because incomplete and then forgotten, they were finally covered with dust and dirt and were destroyed! There are whole classes of books **which** were almost predestined to such a fate. Who would take the trouble to look after the old schoolbooks, the 'Donat-

uses,' the *Doctrinalia* and their countless imitations after
they had served a few generations of school-children
and become shabby and torn? The same fate inevitably
awaited the liturgical books, the missals and breviaries
unavoidably worn out by daily use. The priest was unlikely to think of keeping the half-worn missal of his altar
as a historical document when he could replace it by a
newly, perhaps more suitably printed copy, usually too a
finer and more valuable one. Not a single copy remains of
missals and breviaries which we can prove once existed
in editions of five, six, seven hundred copies. In this very
field however zealous research has made surprising discoveries, bringing to light in church and monastery libraries
imperfect copies of missals and breviaries considered
worthless, which have proved to be long and sadly missed
treasures or entirely unknown books.

Such works were exposed to special danger in the sixteenth century wherever the Reformation was a success. In
these books it saw only the tools of an obnoxious, violently
antagonistic form of worship, and looked upon their destruction as a matter of conscience. Although this passion
for destruction was not often directed against specimens of
early printing, they were at times sacrificed to it, particularly if they were printed on vellum. This precious material was much used by bookbinders in making covers for
documents and books, or to paste upon the covers of their
wooden bindings, a fatal practice of early bookbinders
which has had very important results in incunabula study.

For example, in the Swedish royal archives, the *Manuale Upsalense* printed about 1487, of which not a single copy had been preserved, was successfully reconstructed in one perfect and several imperfect copies, from covers of documents.

23. BINDINGS

This custom of early bookbinders has, as said above, proved of great benefit to incunabula research. It is doubtful whether Italian and French bookbindings contain as many valuable remains of the early printing period as those of Germany, for in those countries paper-making was already more highly developed in the fifteenth century; hence paper was not so expensive a material there that waste was very much used for lining-papers. But as investigation has not been as vigorously carried on in France and Italy as in Germanic countries, surprises there may still be in store for us. In Germany on the other hand the binder's wish to use the more costly material as sparingly as possible has resulted very happily for the study of incunabula. Early binders very often made the pasteboard covers of their bindings by pasting a large number of printed sheets together. Since the outer cover was usually of leather and the inside was as a rule covered with a sheet blank at least on one side, nothing is recognizable outwardly, and it often happens that after painstakingly tearing the pasteboard apart the contents disclosed are quite worthless. Yet just such pasteboards have been known to comprise leaves from an incunabulum which had entirely disappeared, or from an absolutely unknown pamphlet of the time of the Reformation; hence I hereby recommend that students of incunabula spare no pains in examining such covers with the utmost care.

Pasteboard bindings, which were almost unknown in the early period and were still little used in the sixteenth century, are not the only ones which deserve the student's attention. Valuable finds have been made also from wooden covers. Paper or parchment was often used to line the inside covers of the book bound in wood. In this way numerous private documents which were written on parchment have survived, though for the most part badly mutilated. Parchment was also often used for covers, and a wide strip of it also served to attach these to the stitched signatures at the back, part of the strip being fastened to the stitching, the rest pasted to the covers themselves. The earliest binders often put small strips of parchment or paper under the stitching thread of every signature. We find by investigation that even precious parchment leaves printed with the types of the 36- and 42-line Bibles were sacrificed for both these purposes; and that the lining of covers with such leaves was a still more usual procedure. Almost all the fragments of the 'Donatuses' and *Doctrinalia* in the types of the first Dutch printers have survived only because of their use by binders for lining covers.

Paper was used as well as parchment, for such purposes, and has helped us to make valuable discoveries; in fact one whole branch of early literature is now known only because of its use by binders in the ways just described. Yearly almanacs were almost always printed only on one side of the leaf and since they were no longer of value after the year was over for which they were intended, binders chose them whenever they needed linings for covers of large books. Approximately a thousand copies of several hundred different calendars of this kind survive from the days of early printing; but I doubt strongly whether any of them would have come to light in any other way than through their use in book covers. Printers, too, used these leaves with blank versos, putting on them proof- or correction-slips of other books, and some examples of these have been preserved also. When a calendar leaf was decorated with a pleasing astrological woodcut the binder in using it for a lining would occasionally paste it so that the side with the woodcut was left visible. In such a case, however, the verso did not need to be blank; hence here too, leaves could be used on which printers had already experimented.

Very similar to the calendars were the *Prognostica* or *Iudicia* compiled by the astrologers who reckoned the moon's phases and the conjunctions of the stars for the calendars. These little books, which usually consisted of but one to two quarto sheets, often had a woodcut on the title. A binder of Zwickau obviously took pleasure in lining many of his bindings with such calendar-woodcuts, and more than one *Prognosticon* is known to us only because of this practice. The fact that these books too became useless at the end of the year has been of advantage to incunabula study. In the last decades of the fifteenth century binders no longer avoided using printed paper for lining covers. Small books like the *Prognostica* were sold less often by the printers than by binders, and many times if a few copies were unsold these also were used in binding. *Prognostica* printed in all the principal countries have survived, unbound, in large numbers, usually as parts of volumes of miscellanies; but the fact remains that had it not been for its use by bookbinders, we would scarcely be able now to form a true conception of the great extent of this kind of literature.

24. MANUSCRIPT NOTES

Bindings are valuable factors also in still another connection in the study of incunabula. On the covers and preliminary leaves, sometimes on the first and last leaves of the book itself as well, there often occur manuscript notes which merit careful attention. The rubricator had taken over from the scribe the habit of adding at the end of the work a pious expression or occasionally some information as to his name and the place and date of the completion of his task. Thus we learn for example that Johann Bämler worked for years as a rubricator of other printers' productions before he became a printer himself. Information of this kind is especially valuable of course when the book itself gives no sign of its origin.

MANUSCRIPT NOTES

Manuscript notes made by early owners in the books themselves telling how they had acquired them are important, also. As the printed book had not yet become a common occurrence its early owner had more reason even than one in later times to write his name in his book, not only as a means of protection but in order to set down there an account of the circumstances under which it became his. Such entries can be of the utmost importance, as for example the one from which it appears that the earliest Subiaco book, the first printing on Italian soil, was not the *Lactantius,* dated 1465, but the edition of Cicero's *De oratore,* which lacks a colophon. Many hypotheses carefully built with great astuteness on the basis of type-relationships, can be completely overthrown by such an entry. They must of course be used with care and discrimination, for it has happened more than once that over-zealous local patriotism or contemptible greed have falsified such records in order to invest the book with a special value. The names of I. G. Fischer and F. G. H. Culemann have attained sad notoriety in this connection. But wherever critics have been able to establish the authenticity of such information it has proved in many instances strikingly important and valuable for the history of the earliest printing. It is unfortunate that its significance has not been sufficiently taken into consideration by compilers of catalogues of incunabula, and much remains for students to do in this direction.

It is self-evident that the value of these records is not

confined to the book's technical history. They often throw strong light upon its literary history, since they show us who the booklovers were in whose libraries the early books were found, or reveal connections between the authors or editors and the later owners. Carried further, investigation of such notes often allows one to reconstruct the scholarly equipment which a booklover of the fifteenth or sixteenth century knew how to assemble in his study. It is no doubt worth while to trace such owners' marks even if they belong to a later period. The history of early printing has approached definitive treatment through the perfecting of our understanding of incunabula from a bibliographical standpoint; and in attaining this our attention has been directed to the tracing not only of their origin but also of their further fortunes. After the *Gesamtkatalog der Wiegendrucke* has supplied their technical descriptions in an approximately exhaustive manner, the compiler of the catalogue of a single collection will feel more and more how essential is the investigation of the peculiarities of each individual copy, and among these peculiarities an examination of manuscript entries will take first place.

A student of incunabula should give his attention to bindings for their own sake also. It is very evident that the binders' trade must have benefited greatly by the mass-production of books after the invention of printing. Up to that time the centres of book production and collecting had been monasteries, convents and universities which for the most part had their own helpers who clothed their treasured volumes in an enduring garb; and now the multiplication and consequent cheapening of books gave the number of book collectors an unexpected impetus, although most of these new booklovers were not in a position to maintain a binder of their own. Hence the binding trade developed more vigorously than it had had cause to do earlier, although like the printers, the binders of the early period were more truly artists than were later ones. Even if the materials at their disposal for decorating their bindings were modest, they almost always understood how to display a certain originality in every piece of work, and although in some instances we can assemble hundreds of bindings which, judging from the likeness of material must have come from the same bindery, we almost never find two bindings exactly alike.

The history of bookbinding has recently attracted more attention. At one time we were interested only in the costly and splendid bindings of royal booklovers and wealthy monasteries, but now we have begun to consider plainer bindings as well, although the study on the whole is still in its infancy. In this connection the period of early printing has taken on special significance, since it was then that the binders' craft developed. The introduction of binders' rolls and the spread of the use of gilt outside of Italy occurred partly in the early printing period, partly in the age in which incunabula still were the chief constituents of a library; hence in this direction also, incunabula will be the means of giving us a good deal of informa-

tion. It has been assumed that some firms of the incunabula period put their productions on sale already bound,
but even if this is proved false, the study of bindings of
this period promises most interesting results. Bindings by
the Nuremberg Dominican, Konrad Forster, of the years
1434-63 with their printed legends play a very interesting
part in the infancy of printing, and Schwenke's investigations concerning the Erfurt binder Johann Vogel (who
bound among other things a 42-line Bible), and his imitators, have shown how much valuable and interesting
information may be derived from study of this kind. This
study, which is still in its infancy, is an important task for
future cataloguers of incunabula; and the amount of
information which it holds for us cannot be estimated in
advance with any certainty.

25. LITERARY HISTORY

According to a statement in the *Cologne Chronicle* of
1499, which cites the Cologne printer Ulrich Zell as its
authority, Gutenberg may have taken the idea of his invention from Dutch 'Donatuses.' Whether this assertion is to
be taken quite literally is uncertain. The Latin textbook of
the fifteenth century in use in the Low Countries was
apparently not so much the 'Donatus' as the *Doctrinale* of
Alexander de Villa Dei, several editions of which appear
to have been issued as early as 1445 and 1451 by a mechanical process *(gettez en molle)*. 'Donatuses' however also
were printed with the earliest Dutch types, and so, nat-

urally, some of these may have been done by a primitive process. In any case this tradition indicates that the aim of the art of printing was from the first to produce in quantity by a mechanical process the most salable books previously issued in manuscript or by primitive reproductive methods. It is well known that a great many 'Donatus' editions were printed with the types of the 36- and 42-line Bibles, and even if they can scarcely all be attributed to the inventor himself, they show, taken collectively, that the purpose of the new art was from the very first the service of the practical needs of daily life, for which it clearly realized its suitability, as is shown in the printing of indulgences. Thousands of these forms were used, and their production in manuscript took much time and labor, so that the advantage which the invention of printing introduced was quickly recognized. No fewer than twelve different forms of letters of indulgence have survived from the years 1454 and 1455, and probably two different workshops were rivals in their production. We have recently come to realize for the first time what an extraordinary and direct significance the indulgence had in the art of printing, but this fact has not even yet been completely appreciated. Two others of the earliest pieces of printing also belong in this category, the *Türkenkalender* and the German Bull of Calixtus III.

Other specimens of early printing show how the practical needs of that period governed the choice of texts for printing. It seems scarcely conceivable to us that its work

could have begun with the great placard of the astronomical calendar for the year 1448, but there is no doubt
that the character of a calendar appeared to offer a fertile
field for printing. Even though the *Cisianus* had nothing
to do with the year 1443 as was at first supposed, but represented a perpetual calendar, it is evident, judging from this
as well as from the astronomical calendar and that for
1458 that these pieces of printing did not owe their origin
to chance but that the new art consciously took that direction. Indeed, the printing of calendars and the predictions
bound with them *(Iudicia, Prognostica)*, played a very
important part in the history of the earliest printing.

The remains which have come down to us give us the
impression that the first printers, working in the beginning in rivalry with the scribes, engaged in these more
modest productions before they ventured to make larger
books by mechanical means. Pfister's work at Bamberg
with his books which were popular in character, and,
many of them, of modest size, is far more a continuation
of the trend recognizable in the calendars and other job
printing than an attempt at more ambitious ends.

Even the editions of the Bible were not undertaken,
properly speaking, for the benefit of scholars but rather to
meet the general needs of a large part of the population,
i.e., the clergy as a whole; and the editions of the Psalter
from Schöffer's workshop were used, though within much
greater limits, by the same circle. The publications of the
earliest German printing-shops maintained this religious

character throughout the following years as well, but turned from editions of the Bible and from learned theological disquisitions to Canon law. Schöffer's *Cicero* of 1465 was the first appearance in print of a classic, a work of liberal scholarship. The turning toward humanistic scholarship came about very slowly in Germany. Strassburg, Cologne and Basel printing were still very decidedly theological in character as late as the beginning of the seventies, but Roman law gradually attained a certain consideration along with Canon law.

Matters developed quite differently in Italy. The 'Donatus' of Sweynheym and Pannartz was merely a reminiscence of home, although as a matter of fact other German printers in Italy began their work with a 'Donatus'; but the editions of this grammatical schoolbook which were legion in Germany were the exception in Italy. The first larger book from the Subiaco press was in all probability a Cicero, *De oratore,* and of the twenty-eight works printed by Sweynheym and Pannartz at Rome before their petition to Pope Sixtus IV, no fewer than seventeen were classical texts and but nine were theological. The first book printed by Johann of Speyer at Venice was a *Cicero* also, while Johann's brother maintained this policy of printing classics, a policy followed for a few years more by his competitors in Venice, in consequence of which the Italian book market was overloaded as early as the year 1473 with editions of the classics. The result was stagnation, and, naturally, their production ceased.

[219]

Theology in all its branches long remained the chief concern of the German book market, constituting almost half the entire book production of the country, second place being held by schoolbooks, many of their editions of the classics even, being school-texts. Very decided changes were the rule in the Italian market, however. In Italy the period of the printing of classics was followed by one of the literature of jurisprudence, with which within a few years the market was scarcely less overstocked than it had formerly been with classical texts. At the same time however, Italy began to take account of the literature of her national poets. As early as the year 1470 Petrarch's Italian poems were printed, and in 1472 three editions of Dante's *Divina Comedia* appeared, while the *Jacopone da Todi,* the *Giustiniani* and others quickly followed. In Germany, Mentelin's *Partzifal* and *Titurel* of 1477 were quite isolated instances and were not republished during the entire incunabula period, while at no time did the German presses consider the literary works of the Middle Ages worth printing.

Literary development in France was very much like that of Italy. The first press in Paris was influenced by the humanists Fichet and Heynlin, and theological works received almost no consideration. Ulrich Gering and his associates made up for this however, as soon as they became independent, by publishing practically nothing but theology. Paris book production first became more diversified when Keysere and Stoll, to whom we owe the first books printed in French, entered into competition with Gering and his partners. In Lyons, Guillaume Le Roy began his printing with a few theological works. There too however the printing of books in the vernacular was carried on with unusual vigor, some of these, translations from the Latin, with about the same number of original French texts. Soon, however, the Lyons press plunged into international competition, with the object of rivalling Italian editions of works of jurisprudence as well as the theological works of German presses.

About the year 1480 a general calm came over the book market. Supply and demand seemed to have attained a certain balance, and there was no longer a feeling that rival printers would pitch exclusively upon first one and then another kind of literature. The most varied kinds of both scholarly and popular books were printed and every branch of literature came in for more or less attention according to its general importance. The number of those interested in reading now increased materially under the influence of printing, while the relation between scholarly Latin works and books written in the vernacular began to shift gradually in favor of the latter, a development which reached a decisive turning-point however only under the influence of the Reformation movement.

It is not so easy to determine where and when the first book originated which was intended expressly for printing. We have already seen that printing was used immediately to serve the practical needs of daily life; but writings such as the Letter of Indulgence of Calixtus III were not literature, and even though its German translation was probably made with the direct purpose of printing it, printing in this and similar cases was only a means to an end quite other than that of spreading the piece of print- ing itself. The earliest books of travel, as Tucher, Brey- denbach and others were printed very soon after they were written; and in the fact that the Breydenbach travel- ling party was accompanied by an artist, Erhard Reuwich, who was to commemorate noteworthy moments of the journey in his woodcuts, we see that a literary project was in mind from the first. As Reuwich had a share in the printing, this book fulfils almost all the conditions which we are looking for, in order to characterize it as a piece of literature written expressly to be printed. Nevertheless all these travel books were apparently first circulated in manuscript, and printing was not a decisive factor in their composition. Still less can we allow the treatises of the Italian jurists to count in this connection, although they often tell us that they were circulated in print almost im- mediately after their compilation. As they were practically without exception first designed for an academic audi- ence, their circulation in print was not the prime object of their author.

The case was different with the polemical treatisēs of the Italian humanists and the imitations which this form of literary combat inspired in German scholars. In them we must seek the beginnings of the lierature written ex- pressly for printing. Writing of a purely literary sort had not yet a wide enough circle of readers to serve as a profitable subject for the art; and while popular literature availed itself of printing to a considerable extent, this is far more often found to be a literary survival from an early time than a production made expressly to be circulated in print. A literature intended primarily for printing originated only with the disputes of the humanists over the scholasticism of the old and the various movements of the new schools, for these naturally generated a keen desire to bring their arguments before the general public. From this spirit sprang such works as Sebastian Brant's *Narrenschiff,* the writings of the Heidelberg and Leipzig academicians, the Sodalitas of Conrad Celtes and others. Among these the Academia of Aldus Manutius in Venice deserves a word of mention, for it also was anxious to serve as a centre of printing, although its purpose was not so much to create and disseminate a new literature, as to bring to light forgotten treasures of the Greek classics in order to make them better known to its contemporaries.

Earlier incunabula study has in general stressed the literary aspect of early printing, keeping the significance of the history of the printed book quite unduly in the background; yet it is even yet impossible to write a literary history of the incunabula period. Naturally it is of real importance for such a history that Cicero's *Epistolae* went through more than seventy-five editions in the fifteenth century in about thirteen workshops, and that Thomas Aquinas alone is represented by nearly as many incunabula as Augustine and Hieronymus together. But the mere enumeration of countless editions, after the manner of Hain and his followers, is not adequate treatment for figures of such literary significance as these early monuments of printing. It is equally important to trace the conditions under which law books alone were printed in one place while another favored theology and a third medical works; and to unravel the threads linking the many editions of books of Roman and Canon law with the scholarly groups of the places where these books originated. We must trace to the utmost detail the editorial work of the numerous scholarly correctors of the press of the fifteenth century; and must seek to comprehend their literary significance for their own time, from an unprejudiced viewpoint, from the praise of their friends and the abuse of their enemies. In this field alone there is still an endless amount of work for incunabula research to do, while a more exhaustive study of the contents of the early monuments of printing promises the investigator still further rewards. Individual attempts in this direction have already been made. What Bauch accomplished for the learned world of Leipzig, Pauli for the Raimundus Peraudi, Bömer for the Paulus Niavis (and many other similar contributions which I must leave unmentioned here in order not to make this an exhaustive inventory), deserve of course the highest appreciation and are to be considered of the greatest importance in incunabula study. These are but beginnings, however, and we need another equally notable series of individual investigations like these before we can make a comprehensive estimate. Moreover literary history is not the only promising field for incunabula study. Each individual branch of learning has left its impress upon the products of early printing, revealing evidences of development and of the stage which each branch had reached by the end of the Middle Ages; and the significance of incunabula for history, above all for the history of civilization, is not even yet thoroughly understood. Down to the present time, no one has succeeded in producing a clearly outlined history of printing which can universally be acknowledged as true as regards its general features.

The eighteenth and nineteenth centuries were not more one-sided in dealing with incunabula by a literary-bibliographical method than have been the past fifty years in insisting upon the typographical aspect. We have doubtless gained a good deal of information as to incunabula by both methods, and this knowledge must not be underestimated, notwithstanding the fact that it has not yet led to satisfactory results in the form of a definitive history. Incunabula however will receive full justice only from the scholar who combines with the keen eye indispensable for detailed investigation and for peculiarities of form, the far-reaching vision which comprehends all the countless associations which link each incunabulum with its time and the scholarship of that time. No matter how varied a student's interests may be, this study offers him a correspondingly varied field of work which is both wide and fertile; and it is a real source of pleasure to know that the manifold and diverse problems offered by incunabula are being subjected to a searching examination from the most varied scientific standpoints. However small the sphere of incunabula may seem at first glance, it is so closely bound up with every department of our knowledge that we can as little assign exact limits to it as to the field of knowledge itself.

INDEX

INDEX

INDEX

INDEX

[231]

INDEX

INDEX

INDEX

INDEX

INDEX

INDEX

INDEX

INDEX

INDEX

INDEX